T0271013

Analysis and Optimization of Sheet Metal Forming Processes

Analysis and Optimization of Sheet Metal Forming Processes comprehensively covers sheet metal forming, from choosing materials, tools and the forming method to optimising the entire process through finite element analysis and computer-aided engineering.

Beginning with an introduction to sheet metal forming, the book provides a guide to the various techniques used within the industry. It provides a discussion of sheet metal properties relevant to forming processes, such as ductility, formability, and strength, and analyses how materials should be selected with factors including material properties, cost, and availability. Forming processes including shearing, bending, deep drawing, and stamping are also discussed, along with tools such as dies, punches, and moulds. Simulation and modelling are key to optimising the sheet metal forming process, including finite element analysis and computer-aided engineering. Other topics included are quality control, design, industry applications, and future trends.

The book will be of interest to students and professionals working in the field of sheet metal and metal forming, materials science, mechanical engineering, and metallurgy.

Analysis and Optimization of Sheet Metal Forming Processes

Edited by Amrut Mulay,
Swadesh Kumar Singh, and
Andrzej Kocanda

CRC Press
Taylor & Francis Group
Boca Raton London New York

CRC Press is an imprint of the
Taylor & Francis Group, an **informa** business

Designed cover image: www.shutterstock.com

First edition published 2024
by CRC Press
2385 NW Executive Center Drive, Suite 320, Boca Raton FL 33431

and by CRC Press
4 Park Square, Milton Park, Abingdon, Oxon, OX14 4RN

CRC Press is an imprint of Taylor & Francis Group, LLC

ISBN: 978-1-032-57941-2 (hbk)
ISBN: 978-1-032-57946-7 (pbk)
ISBN: 978-1-003-44175-5 (ebk)

DOI: 10.1201/9781003441755

Typeset in Times
by Apex CoVantage, LLC

Lord Sri Ram
Our beloved parents,
and
Late Mrs Chitra Mulay, my mother (first editor), guided
me to study engineering at the right moment during a
challenging financial period.

Contents

Chapter 7   Applications of Incremental Sheet Forming 128

*Sherwan Mohammed Najm, Valentin Oleksik, and
Tomasz Trzepieciński*

Chapter 8   Finite Element Analysis of the Incremental Forming
Process 147

Ömer Seçgin and Vedat Taşdemir

Chapter 9   The Incremental Sheet Forming of Light Alloys 157

G. Karthikeyan, D. Nagarajan, and B. Ravisankar

Foreword

Sheet metal forming is a fundamental pillar of modern manufacturing as it has applications in modern automobiles and in the aerospace and marine sectors. This intricate process, which transforms raw metal sheets into finely detailed, functional components, lies at the core of industries from automotive to aerospace. The scientific developments in *Analysis and Optimization of Sheet Metal Forming Processes* help readers understand the modern sheet-forming processes. Experimental results are discussed whenever appropriate.

Readers will learn about recent trends on sheet metal forming and its applications within this meticulously prepared volume. *Analysis and Optimization of Sheet Metal Forming Processes* is not merely a compilation of methodologies and techniques but an exhaustive guide that seamlessly integrates theoretical foundations with pragmatic applications. Sheet metal parts have the advantage of high elastic modulus and yield strength, allowing the parts produced to be stiff and have an excellent strength-to-weight ratio.

This book navigates the complex interplay of material behaviour, tooling design, and process refinement from fundamental principles to advanced simulations. Each chapter offers a panoramic view of a specific facet, empowering readers with the knowledge to confront challenges in sheet metal forming with assurance and precision. The esteemed authors draw upon a wealth of expertise from their collective academic and industry backgrounds. Through their valuable insights, readers are guided into a domain where intricate physical phenomena are elucidated, laying bare the fundamental principles governing the behaviour of metals subjected to the stresses and strains of the forming process.

This book, a testament to its academic rigour and practical significance, will undoubtedly secure a revered place on the bookshelves of students, researchers, and practitioners alike. It constitutes an invaluable resource for young sheet metal-forming researchers and for industry professionals striving to elevate their processes and products. The authors establish a bridge between theory and practice in these pages, urging readers to start their journey in sheet metal forming. The book is a call to explore the depths of sheet metal forming, where creativity meets scientific investigation and where this discipline's boundless potential continues to influence our industrial world.

I congratulate the book's editors, Dr. Amrut Mulay, Dr. Swadesh Singh, and Dr. Andrzej Kocanda, for their editorial efforts in bringing such a book to publication. The book will not only give insights about the scientific developments in the field but also empower graduates and young researchers.

Dr. Anupam Shukla
Director
Sardar Vallabhabhai National Institute
of Technology, Surat, Gujarat

Foreword

Sheet metal forming developed as a cornerstone of modern manufacturing in an era where precision engineering meets the cutting edge of innovation. The book *Analysis and Optimization of Sheet Metal Forming Processes*, a thorough book that elegantly integrates fundamental concepts with real-world applications, exemplifies this domain.

In recent years, considerable progress has been achieved in the forming of complex parts using finite element methods; many engineers are now employing these systems to form intricate sheet metal parts. Indeed, metal forming is an integral part of modern industry. Stamping, drawing, forging, bending, extrusion, deep drawing, etc. are the most widely used metal-forming techniques, applied in processes such as stamping out automobile body parts and drawing metal utensils, but all these processes require product-specific tooling. Every new product requires a separate die with a high cost. But in this dynamic market, where products are changing rapidly, it is becoming increasingly difficult to make huge investments in the fabrication of specialized dies and tools. These tools serve no purpose once the product changes.

The prominent writers of this volume take readers on a journey into the complex subject matter of sheet metal forming using accurate expertise in a simplified way to describe how the contemporary challenges are being addressed. They reveal the underlying principles that control the behaviour of metals under the stress and strain of forming processes.

This book navigates the world of material behaviour, tooling design, and process improvement, from fundamental concepts to advanced simulation. Each chapter provides a comprehensive overview of a specific dimension, allowing readers to confidently and precisely handle challenges in sheet metal forming.

I acknowledge the efforts of the authors and editors in bringing out this book for the readers. The book will make an effective framework for prospective studies in sheet metal forming processes.

Prof. Harit K. Raval
Professor (HAG)
Department of Mechanical Engineering
Sardar Vallabhbhai National Institute
of Technology, Surat, Gujarat, INDIA

Preface

In sheet metal forming, a sheet blank that has a simple shape is plastically formed between tools to obtain a part with relatively complex geometry with desired tolerances and properties. Sheet metal forming usually produces little scrap and generates the final part geometry in a very short time, usually in one or a few strokes of a press. As a result, sheet forming offers potential savings in energy and material, especially in medium and large production quantities, where tool costs can be affordable.

The ever-increasing costs of materials, energy, and manpower require that sheet metal forming processes and tooling be designed and developed with minimum amount of trial and error with the shortest possible lead times. Therefore, to remain competitive, the cost-effective application of computer-aided technologies, i.e. CAD, CAM, CAE, and especially finite element analysis (FEA), computer-based simulation is an absolute necessity today. Thus, process modelling using FEA has been discussed in all appropriate chapters.

The practical and efficient use of these technologies requires a thorough knowledge of the principal variables of the sheet metal forming processes and their interactions: the flow behaviour and formability of the formed sheet material under processing conditions; die geometry, materials and coatings; friction and lubrication; deformation mechanics, i.e. strains, stresses and forces; characteristics of the sheet metal forming presses and tooling; geometry, tolerances, surface finish and mechanical properties of the formed parts and the effects of the process on the environment.

In *Analysis and Optimization of Sheet Metal Forming Processes*, Chapter 1 covers principles and processes of roll forming in sheet metal forming. Chapter 2 reviews the rotary piercing process. Chapter 3 discusses FEA in metal spinning, and Chapter 4 discusses high-energy rate forming. Chapter 5 reviews the formability of aluminum and its alloy sheet metals. Chapter 6 presents the analysis of deep drawing quality steel using incremental hole flanging with different pre-cut hole diameters. Chapter 7 briefly discusses incremental sheet forming and its applications, and Chapter 8 presents FEA of the incremental forming process. Chapter 9 discusses incremental sheet forming of light alloys, and Chapter 10 discusses the formability and surface integrity of incremental sheet metal parts. Chapter 11 presents the investigation and optimization of process parameters in a novel single-point incremental sheet forming (SPIF) process to reduce excessive thinning for aluminum alloys. Chapter 12 presents a comprehensive study of the impact of generatrix radius variation and temperature on spifability in warm incremental forming.

Chapter 13 presents numeric investigations to improve the final sheet thickness in the SPIF process of DC04 sheets with novel process parameters. Chapter 14 provides the state-of-the-art developments and future trends in hydroforming. Chapter 15 covers a comparative study of adaptive and Arlequin meshing methods applied to tensile testing and various metal forming processes. Chapter 16 discusses the role of edge formability in sheet metal forming. Chapter 17 presents an FEA of bimetallic sheet AA5754/AA6061 using constrained groove pressing. Chapter 18 provides a technical note on fracture limits in sheet metal forming, and Chapter 19 discusses the role of lubrication in sheet metal forming processes.

The preparation of this book was possible through extensive efforts by friends, associates and students of the editors who authored and co-authored many of the chapters. We want to thank them all for their precious contributions. We would also like to thank our families, who offered us enormous support and encouragement throughout the preparation of this book. Thanks are also due to the publishers for bringing out the book in a short span of time. Suggestions for the improvement of the book will be highly appreciated.

MATLAB® is a registered trademark of The Math Works, Inc. For product information, please contact:

The Math Works, Inc.
3 Apple Hill Drive
Natick, MA 01760-2098
Tel: 508-647-7000
Fax: 508-647-7001
E-mail: info@mathworks.com
Web: http://www.mathworks.com

Acknowledgements

We want to express our heartfelt gratitude to all individuals and organizations for their valuable contributions and support in writing of this book.

We deeply thank my research advisors for their guidance, encouragement, and mentorship. Your feedback has been invaluable in enhancing the quality of this chapter. We sincerely appreciate all authors and colleagues collaborating with us on this project. Your insights, expertise, and dedication were instrumental in shaping the content of this chapter. We thank the authors for their time and willingness to contribute to the book. Without their involvement, these chapters would not have been possible.

We want to thank Prof. Anupam Shukla, Director and Dr. A.A. Shaikh, Head, Department of Mechanical Engineering, Sardar Vallbhai National Institute of Technology, Surat, for providing a conducive research environment and access to resources and facilities that were indispensable in the completion of this work.

The input and feedback received from Shatakshi Singh, editorial assistant, mechanical engineering; Nicola Sharpe; Katya Porter, Editor, CRC Press; and reviewers of this book have played a crucial role in refining the book's content, structure, and overall quality. The technical assistant, Mr. Gajendra Panchal, and research scholars Mr. Sudarshan Kumar, Mr. Rahul Ramlal Gurpude, and Mr. Hitesh Dnyaneshwar Mhatre, who assisted in data collection and analysis and administrative tasks, deserve special recognition for their hard work and dedication.

We also want to acknowledge the unwavering support and understanding of my family and friends. Your encouragement and patience throughout this endeavor have been a source of motivation.

We are grateful to my peers and colleagues for their stimulating discussions and insights and for fostering a collaborative research atmosphere.

We are sincerely thankful to every individual and institution mentioned above for their support and contributions, without which this book volume would not have been possible. This book is a result of the collective effort of many, and we are profoundly thankful for the support and collaboration that made it a reality.

.

Editors

Dr. Amrut Mulay is an assistant professor, Department of Mechanical Engineering, Sardar Vallabhbhai National Institute of Technology, Surat (SVNIT; the institute has national importance under the Ministry of Education, Government of India). Before joining the institute, he earned a PhD from the National Institute of Technology, Warangal, India. He was one of the few PhD students from NIT Warangal to be selected for the prestigious Interweave program under the Erasmus Mundus (EMA2 strand1) project fully funded by the European Union. He carried out part of his research work at Warsaw University of Technology, Poland. He completed postdoctoral research at the prestigious Indian Institute of Technology (IIT), Bombay. He has published a good number of articles in reputed journals, books, and conferences. He is associated with many societies and chapters in India. He has worked on sponsored research projects funded by SERB, GoI, SVNIT Surat, among many others.

His major research interests are incremental forming, uniaxial and biaxial cruciform stress–strain field, digital image co-relations, optimization of manufacturing processes, and finite element studies in manufacturing.

Dr. Swadesh Kumar Singh holds a master's degree and a PhD from IIT Delhi, India. He received a gold medal for his bachelor's degree at AMU. He is presently working as a professor in the Department of Mechanical Engineering at GRIET, Hyderabad, India. Prior to joining GRIET he served as assistant executive engineer at Indian Engineering Services and also worked with BHEL for a brief period of time. He was among the 2% top researchers in mechanical engineering worldwide in 2020. Dr. Singh has been involved in research on numerical simulation and experimental studies on sheet metal forming at room and elevated temperatures and the characterization of metals and biocomposites. He has conducted extensive research in characterizing low-carbon and austenitic stainless steels, titanium grade-5, dual-phase steel, and zircaloy. He has received the Young Scientist award from the Department of Science and Technology (DST), Government of India and the Young Teacher award from the All India Council of Technical Education, Government of India for his contribution in research.

He has authored over 175 international journal articles and conference papers and received sponsored project grants from various agencies. He has an h-index of 22 on Scopus and has over 2000 citations. His articles appear in high impact factor journals such as *Materials & Design*, *Journal of Material Processing Technology*, and CIRP. He has received research funding from the DST, the All

India Counsel of Technical Education, the Department of Atomic Energy, and the Aviation Research and Development Board and several industry consultancies. In addition to GRIET, he is also associated with the Institute for Sustainable Industries & Livable Cities, Victoria University. At VU he is working on recycling of used tyres and making activated carbon for filtering different types of waters.

 **Prof. Andrzej Kocanda** completed his PhD and DSc theses at the Warsaw University of Technology in Poland. He received the scientific title of professor in 1996. His scientific, research, and teaching interests include fracture mechanics, fatigue of metals, manufacturing processes with particular emphasis on sheet metal forming, and cold and warm forging of metals, as well as technology transfer and computer modelling of processes.

His professional activities include also consulting and chairmanship in numerous industrial and scientific societies as well as serving on editorial boards of journals and being a reviewer for many international journals. He has published many articles and conference papers as well as several books. He supervised a large number of MSc and PhD theses and was also a reviewer of numerous applications for professorships.

His basic research and teaching work was mostly related to the Warsaw University of Technology, but he also worked for many years at the Kielce University of Technology (Poland). He has also participated in international research projects and given lectures in many countries including the Institute of Physical and Chemical Research (Japan), the Technical University of Denmark, the University of Moratuwa (Sri Lanka, UNIDO expert), the University of Strathclyde (Great Britain), and institutions in Mexico, Romania, France, and Germany.

Contributors

S Abdul Azeez
G. Pullaiah College Of Engineering
And Technology
Kurnool, Andhra Pradesh, India

C Anand Badrish
Defence Research & Development
Laboratory
Hyderabad, India

Ravisankar Balasubramanian
Department of Metallurgical and
Materials Engineering
National Institute of Technology
(NIT) Tiruchirappalli
Tiruchirappalli, Tamil Nadu, India

Raj Ballav
Department of Production &
Industrial Engineering
NIT Jamshedpur
Jamshedpur, Jharkhand, India

Din Bandhu
Manipal Institute of Technology
Bengaluru, Karnataka, India

Tushar Banerjee
Production and Industrial Engineering
Department
NIT Jamshedpur
Jamshedpur, Jharkhand, India

Vishal Bhojak
Mechanical Engineering Department
Malaviya NIT
Jaipur, India

Sudarshan Kumar Choudhary
Department of Mechanical
Engineering
Sardar Vallabhbhai National Institute
of Technology (SVNIT)
Surat, Gujarat, India

Suman Deb
School of Mechanical Sciences
Indian Institute of Technology
Bhubaneswar, Odisha, India

Nagarajan Devarajan
Department of Metallurgical and
Materials Engineering
NIT Tiruchirappalli
Tiruchirappalli, Tamil Nadu,
India

Baloji Dharavath
Department of Mechanical
Engineering
KG Reddy College of Engineering &
Technology
Hyderabad, India

Praveen Kumar Gandla
G. Pullaiah College of Engineering
and Technology
Kurnool, Andhra Pradesh, India

Karthikeyan Gunasekaran
Department of Metallurgical and
Materials Engineering
NIT Tiruchirappalli
Tiruchirappalli, Tamil Nadu, India

Rahul Ramlal Gurpude
Department of Mechanical
  Engineering
SVNIT, Ichchhanath
Surat, Gujarat, India

Jagadeesha T
Mechanical Engineering Department
NIT Calicut
Kerala State, India

Sonanki Keshri
Department of Chemistry
Jyoti Nivas College Autonomous
Bangalore, Karnataka, India

Amaresh Kumar
Department of Production &
  Industrial Engineering
NIT Jamshedpur
Jamshedpur, Jharkhand, India

Gautam Kumar
Mechanical Engineering Department
NIT Patna
Patna, Bihar, India

Jinesh Kumar Jain
Mechanical Engineering Department
Malaviya NIT
Jaipur, India

Luana de Lucca de Costa
Metalforming Innovation Center
Federal University of Rio Grande do
  Sul
Porto Alegre, Brazil

Kuntal Maji
Production and Industrial Engineering
  Department
NIT Jamshedpur
Jamshedpur, Jharkhand, India

Ravi Kumar Mandava
IIITDM Kurnool
Kurnool, Andhra Pradesh, India

Viren Mevada
Department of Mechanical
  Engineering
SVNIT
Surat, Gujarat, India

Hitesh Dnyaneshwar Mhatre
Department of Mechanical
  Engineering
SVNIT
Surat, Gujarat, India

Amrut Shrikant Mulay
Department of Mechanical
  Engineering
SVNIT
Surat, Gujarat, India

Balaji Nallusamy
Department of Metallurgical and
  Materials Engineering
NIT Tiruchirappalli
Tiruchirappalli, Tamil Nadu, India

Sherwan Mohammed Najm
Kirkuk Technical Institute
Northern Technical University
Kirkuk, Iraq

R. Narayanasamy
Department of Production
  Engineering
NIT Tiruchirappalli
Tiruchirappalli, Tamil Nadu, India

Valentin Oleksik
Faculty of Engineering
Lucian Blaga University of Sibiu
Sibiu, Romania

Jigar Patel
Department of Mechanical Engineering
SVNIT
Surat, Gujarat, India

Pravin Pawar
Department of Mechanical Engineering
NIT Goa
Ponda, Goa, India

Balaji Krushna Potnuru
Malla Reddy Engineering College
Hyderabad, India

Harit K. Raval
Department of Mechanical Engineering
SVNIT
Surat, Gujarat, India

André Rosiak
Metalforming Innovation Center
Federal University of Rio Grande do
Sul
Porto Alegre, Brazil

C. Sathiya Narayanan
Department of Production Engineering
NIT Tiruchirappalli, Tiruchirappalli
Tamil Nadu, India

Lirio Schaeffer
Metalforming Innovation Center
Federal University of Rio Grande do Sul
Porto Alegre, Brazil

Ömer Seçgin
Sakarya University of Applied
Sciences Sakarya, Türkiye

Nuri Şen
Düzce University
Düzce, Türkiye

Tejendra Singh Singhal
Mechanical Engineering Department
Malaviya NIT
Jaipur, India

Swadesh Kumar Singh
Department of Mechanical
Engineering
Gokaraju Rangaraju Institute of
Engineering and Technology
Hyderabad, India

K. Sivaprasad
Department of Metallurgical and
Materials Engineering
NIT Tiruchirappalli
Tiruchirappalli, Tamil Nadu, India

Sudha Suriyanarayanan
Department of Chemistry
Jyoti Nivas College Autonomous
Bangalore, Karnataka, India

Vedat Taşdemir
Kütahya Dumlupınar University
Kütahya, Türkiye

Tomasz Trzepieciński
Department of Manufacturing
Processes and Production
Engineering, Faculty of Mechanical
Engineering and Aeronautics
Rzeszow University of Technology
Rzeszów, Poland

S. Vigneshwaran
Department of Mechanical Engineering
SRM Institute of Science and
    Technology Ramapuram
Chennai, Tamil Nadu, India

Gunda Yoganjaneyulu
Production and Industrial Engineering
    Department
NIT Jamshedpur
Jamshedpur, Jharkhand, India

1 Roll Forming in Sheet Metal Forming
Principles and Processes

Suman Deb

1.1 INTRODUCTION

Roll forming is a sheet metal forming process in which flat metal strips are gradually bent into desired cross-sections under rotating rolls (Deb, Panigrahi and Weiss, 2021). A roll-forming mill usually consists of several roller stations in series as shown in Figure 1.1. The sheet metal is continuously fed from a coil using an uncoiler during the operation. The sheet then undergoes flattening and leveling before being directed via the entrance guide/feeder, which ensures appropriate alignment as it feeds the sheet into the roll-forming stations (Troive and Ingvarsson, 2008). The deformation occurs at each station. The roll gap is maintained at the same as the thickness of the sheet to be roll formed. (Bhattacharyya et al., 1984)

Roll forming is highly versatile, used in multiple industries, especially in automotive industries, where parts contain high structural complexity and costs and investments are substantial. Roll forming is a flexible and cost-effective solution for such products (Sweeney and Grunewald, 2003). Longitudinal components can be roll formed to produce cross-sections of different geometries, starting from simple V, U or C cross-section shapes to the more complex shapes shown in Figure 1.2.

Lighter and safer components at competitive prices using high-strength steel are in great demand in the automobile industry. Various roll-formed parts are used in automobiles such as bumpers, side impact beams, pillars, seat components, frame rails, and bottom sill reinforcements (Troive and Ingvarsson, 2008). Along with this, roll forming is used in a variety of other applications, including roofing, siding, shelving, storage racks, electrical appliances, refrigeration units, heating and ventilation systems, railway carriages, power generation facilities, doors and windows, bicycle wheels, fireplaces, other household items, aircraft and spaceships. Some of the reason for the popularity of the roll forming process are its versatility in producing complex components, high productivity and repeatability, ability to produce components with close tolerances and low capital investment as compared to the stamping or hydroforming. Further, the process can be easily automated and integrated with other secondary processes such as cutting, welding and punching (Kiuchi, 1989).

DOI: 10.1201/9781003441755-1

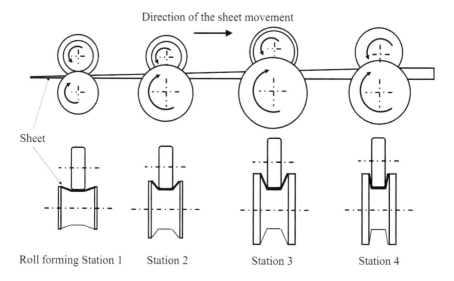

FIGURE 1.1   Schematic representation of the roll forming process.

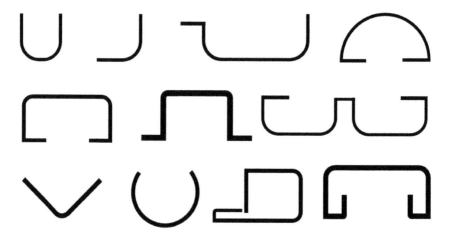

FIGURE 1.2   Different possible roll forming sections.

The deformation steps in roll forming are represented using a flower diagram or flower pattern (Deb, Panigrahi and Weiss, 2021). The flower pattern for a simple U section is shown in Figure 1.3. In this diagram, the deformation at each forming station is superimposed. A variety of approaches can be considered for the same cross-section, but the flower pattern changes based on the required cross-section and becomes more complex in case of complex geometries due to the increased number

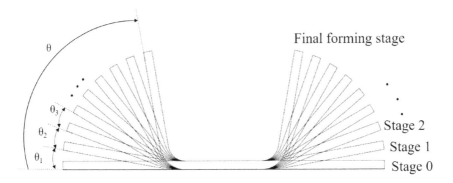

FIGURE 1.3   Flower pattern of a bending sequence in roll forming.

of forming stations. This increases the tooling cost. However, if there are only a few forming stations, the final product may be distorted because of higher stresses generated during deformation (Panton, Zhu and Duncan, 1994). Therefore, the number of steps to bend the sheet into the final product must be optimized. The section depth (maximum vertical dimension of the cross-section) also influences the number of steps required for roll forming. A deeper section needs more forming stations.

The current chapter focuses on the roll forming process in brief and analyzes the deformation behavior during roll forming. Development of longitudinal strain and redundant deformation are discussed in light of the shape defects on the final roll-formed parts. The influence of the processing parameters, geometry of the rollers and forming stations and material properties are also analyzed with relevant examples. Finally, current developments in roll forming are also discussed with case studies.

1.2   DEFORMATION MODES IN ROLL FORMING

The primary mechanism of deformation in roll forming is transverse bending (Panton, Zhu and Duncan, 1994). The sheet metal passes through the series of roll forming stations and incrementally bends into the final cross-section (Figure 1.1). A cross-section of a roll-formed part contains two major elements: a curved or bent part and a straight part. (Brunet, Lay and Pol, 1996) The bending can occur either with constant arc length or constant radius or mixture of both as shown in Figure 1.4. In the process of constant arc length forming, the length of the neutral line remains unchanged, but bend radius reduces, whereas in constant radius forming, one or several segments are bent at final radius in each pass.

During constant length bending, the entire bending length is plastically deformed in each stage; thereby the microstructure strain—and consequently the material—becomes harder after each subsequent stage. In contrast, in constant radius forming, the bend zone plastically deforms only once. Experiments have

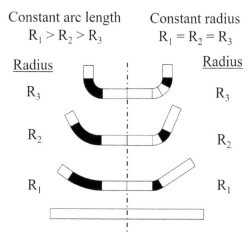

FIGURE 1.4   Constant arc length vs constant radius bending.

shown that longitudinal strain is greater with constant arc length forming than constant radius forming because the outer fiber of the sheet elongates longitudinally at a higher degree with constant arc length roll forming (Badr, Rolfe and Weiss, 2018).

Roll forming utilizes a more complicated deformation mechanism than simple bending (Panton, Zhu and Duncan, 1992). In addition to the transverse bending, other deformation modes are also activated during roll forming that affect the quality of the final products (Panton, Zhu, and Duncan, 1994) (Panton, Duncan and Zhu, 1996). These deformation modes are called redundant deformations and can be listed as follows:

- longitudinal elongation and/or shrinkage
- longitudinal bending and bending back
- transverse elongation and/or shrinkage
- shear along plane of the sheet
- shear along the thickness of the sheet

These redundant deformations often overlap with the transverse bending and result in various shape defects such as longitudinal bow, twist, center wave, edge wave, edge cracking corner buckling, nonuniform springback and end flare.

1.2.1 ORIGINS OF REDUNDANT DEFORMATIONS

1.2.1.1 Longitudinal Elongation and/or Shrinkage
During roll forming, metal strips or longitudinal sheets are gradually bent in the transverse direction, and the height of the transverse cross-section increases while the width decreases according to the product dimension; the sheet deforms

along the horizontal and vertical direction due to the 3D nature of the roll forming process (Bhattacharyya et al., 1984). The edge of the sheet moves verti-
cally upward and horizontally inward toward the center, causing it to elongate
in the longitudinal direction, in contrast with the center of the sheet, which
moves horizontally and experiences longitudinal shrinkage, although the change
in thickness is often small, especially when the sheet is not significantly wide
(Abeyrathna et al., 2016).

The deformation of a flange in U-profile roll forming is shown in Figure 1.5a.
The roll forming is done at the three stations at B, C and D. Assuming the width
of flange (*a*) remains unchanged, i.e., the flange will rotate uniformly, the edge
will lie on a smooth arc ABCD, as shown in Figure 1.5a. The arc is clearly lying
on a cylindrical surface having radius *a* and axis OZ. The process is evaluated
at the very minute where the edge has shifted from point P to point P' along the
z axis at a distance *dz* and along the cylinder *dg* (Figure 1.5b). Now,

$$dg = ad\theta \tag{1.1}$$

and

$$ds = \sqrt{(ad\theta)^2 + dz^2} \tag{1.2}$$

where *ds* is the original length of the flange. At the sheet's mid-surface, the longitudinal strain is expressed as

$$\varepsilon_m = \ln\frac{ds}{dz} = \ln\left[\left(\frac{ad\theta}{dz}\right)^2 + 1\right]^{1/2} \approx \frac{1}{2}\left(\frac{ad\theta}{dz}\right)^2 \tag{1.3}$$

This indicates that the longitudinal strain increases with increasing flange length
and rate of change in the bend angle. Equation 1.3 is derived purely based on
the geometry of the sheet; the material properties, geometry of the roller and

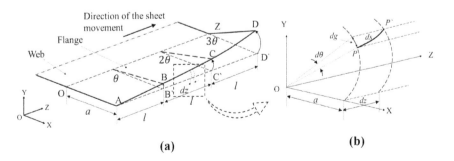

(a)   **(b)**

FIGURE 1.5 (a) Schematic representation of the flange movement during roll forming
and (b) magnified view of the incremental flange length.

thickness of the sheet are not taken into consideration. Experiments have demonstrated that the forming length depends on the thickness of the sheet and the material qualities, whereas the longitudinal edge strain depends on the diameter of the forming rollers. However, equation 1.3 can be used for the initial optimization of the roll forming sequence (flower pattern) and minimizing the longitudinal strain.

1.2.1.2 Longitudinal Bending and Bending Back

Upon entering the roll gap, the metal sheet undergoes longitudinal bending at its edge, as shown in Figure 1.6a. This makes the deformation along the edge nonuniform in nature. This additional local deformation at the region, adjacent to the roll, changes the longitudinal strain distribution. The longitudinal strain at the top and bottom surfaces is determined by the combined effects of the mid-surface strain (ε_m) and bending strain $\left(\varepsilon_b = \pm \dfrac{t \times \beta}{2h} \right)$, which arises due to the curvature of the flange (Abeyrathna et al., 2017). If the top surface is concave (as in Figure 1.6a), ε_b is compressive, and at the bottom surface, ε_b is tensile as indicated in Figure 1.6b. Now the total strain of the surfaces can be expressed as

$$\varepsilon_{top} = \varepsilon_m - \varepsilon_b \tag{1.4}$$

and

$$\varepsilon_{bottom} = \varepsilon_m + \varepsilon_b \tag{1.5}$$

The longitudinal strain experienced at both the top and bottom surfaces can be experimentally derived using a strain gauge fitted to both the surfaces as indicated in Figure 1.7a, and the measured strain usually shows the distribution shown in Figure 1.7b. Experimentally measured total strain at the top surface is very small in between the forming stations and then becomes compressive or negative as the sheet moves towards the roll gap. This is due to the formation of concave upward curvature as the sheet starts to climb the roll.

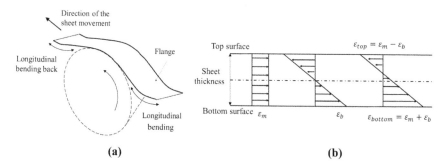

(a)   **(b)**

FIGURE 1.6 (a) Longitudinal bending and bending-back of the flange over the roll and (b) distribution of the longitudinal and bending strain across the thickness.

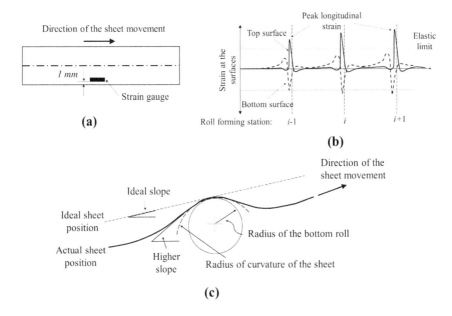

FIGURE 1.7 (a) Location of the strain gauge on the sheet, (b) longitudinal strain distribution at the top and bottom surfaces of the sheet was measured utilizing strain gauges, (c) deviation of the flange from the ideal path while passing over the roll.

Next, when the sheet passes over the roll, it bends positively to form a downward concave curvature, resulting in a high positive peak in the longitudinal strain. Subsequently, a diminished adverse deformation (characterized by an upward concave shape) arises as a consequence of the springback phenomenon. Furthermore, while the sheet goes through the forming station, it does not entirely wrap around the bottom roller. Thus, the actual radius of curvature of the sheet is much larger than the radius of the bottom roll, and the forming path of the sheet near the roller deviates from the helical path, as described in Figure 1.7c. Though the deviation of the path is small, the change in slope is very high, indicating a significant increase in peak strain. The occurrence of shape defects in the roll forming process is strongly correlated with the peak longitudinal strain. Increasing the diameter of the bottom roll can effectively mitigate the magnitude of the peak strain. In practice, the scope for increasing the diameters of rolls is limited.

1.2.1.3 Transverse Elongation and/or Shrinkage and Shear Along the Plane of the Sheet

This type of redundant deformation often occurs when sheets with wide cross-sections are roll formed. When the sheet passes through the forming stations and deforms through transverse bending, the edge of the sheet is pulled toward the center by the rotating rolls along the transverse direction. This results in shear deformation along the plane of the sheet (Panton, Duncan and Zhu, 1996). If the sheet is very wide, a large tensile force is generated along the transverse direction to pull the edge toward the center. The application of a significant transverse tensile force results in the elongation of the sheet metal in the transverse direction. Moreover, if the convex roll pushes the surface of the sheet with considerable force, the thickness of the sheet may reduce locally and cause transvers elongation at the bend line.

1.3   ANALYSIS OF THE ROLL FORMING PROCESS

1.3.1   DEFORMATION LENGTH AND BEND ANGLE

The analysis of the roll forming process is predominantly empirical or semi-empirical in nature, primarily because of the intrinsic complexity that is involved with this particular manufacturing process. The deformation of the material occurs up to a specific distance before reaching the forming rollers as shown in Figure 1.8a. This distance is called deformation length, and it is an essential factor in defining the gap between two forming stations (Bhattacharyya et al., 1984).

While passing through a forming station, the sheet bends as well as stretches. Assuming the sheet is rigid and perfectly plastic and the bending occurs only along the axis parallel to the longitudinal direction (z-axis in Figure 1.18b, neglecting any chances of out-of-plane bending), the plastic deformation per unit length of a bend can be mathematically represented as

$$W_{bend} = \frac{1}{4}\sigma_{YP}t^2\theta \tag{1.6}$$

where σ_{YP}, t and θ are the yield strength of the sheet, thickness of the sheet and bend angle, respectively. Plastic work per unit length for stretching can be represented as

$$W_{stretch} = \frac{1}{6}\sigma_{YP}at^3\left(\frac{d\theta}{dz}\right)^2 \tag{1.7}$$

where a is the width of the flange. Therefore, the total work done per bend (W_{total}) can be evaluated by integrating ($W_{bend} + W_{stretch}$) over the deformation length (L). L can be derived by reducing the energy needed for bending and stretching as

$$L = h\sqrt{8a\theta_f / 3t} \tag{1.8}$$

where θ_f is bend angle in one forming station. According to equation 1.8, L is only determined by geometric characteristics, including flange width, sheet thickness, and bend angle, and is not influenced by material properties. This may not be true in practice as elastic buckling, out-of-plane bending and elastic recovery/

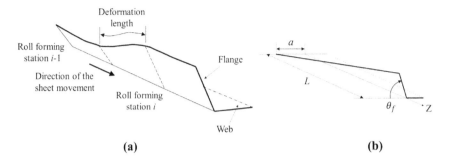

FIGURE 1.8   (a) Schematic representation of the deformation length in roll forming and (b) geometrical representation of the deformation length.

springback of the flange were neglected in this analysis. Nevertheless, equation 1.8 has been widely accepted for estimating roll forming station distance.

Considering an intermediate forming station where the bend angle changes from θ_1 to θ_2, the distribution of the bend angle should be as represented in Figure 1.9a, but the roll imposes geometrical restriction on the bend angle (Panton, Zhu and Duncan, 1992). Therefore, the curve representing bend angle distribution has three distinct zones. In region I, the angle of the bend remains constant, whereas in region II, the distribution of the bend angle may be described by the given equation 1.9 and in region III it follows the geometry of the roll as indicated in Figure 1.9b (Bhattacharyya et al., 1984):

$$\theta(z) = \theta_1 + \frac{3t}{8a^3}(z - z_1)^2 \qquad (1.9)$$

Hence, the deformation length in reality has at least two distinct regions. It is evident from Figure 1.9b that in region II, the curve exhibits concave upward behavior, while in region III, it has the opposite characteristic. Therefore, the maximum of $\dfrac{d\theta}{dz}$ always occurs at the boundary of regions II and III, and the peak longitudinal tensile strain appears at the boundary (from equation 1.3) where the sheet first touches the roll.

1.3.2   EFFECTS OF GEOMETRIC VARIABLES

Numerous geometric factors, including flange width, bend angle increment, sheet thickness and roll diameter, have critical impacts on the roll forming process (Zhu, Panton and Duncan, 1996) (Lindgren, 2007). The overall impact of these factors is responsible for determining the magnitude of peak longitudinal strain, which subsequently affects the quality of the finished product (Han et al., 2001). Typically, there is a positive correlation between the magnitude of peak longitudinal strain and the corresponding rise in flange width (a). This is also evident from equation 1.3. However, when the flange width increases, the slope of the

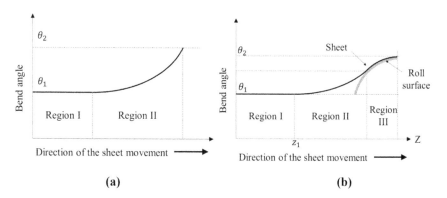

FIGURE 1.9 Distribution of the bend angle (a) without geometrical restriction and (b) with the geometric limitation that the roll imposes.

bend angle $\left(\dfrac{d\theta}{dz}\right)$ at the boundary between region II and region III decreases (from equation 1.9). Hence, the ultimate impact of an increase in flange width on the peak longitudinal strain is determined by the combined influence of these two opposite factors.

In practice, the peak longitudinal strain exhibits an initial rise in response to an augmentation in flange width, reaches a maximum value at the critical flange width and then decreases. Again, for a fixed a, the slope, $\dfrac{d\theta}{dz}$ varies linearly with z and sheet thickness t. If the sheet thickness increases the slope $\dfrac{d\theta}{dz}$ as the boundary increases, the peak longitudinal strain exhibits an increase. Moreover, an increase in $\dfrac{d\theta}{dz}$ shortens region II as the bend angle curve, represented by equation 1.9, now touches the roll surface earlier. Hence, deformation length becomes shorter for thicker sheets, and the distance between the stations can be reduced. The increment in bend angle between the two forming stations also adversely impacts the quality of product, and the effect is more pronounced if the bend angle increment is higher.

Conversely, an increase in the roll diameter leads to a reduction in the peak longitudinal strain. As noted in Figure 1.9b, the geometric restriction provided by the roll reduces if its diameter increases. The appearance of the roll surface becomes flatter, and $\dfrac{d\theta}{dz}$ decreases. Therefore, the severity of the deformation is diminished, and peak longitudinal strain drops. It should be noted that increasing the diameter of the rolls alone is neither very economical nor very effective in reducing the peak longitudinal strain as multiple factors are involved in complex combinations in order to define the final product quality.

1.3.3 DETERMINATION OF ROLL-FORMING LOADS

The roll load or torque is another crucial element in roll forming. Prior under-standing of the roll load is required for smooth functioning of the entire pro-cess. Broadly, the roll load can be classified into three elements: (a) load due to transverse bending and longitudinal stretching (P_{tb}), (b) load due to longitudi-nal reverse bending (P_{lb}) and (c) clamping load (P_c) due to biting of the sheet by the top and bottom rolls at the forming station, which can set by adjusting the clearance between the rolls (Bhattacharyya et al., 1987). Assuming the process is quasi-static, the pressure distribution along the roll axis exhibits a linear variation from the entry point to the plane that connects the roll centers and the height of the bend rises linearly, the forming load for transverse bending and longitudinal stretching can be calculated from equation 1.10:

$$P_{tb} = \sigma \sqrt{\frac{2t^3\theta^3 a}{3\sin^2\theta}} \tag{1.10}$$

Longitudinal bending occurs due to the downward deflection of the partially formed sheet located between the stations, as shown in Figure 1.10. This down-ward deflection is the result of peak longitudinal strain followed by partial recov-ery. Therefore, a vertical lift is required for the semi-formed part to enter the next forming station. Considering the semi-form sheet as cantilever of length $(D-x)$, fixed at the forming station, vertical lift (h) can be determined as

$$h = \frac{D(D-2x)}{2(R-r)} \tag{1.11}$$

where D, R and r are the distance between two forming stations, radius of cur-vature of the deformed sheet and radius of the roll, respectively. This h causes reverse longitudinal bending and puts an additional load on the rolls, which can be expressed as

$$P_{lb} = \frac{3hEI}{(D-x)} \tag{1.12}$$

where E is the Young's modulus of the sheet and I is the second moment of inertia.

Therefore, the load required for the reverse bending increases with the decrease in cantilever length at the point of contact, $(D-x)$, and with the increase in h, which itself is a function of roller diameter, distance between two forming stages and material properties.

Theoretical load calculation yields a conservative approximation of the origi-nal roll load. In practice, the yield strength of the material at the bend increases while passing through a roll due to strain hardening. To eliminate further com-plexity in calculating roll load, designers often consider a higher yield strength based on the strain-hardening characteristics of the sheet. Despite this, several factors such as bending of flange, buckling and other redundant deformation are

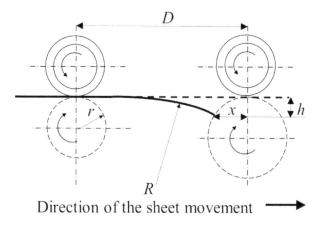

Direction of the sheet movement $\longrightarrow$

FIGURE 1.10   Longitudinal bending of the semi-formed sheet.

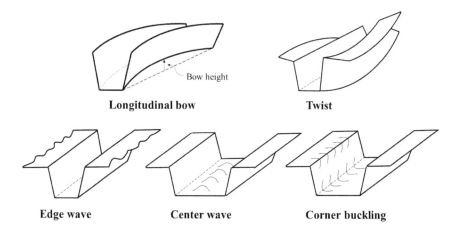

FIGURE 1.11   Various possible defects in roll-formed parts.

still neglected in the calculation. So, the equations 1.10 and 1.12 loads should be used with caution, especially when dealing with sheets having higher thickness and/or shorter flange width.

1.4   DEFECTS IN ROLL FORMING

The presence of redundant deformations, resulting from the three-dimensional attributes of the roll forming process, induces undesirable strain in the sheet metal. Consequently, this leads to the occurrence of imperfections in the ultimate roll-formed components including longitudinal bow, twist, edge and center wave, thinning, cracking and springback (Figure 1.11). The generation of these defects is further discussed in the current section.

1.4.1 LONGITUDINAL BOW AND TWIST

The occurrence of these defects can be attributed to the nonuniform distribution of longitudinal strain in the transverse direction, arising due to longitudinal elongation and shrinkage of the sheet as described in Section 1.2.1.1. These are the most typical defects seen in roll-formed components (Bui and Ponthot, 2008). The highest longitudinal strain occurs at the deformation zone close to the roll gap as shown Figure 1.7b. In the event that the peak longitudinal strain surpasses the elastic limit (marked in Figure 1.7b), permanent longitudinal elongation occurs.

The longitudinal peak strain exhibits variation in the flange-to-center direction as a consequence of the incremental characteristics inherent in the roll forming process, rendering its avoidance unattainable. Nevertheless, the longitudinal strain magnitude and nonuniformity may be mitigated by the implementation of appropriate design strategies in the roll forming process. One way is to increase the number of forming stations, thereby reducing the severity of the deformation in each station. However, an excessive number of passes raises the capital cost. Alternatively, the roll profile and its position can be adjusted to ensure the smooth flow of the material. The entry guides and intermediate guides can be utilized in order to control the special length of the flow line and provide additional support to the flange. This can effectively reduce the longitudinal strain.

1.4.2 EDGE WAVE

Edge waves are generated due to the elastic or elastoplastic buckling of the sheet metal. This is also a very common defect in roll forming, especially for thin sheets. As discussed before, the occurrence of longitudinal elongation near the edge of the sheet is unavoidable. Additionally, the sheet can elongate while inside the roll gap. This elongated edge should shrink after leaving the roll gap in order to keep the product straight.

In the case of very high localized elongation at the edge, sufficient shrinking is often difficult to achieve. Hence, in the presence of significant compressive stress, buckling may manifest around the edge of the sheet. The edge wave can be reduced by reducing the peak longitudinal strain. Local heating can also be employed to reduce the high plastic deformation of the edge. Preforming the sheet by using a preforming roller set prior to the roll forming can introduce longitudinal elongation toward the center of the sheet. This longitudinal elongation can compensate for the nonuniform longitudinal elongation that occurs during roll forming, thereby reducing the chances of edge waves.

1.4.3 CENTER WAVE AND CORNER BUCKLING

Center waves or pocket waves develop due to the nonuniform transverse tensile stress during roll forming. When the sheet metal traverses the roll gap, transverse tensile stresses are generated as described in section 1.2.1.3. The magnitudes of the transverse tensile stresses are greater in the central region than at the edges.

If the bend lines are situated close to the center, they can significantly elongate along the transverse direction but shrink along the longitudinal direction. As a result, compressive stresses are generated on flat portions of the component such as webs and flanges. When the compressive stresses are too high, elastic buckling occurs, resulting in the formation of center waves in the roll-formed component. Center wave occurs when the stiffness of the flat region is less than that of the corners. The corners have low stiffness, the sheet has low thickness, buckling takes place at the corner and the wave at the corners appears as herringbone. Reducing the transverse tensile stress by roll forming sequence optimization can prevent the corner buckling and the center wave.

1.4.4 EDGE CRACKING AND SPLITTING

When the transverse elongation exceeds the critical limit, cracking and splitting take place. This is more severe for materials with low formability and low thickness and if the product has multiple sharp corners.

1.4.5 SPRINGBACK

Springback refers to the phenomenon of elastic recovery shown by a material after the removal of forming stresses (Bui and Ponthot, 2008). In roll forming, deformation in sheet metal is mostly attributed to transverse bending, along with the complex action of other redundant deformations. The shape of the final roll formed part changes as a result of springback. Figure 1.12a schematically represents the springback at the cross-section of a roll formed part. In equation 1.13, φ is the actual angle of the final product, while φ' is the desired angle defined by the geometry of the rolls. The springback can be estimated as

$$\text{Springback} = \varphi' - \varphi \qquad (1.13)$$

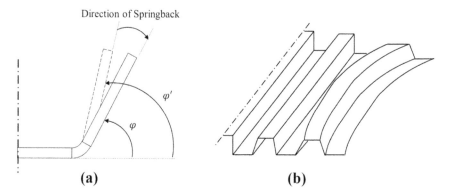

(a)   **(b)**

FIGURE 1.12   (a) Springback in a roll-formed part and (b) nonuniform springback in a corrugated sheet.

The simplest way to overcome springback is to overbend the sheet during roll forming at an angle similar to the springback.

The estimation of springback is more complex than equation 1.13. For example, when the sheet is very wide, transverse tensile stress is higher at the center than at the edges. Therefore, there is more springback at the bend situated at the edges than in the center part of the sheet. That is, nonuniform springback takes place along the transverse direction of the sheet (Figure 1.10b). This phenomenon often occurs in the case of corrugated sheets and results in additional defects such as cross bow. Roll profiles should be optimized in order to nullify the effect of the springback.

1.4.6 END FLARES

Additional springback is often experienced by the roll-formed part's entrance and exit cross-section. The entrance and exit of each component exhibit extra distortion when a continuous roll formed object is sliced into parts. This is due to the distribution of residual shear stress along the direction of the sheet thickness. The lead-in portion of the concave roll directs the whole sheet into the roll gap as it approaches it. The sheet travels along the flow lines that follow the contour of the concave roll because its width is greater than the roll gap.

The sheet's inner layer, which is closest to the convex roll's surface, travels more slowly than the layer in touch with the concave roll when it enters the roll gap. However, it is essential that all layers depart the roll gap simultaneously in the longitudinal direction, resulting in the occurrence of inverse shear deformation during the exit process. The act of shearing and then reversing the shearing of a sheet throughout its thickness induces residual shear stress inside the sheet metal, with a particular emphasis on the edges. During cut off, these residual stresses are released, and flare deformation takes place. The entry edge leans to deform in a way that closes the cross-section, while the exit edge opens it (Figure 1.13). The

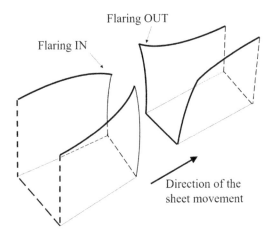

FIGURE 1.13   Schematic representation of End flare

end flare can be reduced by providing over-bending and bending back before the final forming station in order to nullify the residual shear stresses accumulated in the preceding stations.

1.4.7 EFFECTS OF RESIDUAL STRESSES ON ROLL-FORMED PRODUCT QUALITY

Sheet metals are produced from metal blocks by thickness reduction using rolling. Frequently, these sheets undergo temper rolling, roller leveling or tension leveling, and these processes generate residual stresses inside the sheet metal. The process of rolling typically decreases thickness, which in turn induces tensile residual stresses on the surface of the material and compressive residual stresses in the core of the sheet along the thickness direction (Abvabi et al., 2015). The magnitude of residual stresses is directly proportional to the extent of thickness decrease.

As a consequence of strain hardening, the material yield strength increases. As previously mentioned, the longitudinal strain experienced during the roll forming process exhibits variations over the width, transitioning from tensile strain at the flange to compressive strain at the center. The presence of previous tensile residual stress during roll forming increases the overall tensile stress experienced at the surface of the sheet. Consequently, the permanent longitudinal strain on the surface becomes more pronounced and exhibits a tensile nature throughout the width of the sheet.

Furthermore, the amount of plastic deformation brought on by the longitudinal strain during roll forming is decreased by the sheet's increased yield strength. Increased material strength and residual stresses work together to lessen the discrepancy between the stresses at the sheet's edge and center, resulting in a more even distribution of stresses across the sheet's width. This leads to decrease in the height of the longitudinal bow in a roll-formed part. This effect is quite opposite in case of springback, which is basically the elastic recovery after roll forming. Moreover, any preexisting residual stresses increase the net residual stress at the front and end portion of the roll formed part, thereby increasing the end flare.

Researchers have found that materials with low strain hardening can be easily roll formed (Deb, Panigrahi and Weiss, 2021) (Badr et al., 2015) (Marnette, Weiss and Hodgson, 2014) (Deole, Barnett and Weiss, 2018). The enhanced formability seen in roll forming may be attributed to the progressive character of the process, which distinguishes it from simple bending deformation. Moreover, the rollers provide geometrical constraint, thereby restricting the formation of diffused neck due to the geometric limitations imposed by the top and bottom rolls. This allows the roll forming of materials with lower formability and ductility. Low ductile materials such as ultra-high-strength steel, titanium alloys and cryo-rolled aluminum alloys are being effectively roll formed to different cross-sections and profile radii.

1.5   RECENT DEVELOPMENTS IN ROLL FORMING

1.5.1   FLEXIBLE ROLL FORMING

One type of roll forming, flexible forming, allows for the manufacturing of longitudinal components with variable widths and depths, while traditional roll forming can only produce longitudinal components with fixed cross-sections (Sreenivas et al., 2022). These profiles are useful for automotive applications to produce bumpers, cross members and frame rails. The variations are achieved through introducing curvature to the flange and/or web by providing additional rotation and translation into the rolls. As a result, the flange experiences longitudinal elongation in certain length and longitudinal compression in others. If the longitudinal deformation is not enough to accommodate the stretching and compression of the flange during transverse bending, then the web tends to bend and causes web warping; web warping can be eliminated by constraining the web using blank holder dies. As a result, unbalanced compressive stress may generate in the flange, which causes buckling or wrinkling of the edge (Woo et al., 2019). The wrinkling becomes severe with increasing flange width and decreasing sheet thickness.

1.5.2   MICROROLL FORMING

Microroll forming is a very efficient method for producing microchannels. Microchannels are basically used as bipolar plates in fuel cells. The existing process in industry is to produce microchannels through micro-EDM, micro-ECM, microstamping and hydroforming. These processes either are costly or cause excessive thinning of the sheets. Hence, microroll forming is a promising solution for producing bipolar plates (Abeyrathna et al., 2019). This process enables roll forming of sheets with thickness of 0.1 mm. Special attention should be given to the alignment of rolls as a small misalignment may affect the final product quality. Local thinning, uneven profile and longitudinal bow are among the often-seen shape defects that typically manifest in microroll-produced components. These defects are generally the result of roll inaccuracies, positioning and surface roughness. High-precision manufacturing processes are required to produce more accurate tools with better surface finish. Thus, the overall tooling cost will be improved.

1.5.3   BRANCHING IN ROLL FORMING

Branch profiles can be produced by roll forming through the principle of linear flow splitting (Groche, Vucic and Jöckel, 2007). This is a modification of conventional roll forming in which the splitting rolls have obtuse angles, and additional supportive rolls are introduced to cause splitting of the edge of the sheet. Localized generation of high compressive hydrostatic stress results in surface enlargement of the sheet and leads to splitting. The generated profile has one web but two flanges. This process produces a whole structure using sheet metal without the assistance of any joining process. The roll-formed properties of the branched sheet exhibit distinct features from those of a flat sheet. Recent research has shown modest variations in the magnitude and location of the maximum longitudinal tensile strain between branched sheets and flat sheets (Görtan et al., 2009). However, fundamental studies are still required in order to establish the process in industry to produce more intricate product geometry.

1.6   SUMMARY AND CONCLUSIONS

The present chapter provides an overview of the basic principles underlying the roll forming process, with a particular focus on the primary aspects of the operation. Roll forming is a highly adaptable manufacturing technique in which a longitudinally oriented sheet undergoes progressive bending as it passes through a series of consecutive forming stations, and it is capable of producing various cross-sections. Transverse bending is the primary mode of deformation, which limits the development of a diffused neck. This characteristic allows for the roll forming of materials with greater strength but lower ductility, such as ultra-high strength steels, titanium alloys, and ultrafine-grained aluminum. Major deformation is restricted at the bend region.

During the process of roll forming, it is essential to ensure that the spacing between the rollers remains consistent with the thickness of the sheet being used. Additionally, the determination of the number of roll forming stations is contingent upon the intricacy of the cross-sectional design of the final product. Moreover, the activation of redundant deformation modes is seen in roll forming as a result of the inherent three-dimensional characteristics of the process. These deformations often overlap with the transverse bending and result in various shape defects such as longitudinal low, twist and waves. Peak longitudinal strain has been shown to be a significant factor in determining the quality of the product. Analytical models were presented in this chapter of longitudinal strain, the development of flower pattern and roll load to comprehend the deformation behavior during roll forming. The effects of the geometric parameters and materials condition on the shape defects were explained. The chapter concludes by discussing variations in the conventional roll forming process and the evolution of related technologies.

REFERENCES

Abeyrathna, B. *et al.* (2016) 'An extension of the flower pattern diagram for roll forming', *The International Journal of Advanced Manufacturing Technology*, 83(9–12), pp. 1683–1695. doi:10.1007/s00170-015-7667-0.

Abeyrathna, B. *et al.* (2017) 'Local deformation in roll forming', *The International Journal of Advanced Manufacturing Technology*, 88(9–12), pp. 2405–2415. doi:10.1007/s00170-016-8962-0.

Abeyrathna, B. *et al.* (2019) 'Micro-roll forming of stainless steel bipolar plates for fuel cells', *International Journal of Hydrogen Energy*, 44(7), pp. 3861–3875. doi:10.1016/j.ijhydene.2018.12.013.

Abvabi, A. *et al.* (2015) 'The influence of residual stress on a roll forming process', *International Journal of Mechanical Sciences*, 101–102, pp. 124–136. doi:10.1016/j. ijmecsci.2015.08.004.

Badr, O.M., Rolfe, B. and Weiss, M. (2018) 'Effect of the forming method on part shape quality in cold roll forming high strength ti-6al-4v sheet', *Journal of Manufacturing Processes*, 32, pp. 513–521. doi:10.1016/j.jmapro.2018.03.022.

Badr, O.M. *et al.* (2015) 'Forming of high strength titanium sheet at room temperature', *Materials & Design*, 66, pp. 618–626. doi:10.1016/j.matdes.2014.03.008.

Bhattacharyya, D. *et al.* (1984) 'The prediction of deformation length in cold roll-forming', *Journal of Mechanical Working Technology*, 9(2), pp. 181–191. doi:10.1016/0378-3804(84)90004-4.

Bhattacharyya, D. *et al.* (1987) 'The prediction of roll load in cold roll-forming', *Journal of Mechanical Working Technology*, 14(3), pp. 363–379. doi:10.1016/0378-3804(87)90019-2.

Brunet, M., Lay, B. and Pol, P. (1996) 'Computer aided design of roll-forming of channel sections', *Journal of Materials Processing Technology*, 60(1–4), pp. 209–214. doi:10.1016/0924-0136(96)02331-x.

Bui, Q.V. and Ponthot, J.P. (2008) 'Numerical simulation of cold roll-forming processes', *Journal of Materials Processing Technology*, 202(1–3), pp. 275–282. doi:10.1016/j. jmatprotec.2007.08.073.

Deb, S., Panigrahi, S.K. and Weiss, M. (2021) 'Understanding material behaviour of ultrafine-grained aluminium nano-composite sheets with emphasis on stretch and bending deformation', *Journal of Materials Processing Technology*, 293, p. 117082. doi:10.1016/j.jmatprotec.2021.117082.

Deole, A.D., Barnett, M.R. and Weiss, M. (2018) 'The numerical prediction of ductile fracture of martensitic steel in roll forming', *International Journal of Solids and Structures*, 144–145, pp. 20–31. doi:10.1016/j.ijsolstr.2018.04.011.

Görtan, M.O. *et al.* (2009) 'Roll forming of branched profiles', *Journal of Materials Processing Technology*, 209(17), pp. 5837–5844. doi:10.1016/j. jmatprotec.2009.07.004.

Groche, P., Vucic, D. and Jöckel, M. (2007) 'Basics of linear flow splitting', *Journal of Materials Processing Technology*, 183(2–3), pp. 249–255. doi:10.1016/j. jmatprotec.2006.10.023.

Han, Z.W. *et al.* (2001) 'The effects of forming parameters in the roll-forming of a channel section with an outer edge', *Journal of Materials Processing Technology*, 116(2–3), pp. 205–210. doi:10.1016/s0924-0136(01)01041-x.

Kiuchi, M. (1989) 'CAD system for cold roll-forming', *CIRP Annals*, 38(1), pp. 283–286. doi:10.1016/s0007-8506(07)62704-8.

Lindgren, M. (2007) 'An improved model for the longitudinal peak strain in the flange of a roll formed u-channel developed by Fe-Analyses', *Steel Research International*, 78(1), pp. 82–87. doi:10.1002/srin.200705863.

Marnette, J., Weiss, M. and Hodgson, P.D. (2014) 'Roll-formability of cryo-rolled ultrafine aluminium sheet', *Materials & Design*, 63, pp. 471–478. doi:10.1016/j. matdes.2014.06.036.

Panton, S.M., Duncan, J.L. and Zhu, S.D. (1996) 'Longitudinal and shear strain development in cold roll forming', *Journal of Materials Processing Technology*, 60(1–4), pp. 219–224. doi:10.1016/0924-0136(96)02333-3.

Panton, S.M., Zhu, S.D. and Duncan, J.L. (1992) 'Geometric constraints on the forming path in roll forming channel sections', *Proceedings of the Institution of Mechanical*

Engineers, Part B: Journal of Engineering Manufacture, 206(2), pp. 113–118. doi:10.1243/pime_proc_1992_206_063_02.

Panton, S.M., Zhu, S.D. and Duncan, J.L. (1994) 'Fundamental deformation types and sectional properties in roll forming', *International Journal of Mechanical Sciences*, 36(8), pp. 725–735. doi:10.1016/0020-7403(94)90088-4.

Sreenivas, A. *et al.* (2022) 'Longitudinal strain and wrinkling analysis of variable depth flexible roll forming', *Journal of Manufacturing Processes*, 81, pp. 414–432. doi:10.1016/j.jmapro.2022.06.063.

Sweeney, K. and Grunewald, U. (2003) 'The application of roll forming for automotive structural parts', *Journal of Materials Processing Technology*, 132(1–3), pp. 9–15. doi:10.1016/s0924-0136(02)00193-0.

Troive, L. and Ingvarsson, L. (2008) 'Roll forming and the benefits of ultrahigh strength steel', *Ironmaking & Steelmaking*, 35(4), pp. 251–253. doi:10.1179/174328108x301714.

Woo, Y.Y. *et al.* (2019) 'Shape defects in the flexible roll forming of automotive parts', *International Journal of Automotive Technology*, 20(2), pp. 227–236. doi: 10.1007/s12239-019-0022-y.

Zhu, S.D., Panton, S.M. and Duncan, J.L. (1996) 'The effects of geometric variables in roll forming a channel section', *Proceedings of the Institution of Mechanical Engineers, Part B: Journal of Engineering Manufacture*, 210(2), pp. 127–134. doi: 10.1243/pime_proc_1996_210_098_02.

2 A Systematic Review of the Rotary Piercing Process

Pravin Pawar, Amaresh Kumar, and Raj Ballav

2.1 INTRODUCTION

Welded pipes and seamless pipes are the two types of metal pipes. Metal sheets are bent and welded to create welded pipes, whereas piercing is used to create seamless pipes [1–3]. It is widely accepted that seamless pipe has more advantages than welded pipe, including higher pressure ratings; uniformity of shape; and better material qualities, structural strength, and fatigue capacity under load [4]. Seamless pipe has greater dependability, structural rigidity, and prolonged service life under high pressure or repetitive loads, and therefore, seamless pipes are more frequently employed in the oil and gas sectors than welded pipes [1, 5].

Because of their exceptional circumferential uniformity and excellent resistance to internal pressure and torsion, seamless steel tubes are a highly desirable material [5]. In rotary piercing, two skewed rolls rotating in the same direction feed a heated cylindrical workpiece into a plug. The rolls provide the workpiece with both rotational and translation movement since they are inclined and positioned on opposite sides of the workpiece [1, 4]. Revolving piercing is another term for Mannesmann piercing, which was invented by the Mannesmann brothers in the year 1886 [6, 7]. More than a century later, Sumitomo Metal Industries, Ltd. created a cone-type piercing mill with a steep toe angle [8]. The following three methods of rolling are typically used to make seamless pipes. To begin with, pierce rolling is used to create thick tubes from bars. Next, elongate rolling is used to create thin pipes from the thick pipes. Reduce rolling is used to create seamless pipes, which are the end products, from the thin pipes. In addition to piercing by the plug, Calmes (1967) introduced push roll piercing, which rolls the material by pressing it in the rolling direction. The material is not prone to shearing deformation brought on by rotary forging because neither the material nor the plug rotates while being pierced [9]. The Mannesmann piercing mill guiding principle is the effective use of the effects of rotary forging. In this piercing process, redundant shear deformations such as longitudinal shear strain, shear strain from the surface twist, and circumferential shear strain are inescapable [6]. Figure 2.1 shows the schematic diagram of the rotary piercing process.

DOI: 10.1201/9781003441755-2

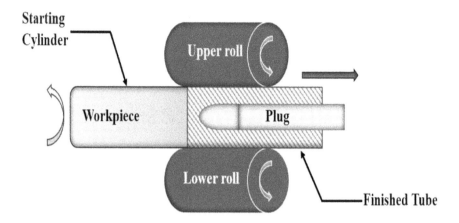

FIGURE 2.1   Schematic diagram of the rotary piercing process.

The Mannesmann effect, which is tensile stress at the front edge of the mandrel that results in a crack or hole owing to concentrative damage, is exploited in Mannesmann roll piercing. The Stiefel supporting or guiding shoe, the Mannesmann guiding roll, and the Diescher guiding disk are only a few examples of the various forms of supporting equipment [3, 10].

Three processes—roll piercing, also known as pierce rolling, elongation rolling, and reduction rolling—make up the fabrication sequence for seamless pipes. Press roll piercing, corn-type rotary piercing, and barrel-type rotary piercing are the three different types of piercing. Mannesmann piercing is a barrel-type rotary piercing in which the roll has a barrel-like shape, in contrast to a cone-like shape in cone-type rotary piercing. [3, 9, 11]. Piercing and extrusion are used to create seamless tubes. Rotary tube piercing involves two or three rollers. Each type features components that are specifically created, such as specific guides at the rolling zones. Additionally, the two basic roller shapes are Diescher and conical [12, 13].

In terms of rolling, there are primarily three-roll rolling mills without guiding devices and two-roll skew rolling mills with two guiding devices (roll and flat) [3]. The rotary tube piercing of seamless tubes is a difficult procedure that requires high temperatures between 1200 and 1300 °C [14]. The primary process variables for rotary tube piercing include the feed angle, the temperature of the billets, the plug insertion depth, the area reduction ratio, the cross angle, the shape of the roll and plug, and the lubrication condition [15]. This chapter's three sections will cover the previous literature on theoretical studies of the rotary piercing process, experimental studies of rotary piercing, and comparative studies of experimental methods versus the finite element method (FEM) of rotary piercing (Figure 2.2).

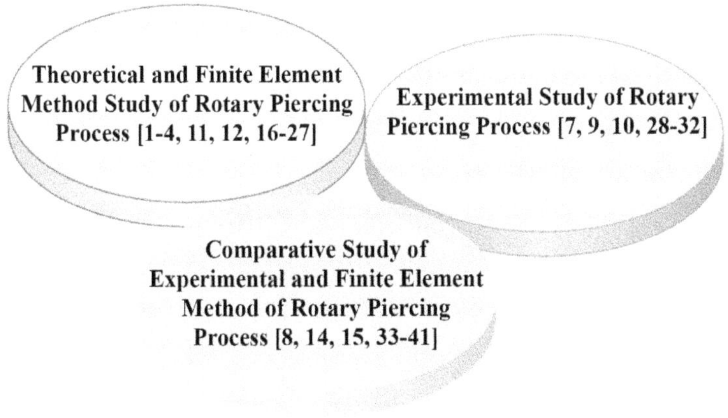

Theoretical and Finite Element Method Study of Rotary Piercing Process [1-4, 11, 12, 16-27]

Experimental Study of Rotary Piercing Process [7, 9, 10, 28-32]

Comparative Study of Experimental and Finite Element Method of Rotary Piercing Process [8, 14, 15, 33-41]

FIGURE 2.2   Different studies on the rotary piercing process.

2.2 THEORETICAL AND FINITE ELEMENT METHOD STUDIES OF THE ROTARY PIERCING PROCESS

Top et al. [1] used LS-Dyna software to create a 3D model of the rotary piercing process and ran numeric simulations using the arbitrary Lagrangian–Eulerian (ALE) formulation. The simulation results were compared with experimental data and demonstrated the efficacy of the fluid-structure interaction approach in modeling the piercing procedure. By running the numerical simulations at three various plug diameters, they looked at the impact of the pipe thickness [2].

Tube rolling in a planetary rolling mill is examined using the FEM approach and the dual-stream function and upper bound method. The rolling load, exit velocity of the workpiece, and energy lost in deformation are all impacted by the offset angle of the roller. The efficiency of the rolling process can be improved by increasing the offset angle, but doing so can also result in an increase in rolling load and energy loss.

Three-dimensional metal-forming issues can be analytically analyzed by combining the upper bound method and dual-stream functions. Cho et al. [3] used FEM to conduct a comparison study of Mannesmann roll piercing with two different guiding tools, Diescher's guiding disks and Stiefel's guiding shoes. For this, they used a rigid-thermoviscoplastic FEM with intelligent tetrahedron remeshing. In terms of material deformation history, ellipticity change at the roll bite, and effective strain, the predictions for the two distinct guidance instruments were compared. The guiding shoes of the Stiefel are comparatively space-efficient, but the Stiefel's shoes require more careful process control and an improved geometry design.

Topa et al. [4] used an ALE framework to carry out three-dimensional numerical piercing simulations using LS-DYNA. They discovered that plug diameter and workpiece temperature have the opposite effect from the piercing velocity, which has a proportional link with the maximum stress. Because of the similarities between the stress–strain behavior of synthetic material and that of steel at high temperatures, these relationships can be applied to the piercing of steel.

Komori and Suzuki [11] analyzed press roll piercing deformation and temperature using a standard 3D rigid-plastic FEM. According to the results of the simulation, mandrel force and roll force both decrease as the plug diameter is reduced. Darki and Raskatov [12] tested and simulated a new input and output guide design for all different types of rollers. Specifically, they considered three instances including conical and Diescher and two sets of rollers, one set with three rollers and another with two. They used FEM to derive temperature distribution, total force, torque, and strain from the simulation output findings. The greater quality of the tube generated by specimen 3 is indicated by the uniform distribution of strain in the tube wall and, in contrast, the uniform distribution of temperature along the tube.

To determine the geometric parameters of the deforming region in a piercing mill, Gulyaev et al. [16] proposed a mathematical model that can be used to correct the calibration parameters of the working rollers and permit industrial adjustment of rotary rolling mills. Kajtoch et al. [17] suggested twisting layers of the material across the deformation zone as the material deforms plastically in a circumferential manner, a novel analytical methodology for calculating the torsion parameters in the material outer layers. The authors established the torsion parameter distributions for both the plastic zone and the final hollow.

The actual geometric and kinematic circumstances in the plastic zone determine the torsion parameters. The direction of the torsion vector in the piercing zone is the opposite of the hollow turning, whereas it changes signs in the wall reduction zone. The feed angle had little effect on the overall torsion parameter values in the piercing zone, but the absolute values of these parameters typically reduced as the feed angle increased in the final hollow. The final hollow layers' torsion direction was consistent with how the hollow turned [17].

Hrycaj et al. [18] described improvement, effective control, and analyses of the manufacture of seamless tubes by applying thermoviscoplastic FEM analysis. They studied the efficiency of the FEM model by considering some important aspects such as strain rate, stress distribution, adiabatic temperature, the influence of viscoplastic coefficient consistency, and the influence of incipient cavity dimensions. According to them, the finite element model developed in ASTRID software could be a significant means for secure management of piercing mills.

Joun et al. [19] studied the Mannesmann effect using a hollow cylinder model. In this simulation, the authors examined the phenomenon of hole enlargement during the piercing of hollow cylinders by two-barrel type rolls without the use of mandrels or guiding tools. They performed the simulation using rigid-thermoviscoplastic FEM and observed that as the original hole diameter decreased, there was an increase in the relative hole expansion ratio between the maximum hole diameter and the beginning hole diameter. Additionally, the hole expansion is connected to the cavity development that occurs immediately.

Mori et al. [20] proposed a simplified three-dimensional method for simulating the rotary piercing of pipes to increase the accuracy of the findings. They introduced twisting shear deformation to generalized plane-strain modeling and found similar results for the deformed workpiece acquired by FEM simulation similar to those of the experimental work for plasticine workpieces. Huang [21] utilized rigid-plastic FEM and performed a numerical simulation for cone-type rotary piercing. According to the simulation's findings, the space between the rolls and plug at the reeling section and the relief section is affected by the plug location.

Urbanski and Kazanecki [22] identified the distribution of strains in the deformation zone using finite element simulation. The material layers closest to the plug experienced significant deformation during the examined process, and the areas closest to the developed inner surface showed the greatest strain gradients. When material flows, the length of the cavity in front of the plug is determined, leading to high values of effective strain. Large cavities lead to material flaws like slivers and uneven surface texture. As a result, the process variables should be changed to ensure that the axial cavity is small. The distribution and level of strain are influenced by the plastic zone's length. However, the feed angle affects the length of the deformation zone.

Pater et al. [23] differentiated between a three-roll and a two-roll Diescher mill's piercing operation using numerical simulations. They found out that three-roll piercing causes less ovalization of the material and therefore increases the manufactured object's angular speed. The tube produced in the three-roll mill

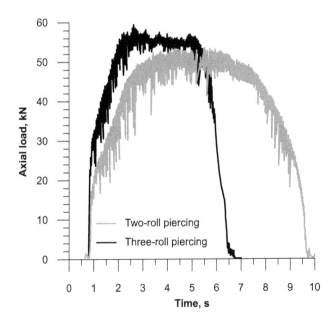

FIGURE 2.3 Distribution of axial load on the piercing plug during the piercing process [23].

has a higher temperature and a little thinner wall, which reduces the amount of energy needed for the subsequent shaping processes. The forming tools are subjected to greater forces with three-roll than with two-roll piercing.

Darki and Raskatov [24] studied the effects of roller geometry in rotary tube piercing and conducted mandrel wear analyses according to the FEM. Comparing the Diescher type with three rollers to the conical type with three rollers, they found that the average force for the Diescher type was 7.9% lower. Additionally, torque and contact surface had a direct relationship: The application of torque was increased by the push-back distribution among the two rollers. Mandrel wear was mostly caused by rotational motion, and the feed angle had an inverse relationship with the torque supplied to the rollers.

Yamane et al. [25] focused on the piercing mill and mandrel mill used in Mannesmann rolling and applied typical numerical simulation approaches. They developed a new method for material testing which can accurately calculate the processing heat. It helps to predict the material flow caused by grain boundary melting. The degree of pipe underfill may now be properly predicted for the mandrel mill using a simulation that accurately reflects the right material qualities. All this substantially aids in the stable rolling of materials made of high alloy steel. Ye et al. [26] investigated four rotary tube piercing processes and simulated them through FEM. The wall thickness was reasonably uniform but only when the feed angle was less than 12°. The wall thickness of the tube billet tended to reduce with the rise in feed angle, and the tube billet's wall thickness homogeneity declined; with the rise in plug advance, the tube billet's wall thickness homogeneity increased. Tube blanks with high roll speeds had homogeneous wall thicknesses. Bogatov et al. [27] conducted an FEM simulation of the rotary piercing method using the Deform-3D program. The values of reduction and ovality were minimum at the entrance to the deformation zone, maximum in the gorge, and again minimum at the exit from the deformation zone. In an axial direction, the surface layers of the billet deform much more severely than the inside layers.

2.3   EXPERIMENTAL STUDIES OF ROTARY PIERCING

Hayashi and Yamakawa [7] studied the effects of cross angle and feed angle on redundant shear deformations and circumferential shear strain. The reduction in circumferential shear strain was higher the greater the cross and feed angles. If the feed angle and cross angle are both adequately large, the circumferential shear strain can be completely removed. Komori and Mizuno [9] used modeling clay to develop a rotary piercing mill model for the plastic deformation of cone-shaped objects. They discovered that when the cross-angle is about 20°, the shearing strain components are virtually nil and the feed angle is inconsequential.

Blazynski and Cole [10] analyzed redundant deformations in rotary piercing. They discovered that when the plug is free to rotate, it is hard to get rid of any redundant strains, and changing the mill settings only slightly reduces redundancy. For this experimental study, two Tufnol guides were applied. The spacing of guides was adjusted to achieve the removal of ovality. Murillo-Marrodán et al. [28] created a cone-type rotational piercing process. The efficiency of feed increases with increasing cross and feed angles and decreases as the expansion ratio increases. When the cross angle is set to a lower value, this tendency stands out.

Khudheyer et al. [29] evaluated the force settings and the results of redundant-macro shear effects in two- and three-roll tube cone rotary piercers for internally-marked wax billets and lead. The testing findings showed that the three-roll process performed better than the process with the two-roll piercer. Al-Dahwi and Blazynski [30] explored rotary two-cone seamless tube roll piercing, specifically the relationships between force and redundant macro shearing parameters connected to a rational pass design. They used a variety of operating circumstances, including decelerated, accelerated, and uniform flow rates; various plug advances; and four feed angle values, to pierce wax and lead billets. When the distance along the operating zone of the pass was extended, the roll force, plug load, and roll torque increased. The billet-gorge diameter ratio increased, and a smaller feed angle was used that in turn increased homogenous strain.

Zhao et al. [31] studied rotary piercing for Ti-26 material and identified small grains with good ductility as well as one metastable phase. According to the experimental findings, the rates of expansion and contraction of the cross-sectional area were 11% and 40%, respectively. Blazynski [32] determined how the redundant shear in the pass is affected by the predefined rate of deformation. They discovered that there is a clear relationship between these strains and the geometry of pass, which causes the rate of deformation. The final product's quality was achieved by maintaining a low feed angle in conjunction with a decelerated rate in the pass's piercing zone and a consistent rate in the elongating zone.

2.4 COMPARATIVE STUDIES OF ROTARY PIERCING WITH EXPERIMENTAL METHODS AND FEM

Yamane et al. [8] used the FEM to perform 3D numeric analysis of rotary piercing with rigid plastic. The plug serves as an internal tool, piercing the heated circular carbon steel C45 billet as it is rotated by the rollers. The authors looked at how deformation occurred during redundant shear deformation and rotational piercing. When both the feed and toe angles were raised, circumferential shear strain reduced. Additionally, Yamane et al. discovered good agreement between the circumferential shear strain output response of numeric three-dimensional analysis and experimental techniques.

Murillo-Marrodán et al. [14] applied an incremental constitutive model for finite element analysis of material dynamic restoration in rotary tube piercing.

They could accurately reproduce the average wall thickness, wall thickness eccentricity, elongation, and longitudinal torsion parameters of the tube geometry using their constitutive model to simulate rotary tube piercing. The precise wall thickness eccentricity findings demonstrated the constitutive model's dependency on temperature. The most important factor in determining the prevailing dynamic restoration process was the material strain rate.

Lee et al. [15] constructed a rotary tube piercing machine. To accurately estimate the internal crack initiation position on a carbon steel billet, they carried out finite element simulations with and without the plug as well as experimental evaluations without the plug. Murillo-Marrodán et al. [33] developed a finite element model of the rotary tube piercing process to study the wall thickness eccentricity imperfection. The experimental tube deformation results were validated with the FE model. The authors found similar and accurate results for the experimental and the FE models. The findings demonstrated that higher tube angular velocity at the roll secondary region created positive tube longitudinal torsion. In the early roll zone, the tube's angular velocity reduced from friction with the lateral Diescher discs.

Ceretti et al. [34] developed an FEM model to investigate the factors affecting the hole creation during the rotary piercing process and used Deform 2D for the FEM analysis. These authors investigated the bar at its breaking point and compared the FEM results with TENARIS-DALMINE experiments. They found that the position of the crack initiation with respect to the mandrel position represented the length of the Mannesmann cone.

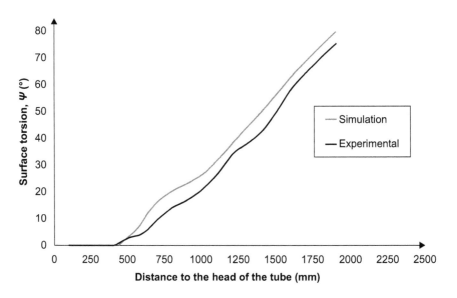

FIGURE 2.4   Comparison of the longitudinal torsion of the tube [33].

Pater et al. [35] performed a 3D FEM analysis of the process parameters and their impacts on the rotary piercing mill. They noted that the exterior and interior diameter variations were affected by the piercing plug's position and shape: While the arrangement of the forward plug tended to reduce internal deviations, the external ones tended to increase due to more intensive material movement before the workspace between the rolls and discs. Greater angle values are advantageous for producing parts with tight tolerances, but this parameter must also be associated with the appropriate plug position and area reduction ratio.

Zhang et al. [36] studied rotary tube piercing and Inconel 718 alloy microstructure evolution using experimental methods and FEM. They found that rotating tube piercing, a mechanism of large plastic deformation, is dominated by the periodic shear strain components and that the piercing procedure is significantly impacted by the reduction rate. Mixed grains and the rolling block phenomenon can result in a lower reduction rate. Additionally, microstructure homogeneity is significantly impacted by the roll speed; with increasing roll speed, the radial variance in average grain size increases.

Zhang et al. [37] utilized computational and experimental data to examine how shear torsion deformation affected the microstructure of the material and the flaws in the separation layer. The shear torsion deformation in the region close to the outside was most influenced by rolling speed. Additionally, the metal around the rolls and plug was severely sheared as a result of the separation layer faults that cause circumferential and longitudinal shear deformation to rise.

Ding et al. [38] used rotary piercing to create AZ31 magnesium alloy seamless tubes in order to get around the shortcomings of the current method. Using the FEM and an experimental approach, they examined the stress, temperature, and strain field distributions. The rotary piercing success was significantly impacted by temperature. At rolling temperatures in the 350–450°C range, the penetration rate of the tube was greater than 90%; however, when the temperature of the billet was lower or higher than these ranges, the penetration rate fell. The tube wall thickness also shrank as the plug diameter and plug advance increased.

Ghiotti et al. [39] investigated a Mannesmann effect to identify the consequences of existing flaws in the working material using a novel damage rule. The created model made precise predictions about the time it would take to cause a fracture as well as where it would occur under external loading. The application of the model was restricted to large batches because of the time-consuming testing, but statistical methods can lessen the influence of the fracture issue on the industrial acceptability.

Zhang et al. [40] discovered that additional tensile stress rises and subsequently reduces as the reduction rate rises and that the maximal tensile stress's corresponding decrease rate is between 10 and 16%. The rolling block phenomenon is significantly impacted by the decrease rate, which should be larger than or equal to 6% in order to guarantee the smooth progress of the piercing process. The outcomes of the simulation demonstrate that the roll speed significantly affects the rise in temperature: The piercing temperature increases with increased roll speed. The acceptable roll speed for creating titanium alloy thick-walled tubes is between 30 and 60 rpm. Zhang et al. [41] compared experimental methods with FEM methods and found that roll speed and reduction rate have considerable impacts on the exterior separation layer defect. The best results obtained for the optimum process parameters are a reduction rate of 13%, roll speed of 35 rpm, Temperature of 1040°C, cross angle of 15°, and feed angle of 8°.

2.5   SUMMARY OF THE ROTARY PIERCING PROCESS

Piercing is widely used for making seamless tubes and therefore has various applications in bridges; buildings; and the oil and gas, automotive, and aerospace industries [1, 5]. The many studies discussed in this chapter demonstrate that piercing processes call for many different input parameters, workpiece materials, and output parameters. Figure 2.5 shows the percentages of different materials used in rotary piercing processing; the figure shows that steel alloy materials are mostly used followed by titanium alloy and wax. Figure 2.6 shows the various rotary piercing output parameters that researchers have investigated, most notably the shape of the workpiece, shear deformation, strain, elongation, redundancy factors, mandrel force, and wall thickness. Table 2.1 shows that the review of past research work based on workpiece material, input process parameters and its output responses of piercing process by various researchers.

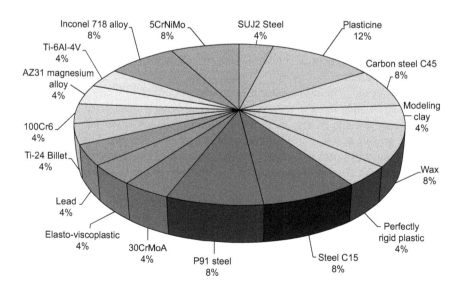

FIGURE 2.5   Percentages of different materials used in rotary piercing processing.

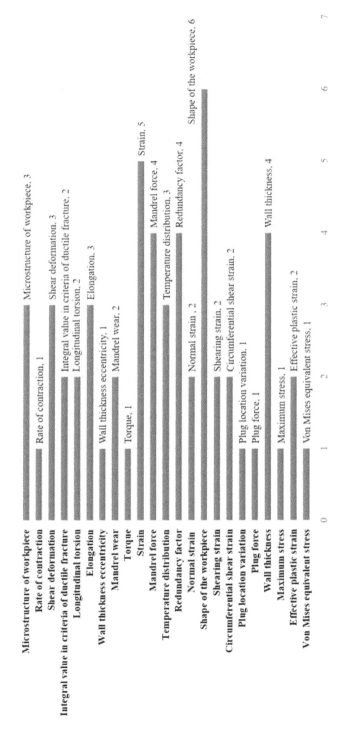

FIGURE 2.6 Numbers of output responses of rotary piercing process investigated by researchers.

TABLE 2.1

Overview of Past Research Work on Rotary Piercing Workpiece Materials and Input and Output Parameters

Sr. No.	Type of workpiece material	Input parameters	Output parameters	Reference No.
1	Plasticine	Distance between roll axes, Initial workpiece diameter, Feed angle, Guide shoe diameter, Maximum plug diameter, Minimum roll gap, Entrance face angle, Exit face angle, Plug advance, Roll velocity	Effective plastic strain, diameter, Von-Mises equivalent stress	[1]
2	SUJ2(100Cr6)	Radius of initial material, Angular speed of work roll, Angular speed of guide roll, Initial temperature of material, Initial temperature of mandrel, Initial temperature of guide roll and work roll	Strain, deformation	[3]
3	Plasticine	Distance between roll axes, Initial workpiece diameter, Feed angle, Minimum roll gap, Workpiece velocity, Entrance face angle, Guide shoe diameter, Maximum plug diameter, Exit face angle, Plug advance	Maximum stress	[4]
4	Carbon structural steel	Feed angle, Exit face angle, Cross angle, Roll diameter, Entrance face angle, Billet diameter, Tube dimension, Minimum roll gap, Plug advance, Rolling speed, Initial temperature of billet, Maximum plug diameter, Relief section width of plug, Spring stiffness	Plug force, Wall thickness, Plug location variation	[5]
5	Carbon steel C45	Toe angle, Feed angle, Diameter expansion ratio, Elongation ratio, Diameter expansion ratio, Billet diameter	Circumferential shear strain	[8]

TABLE 2.1 (*Continued*)

Overview of Past Research Work on Rotary Piercing Workpiece Materials and Input and Output Parameters

Sr. No.	Type of workpiece material	Input parameters	Output parameters	Reference No.
6	Modeling clay	Cross angle, Feed angle, Diameter of specimen, Plug advance against gorge, Temperature of specimen, Distance among two rolls at gorge, Length of specimen	Normal strain component, Shape of the specimen, Shearing strain components	[9]
7	Wax	Plug profile, Roll profile, Guide profile, Roll inclination angle, Roll speed, Plug positions	Longitudinal shear, Redundancy factor, Circumferential shear, Shear because of twist	[10]
8	Perfectly rigid plastic	Friction, Initial material width, Distance between two rolls, Radius of plug top, Cone angle of plug, Initial material height, Plug position, Velocity of pusher, Diameter of caliber, Diameter of plug, Revolutions per minute of roll	Mandrel force, Material shape, Temperature distribution, Roll force	[11]
9	Steel C15 (DIN) AISI-H-13	Feed angle, Cross angle, Roll entrance angle, Roll output angle, Roll diameter, Roll length, Angular velocity	Torque, Total force, Temperature distribution and strain	[12]
10	DIN-C15	Number of rolls, Shaper of rolls, Cross angle, Feed angle, Roll entrance angle, Roll output angle, Roll diameter, Roll length, Roll angular velocity	Mandrel wear, Temperature distribution, Total force, and strain	[13]
11	P91 steel	Roll angular velocity, Cross angle, Diescher diameter, Feed angle, Roll diameter, Diescher angular velocity	Wall thickness eccentricity, Average wall thickness, elongation, Longitudinal torsion, Strain rate	[14]

(Continued)

TABLE 2.1 (*Continued*)

Overview of Past Research Work on Rotary Piercing Workpiece Materials and Input and Output Parameters

Sr. No.	Type of workpiece material	Input parameters	Output parameters	Reference No.
12	Plasticine	Maximum roll diameter, Maximum plug diameter, Initial workpiece diameter, Roll feed angle, Roll face angle, Plug advance, Minimum roll gap	Integral ductile fracture criteria, Elongations in rolling direction, Average outer diameter	[20]
13	30CrMoA	Roll speed, Feed angle, Cross angle, Plug advance	Average wall thickness, Wall thickness variance	[26]
14	Elasto-viscoplastic	Feed angle, Cross angle, Roll angular velocity, Roll diameter, Diescher diameter, Separation between the rolls, Separation between the Diescher discs, Diescher angular velocity	Surface twisting, Shear deformation	[28]
15	wax and lead	Type of rolls, Cone angle, Feed angles, Roll speed, Plug advance	Redundancy factor, Level of force parameters	[30]
16	Ti-24 Billet	Temperature, Rotational speed, Approach angle	Grain size, Deformation degree, Rate of elongation, Rate of contraction	[31]
17	Wax	Deformation rate (uniform, accelerated, feed angle, decelerated), Axial velocity, Roll speed, Ratio of billet diameter to piercing gorge diameter	Total strain with essential stain in elongating pass and piercing pass, Redundancy	[32]
18	P91 steel	Cross angle, Roll angular velocity, Feed angle, Diescher angular velocity, Roll diameter, Diescher diameter	Elongation ratio, Force of Dieschers, Power of the rolls, Average velocity of the tube, Average wall thickness of the tube, Slope of the tube torsion	[33]
19	100Cr6	Angle, Distance, Area reduction ratio, Plug position	Plug dimensions	[35]

TABLE 2.1 (*Continued*)
Overview of Past Research Work on Rotary Piercing Workpiece Materials and Input and Output Parameters

Sr. No.	Type of workpiece material	Input parameters	Output parameters	Reference No.
20	Inconel 718 and 5CrNimo	Longitudinal shear angle, Ratio of billet diameter and plug diameter, Feed angle, Torsion angle, Roll speed, Circumferential shear angle, Reduction rate	Microstructure evolution, Shear-torsion deformation on the separation layer defects	[37]
21	AZ31 magnesium alloy	Temperature, Feed angle, Plug advance, Piercing section roll length, Piercing section roll diameter, Roll gape, Roll rotating velocity, Plug diameter, Radial rolling reduction, Entrance face angle	Tube diameter, Penetration rate, Tube wall thickness, Deformation	[38]
22	Ti—6Al—4V	Roll diameter at gorge, Maximum plug diameter, Plug nose diameter, Exit angle, Entrance angle, Plug length, Billet diameter, Temperature, Feed angle, Guide space, Reduction rate, Cross angle, Plug advance against gorge roll speed	Geometric dimension, microstructural analysis, Tensile stress	[40]
23	Inconel 718 alloy and 5CrNiMo	Feed angle, Billet diameter, Billet length, Temperature, Guide plate space, Roll speed, Reduction rate, Plug advance against gorge	Defect control, Lose stability of mandrel, Temperature, Microstructure distribution, Strain	[41]

2.6 CONCLUSION

Seamless pipes are more commonly used for various purposes due to their unique characteristics such as uniformity of shape, structural strength, fatigue capacity under load, and higher pressure ratings. These pipes are effectively produced by the piercing process. The review of the past literature revealed a number of findings, such as that many researchers have used the finite element method to analyze the most common materials used during rotary piercing; the review also identified the most common input and output parameters that researchers have studied and obtained. The most common workpiece material has been steel alloy, and among the most commonly investigated input parameters have been cross angle, roll speed, feed angle, workpiece diameter, plug diameter, and initial temperature. The most common output parameters have been workpiece shape, shear deformation, strain, elongation, redundancy, mandrel force, and wall thickness.

REFERENCES

1. Topa, A., Kim, D.K., and Kim, Y. (2018) "3D numerical simulation of seamless pipe piercing process by fluid-structure interaction method", *MATEC Web of Conferences*, 203, p. 06016. doi: 10.1051/matecconf/201820306016.
2. Shih, C.K., Hsu, R.Q., and Hung, C. (2002) "A study on seamless tube in the planetary rolling process", *Journal of Materials Processing Technology*, 121(2–3), pp. 273–284. doi: 10.1016/S0924-0136(01)01265-1.
3. Cho, J.M., Kim, B.S., Moon, H.K., Lee, M.C., and Joun, M.S. (2013) "Comparative study on Mannesmann roll piercing processes between Diescher's guiding disk and Stiefel's guiding shoe", *AIP Conference Proceedings*, 1532(1), pp. 843–849. doi: 10.1063/1.4806919.
4. Topa, A., Cerik, B.C., and Kim, D.K. (2020) "A useful manufacturing guide for rotary piercing seamless pipe by ALE method", *Journal of Marine Science and Engineering*, 8(10), pp. 756(1–24). doi: 10.3390/jmse8100756.
5. Hashmi, M.S.J. (2006) "Aspects of tube and pipe manufacturing processes: Meter to nanometer diameter", *Journal of Materials Processing Technology*, 179(1–3), pp. 5–10. doi: 10.1016/j.jmatprotec.2006.03.104.
6. Hayashi, C., Akiyama, M., and Yamakawa, T. (1999) "Advancements in cone-type rotary piercing technology", *ASME: Journal of Manufacturing Science and Engineering*, 121(3), pp. 313–320. doi: 10.1115/1.2832683.
7. Hayashi, C., and Yamakawa, T. (1997) "Influences of feed and cross angle on rotary forging effects and redundant shear deformations in rotary piercing process", *ISIJ International*, 37(2), pp. 146–152. doi: 10.2355/isijinternational.37.146.
8. Yamane, K., Shimoda, K., and Yamane, A. (2017) "Three-dimensional numerical analysis of rotary piercing process", In *Proceedings of the XIV International Conference on Computational Plasticity: Fundamentals and Applications (COMPLAS 2017)*, Barcelona, Spain, 5–7 September, pp. 114–121.
9. Komori, K., and Mizuno, K. (2009) "Study on plastic deformation in cone-type rotary piercing process using model piercing mill for modeling clay", *Journal of Materials Processing Technology*, 209(11), pp. 4994–5001. doi: 10.1016/j.jmatprotec.2009.01.022.
10. Blazynski, T.Z., and Cole, I.M. (1963) "An analysis of redundant deformations in rotary piercing", *Proceedings of the Institution of Mechanical Engineers*, 178(1), pp. 867–893. doi: 10.1177/0020348363178001127.
11. Komori, K., and Suzuki, M. (2005) "Simulation of deformation and temperature in press roll piercing," *Journal of Materials Processing Technology*, 169(2), pp. 249–257. doi: 10.1016/j.jmatprotec.2005.03.017.
12. Darki, S., and Raskatov, E.Y. (2021) "Analysis of the tube piercing process types in terms of final product properties", *IOP Conference Series: Materials Science and Engineering*, 1140, p. 012012.

13. Darki, S., and Raskatov, E.Y. (2021) "Study of the rollers geometry effects in rotary tube piercing and wear analyses of mandrel according to the finite element method", *Proceedings of the Institution of Mechanical Engineers, Part E: Journal of Process Mechanical Engineering*, 235(5), pp. 1676–1684. doi: 10.1177/09544089211015777.

14. Murillo-Marrodán, A., García, E., Barco, J., and Cortés, F. (2020) "Application of an incremental constitutive model for the FE analysis of material dynamic restoration in the rotary tube piercing process", *Materials*, 13(19), p. 4289. doi: 10.3390/MA13194289.

15. Lee, H.W., Lee, G.A., Kim, E.Z., and Choi, S. (2008) "Prediction of plug tip position in rotary tube piercing mill using simulation and experiment", *International Journal of Modern Physics B*, 22(31 and 32), pp. 5787–5792.

16. Gulyaev, Y.G., Shifrin, E.I., Lube, I.I., Garmashev D.Y., and Nikolaenko, Y.N. (2013) "Geometry of the deforming region in rotary-rolling mills", *Steel in Translation*, 43(11), pp. 758–761. doi: 10.3103/S0967091213110089.

17. Kajtoch, J., and Urbanski, S. (1994) "Distribution of metal torsion in a rotary piercing mill", *Steel Research*, 65(9), pp. 382–389. doi: 10.1002/srin.199401181.

18. Hrycaj, P., Lochegnies, D., Oudin, J., Gelin, J.C., and Ravalard, Y. (1988) "Finite element analysis of two-roll hot piercing", In Chenot, J.L., Oñate, E. (eds), *Modelling of Metal Forming Processes*, pp. 329–336, Springer, Dordrecht.

19. Joun, M.S., Lee, J., Cho, J.M., Jeong, S.W., and Moon, H.K. (2014) "Quantitative study on Mannesmann effect in roll piercing of hollow shaft", *Procedia Engineering*, 81, pp. 197–202. doi: 10.1016/j.proeng.2014.09.150.

20. Mori, K.I., Yoshimura, H., and Osakada, K. (1998) "Simplified three-dimensional simulation of rotary piercing of seamless pipe by rigid—Plastic finite-element method", *Journal of Materials Processing Technology*, 80–81, pp. 700–706. doi: 10.1016/S0924-0136(98)00128-9.

21. Huang, H., Wang, W., and Du, F. (2009) "A study of formation mechanism of the spiral mark defect in cone-type rotary piercing process", *Applied Mechanics and Materials*, 16–19, pp. 601–606. doi: 10.4028/www.scientific.net/AMM.16-19.601.

22. Urbanski, S., and Kazanecki, J. (1994) "Assessment of the strain distribution in the rotary piercing process by the finite element method", *Journal of Materials Processing Technology*, 45(1–4), pp. 335–340. doi: 10.1016/0924-0136(94)90362-X.

23. Pater, Z., Łukasz, W., and Walczuk, P. (2019) "Comparative analysis of tube piercing processes in the two-roll and three-roll mills", *Advances in Science and Technology Research Journal*, 13(1), pp. 37–45. doi: 10.12913/22998624/102766.

24. Darki, S., and Raskatov, E.Y. (2021) "Study of the rollers geometry effects in rotary tube piercing and wear analyses of mandrel according to the finite element method", *Proceedings of the Institution of Mechanical Engineers, Part E: Journal of Process Mechanical Engineering*, 235(5), pp. 1676–1684. doi: 10.1177/09544089211015777.

25. Yamane, A., Shitamoto, H., and Yamane, K. (2015) "Development of numerical analysis on seamless tube and pipe process", *Journal of Nippon Steel & Sumitomo Metal Technical Report*, 107, pp. 108–113.

26. Ye, C.Q., Shu, X.D., Lin, F., Wang, J.T., Xia, Y.X., and Zhang, S. (2022) "Effect of process parameters on forming quality of hollow axle blank with rotary piercing", *IOP Conference Series: Materials Science and Engineering*, 1270, pp. 012062. doi: 10.1088/1757-899x/1270/1/012062.

27. Bogatov, A.A., Nukhov, D.S., and Toporov, V.A. (2017) "Simulation of rotary piercing process", *Metallurgist*, 61(1–2), pp. 101–105. doi: 10.1007/s11015-017-0460-6.

28. Murillo-Marrodán, A., García, E., and Cortés, F. (2019) "Modelling of the cone-type rotary piercing process and analysis of the seamless tube longitudinal shear strain using industrial data", *AIP Conference Proceedings*, 2113(1), p. 040003. doi: 10.1063/1.5112537.

29. Khudheyer, W.A., Barton, D.C., and Blazynski, T.Z. (1997) "A comparison between macroshear redundancy and loading effects in 2-and 3-roll rotary tube cone piercers", *Journal of Materials Processing Technology*, 65(1–3), pp. 191–202. doi: 10.1016/S0924-0136(96)02261-3.

30. Al-Dahwi, A.K.A., and Blazynski, T.Z. (1992) "Inhomogeneity of flow, force parameters and pass geometry in rotary cone-roll tube piercing", *Materialwissenschaft und Werkstofftechnik*, 23(1), pp. 29–38. doi: 10.1002/mawe.19920230109.

31. Zhao, H., Yang, Y., Guo, D., Lu, Y., Xi, Z., and Xion, L. (2011) "Study on rotary piercing technology for high strength titanium", *Proceedings of the 12th World Conference on Titanium*, pp. 291–294.

32. Blazynski, T.Z. (1968) "Rate of deformation and redundant shears in the combined rotary piercing-elongating process", *Journal of Strain Analysis*, 3(4), pp. 264–272. doi: 10.1243/03093247V034264.

33. Murillo-Marrodán, A., García, E., Barco, J., and Cortés, F. (2020) "Analysis of wall thickness eccentricity in the rotary tube piercing process using a strain correlated FE model", *Metals*, 10(8), p. 1045. doi: 10.3390/met10081045.

34. Ceretti, E., Giardini, C., and Brisotto, F. (2004) "2D imulation and validation of rotary tube piercing process", *AIP Conference Proceedings*, 712(1), pp. 1154–1159. doi: 10.1063/1.1766684.

35. Pater, Z., Bartnicki, J., and Kazanecki., J. (2012) "3D Finite Elements Method (FEM) analysis of basic process parameters in rotary piercing mill", *Metalurgija*, 51(4), pp. 501–504.

36. Zhang, Z., Liu, D., Yang, Y., Wang, J., Zheng, Y., and Zhang, F. (2019) "Microstructure evolution of nickel-based superalloy with periodic thermal parameters during rotary tube piercing process", *The International Journal of Advanced Manufacturing Technology*, 104, pp. 3991–4006. doi: 10.1007/s00170-019-04126-x.

37. Zhang, Z., Liu, D., Li, Z., Zhang, Y., Zhang, R., Yang, Y., Pang, Y., and Wang, J. (2021) "Study on the shear-torsion deformation of rotary tube piercing process for nickel base superalloy", *Journal of Materials Processing Technology*, 295, p. 117153. doi: 10.1016/j.jmatprotec.2021.117153.

38. Ding, X.F., Shuang, Y.H., Liu, Q.Z., and Zhao, C.J. (2018) "New rotary piercing process for an AZ31 magnesium alloy seamless tube", *Materials Science and Technology*, 34(4), pp. 408–418. doi: 10.1080/02670836.2017.1393998.

39. Ghiotti, A., Fanini, S., Bruschi, S., and Bariani, P.F. (2009) "Modelling of the Mannesmann effect", *CIRP Annals-Manufacturing Technology*, 58(1), pp. 255–258. doi: 10.1016/j.cirp.2009.03.099.

40. Zhang, Z., Liu, D., Yang, Y., Zheng, Y., Pang, Y., Wang, J., and Wang, H. (2018) "Explorative study of rotary tube piercing process for producing titanium alloy thick-walled tubes with bi-modal microstructure", *Archives of Civil and Mechanical Engineering*, 18(4), pp. 1451–1463. doi: 10.1016/j.acme.2018.05.005.

41. Zhang, Z., Liu, D., Zhang, R., Yang Y., Pang, Y. Wang, J., and Wang, H. (2020) "Experimental and numerical analysis of rotary tube piercing process for producing thick-walled tubes of nickel-base superalloy", *Journal of Materials Processing Technology*, 279, p. 116557. doi: 10.1016/j.jmatprotec.2019.116557.

3 Finite Element Analysis of the Metal Spinning Process

N. Balaji, and D. Nagarajan

3.1 INTRODUCTION

Metal spinning is one of the most versatile and useful sheet metal forming processes. The spinning process was developed from the ancient Egyptian craft of pottery making on a potter's wheel. It offers several unique advantages compared with other sheet metal forming methods. One of its key features is incremental forming, which sets it apart from alternative processes. The process can be used for a range of materials for obtaining a variety of shapes with different surface finishes. Recent innovative spinning technologies are used to manufacture complicated geometric components such as axisymmetric, non-axisymmetric, and noncircular cross-sections and oblique bottom filling and tooth gear-shaped parts (Wong et al., 2003; Runge, 1994; Music et al., 2010; Xia et al., 2014).

A schematic of the conventional spinning process is shown in Figure 3.1(a). The roller tool moves along the preprogrammed path and deforms the sheet metal clamped between the mandrel and holder (i.e., tailstock) into the final shape. The sheet metal is deformed plastically through small strain increments beneath the roller tool region, and the final geometry is obtained as a result of these successive incremental plastic deformations. Several process variables such as blank material, blank diameter, sheet thickness, roller nose radius, mandrel speed, feed rate, and roller tool path profile affect the overall plastic deformation attained during the spinning process.

It is expensive, time-consuming, and challenging to conduct extensive experiments with parametric variation for each of the variables that affect the product quality. In the modern industrial spinning process, trial and error is still widely employed to optimize the above process parameters during the creation of new parts, but in many circumstances, it leads to considerable variations and inconsistencies in the final product quality and geometrical aspects. In certain situations, forming defects like wrinkles and fractures also occur, as shown in Figure 3.1(b), due to improper process design and understanding of the material's behavior.

Because it is so costly, time-intensive, and difficult to optimize metal spinning processes, developing analytical and finite element approaches is critical

DOI: 10.1201/9781003441755-3

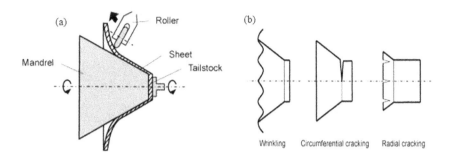

FIGURE 3.1   (a) Conventional spinning process (Music et al., 2010), (b) Defects in conventional spinning (Music et al., 2010).

for fully understanding the material and process behaviors involved. This will thereby reduce the number of necessary experiments and overall process development time and increase the chances of obtaining defect-free products on the first attempt. It is fully possible to understand material deformations involved during conventional spinning processes and analyze how the process parameters affect spun parts through analytical and finite element approaches.

3.2   PRINCIPLES AND PROCEDURES OF METAL SPINNING

The main principle of the metal spinning process is small, incremental deformation. The sheet metal is clamped between the mandrel and holder/tailstock of the computer numerical control lathe machine. During the process, the sheet metal rotates at a high speed along with the mandrel and is wrapped over the mandrel with the help of a roller tool. The predefined tool path generated using computer-aided manufacturing software is used to move the roller tool over the mandrel and wrap the sheet to the final geometry. The entire deformation is focused only in the region where the roller tool comes into contact with the sheet metal and the other areas of the sheet metal remain undeformed during this time. This small, incremental deformation progresses with the movement of roller tool, and the entire plastic deformation is a result of such progressive deformations. This small, incremental deformation helps in easily relaxing the stresses during the deformation and improves the formability of the sheet metal during the spinning process, unlike any conventional volume/bulk deformation processes such as deep drawing or stamping.

Spinning is classified into three processes: conventional metal spinning, shear metal spinning, and tube metal spinning or flow forming. The fundamental difference among these processes is the wall thickness of the produced parts. In conventional spinning, the roller tool progressively deforms the blank over several passes without changing the initial thickness, but the circular blank diameter (D_1) is reduced from its initial diameter (Do), as shown in Figure 3.2 (a). As shown in (Figure 3.2 (b and c)), the wall thickness of the forming element is reduced during shear spinning and tube spinning while the internal diameter (D_1) is held

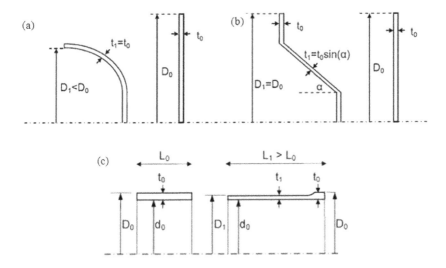

FIGURE 3.2 (a) Conventional spinning part and blank (Music et al., 2010), (b) shear spinning part and blank (Music et al., 2010), and (c) tube spinning preform and formed part (Music et al., 2010).

constant. The final thickness of the component t_1 formed during shear spinning can be calculated using the sine law equation given in (3.1):

$$t_1 = to\ (sin(\alpha)) \qquad\qquad (3.1)\ (Music\ et\ al.,\ 2010)$$

where t_0 stands for the circular blank initial thickness and α is the inclined angle of the mandrel during the process of shear forming.

In tube spinning, the required increase in length of the preformed tube after the process determines the ultimate thickness (Figure 3.2 (c)). Although conventional and tube spinning components can be produced in either single or several passes, shear spinning components are always produced in single passes (Runge, 1994; Music et al., 2010).

3.3 SPINNING PROCESSES

Spinning is considered a compressive draw forming process because of the radial, tensile, and compressive stresses that influence the metal flow in the localized work roller zone and adjacent areas. Spinning processes are classified as conventional, shear, flexible or mandrel free, or hot.

3.3.1 CONVENTIONAL OR TRADITIONAL SPINNING

A sheet blank can be conventionally formed into suitable shapes without changing the wall thickness either along the entire length or by reducing the diameter in certain areas (Figure 3.3 (a)). This can be achieved in a single pass or in several passes by gradually pressing the sheet over the mandrel using the roller tool. The geometries that can be made using the conventional spinning process are shown in Figure 3.3 (b).

Internal tension is created in the sheet when it is deformed by the roller, as shown in Figure 3.4. The formation of this stress differs between the blank's periphery side and mandrel side. When the roller tool moves toward the edge of the blank, it generates tensile radial stress and compressive circumferential stress. If the tensile radial stresses were not balanced by compressive circumferential

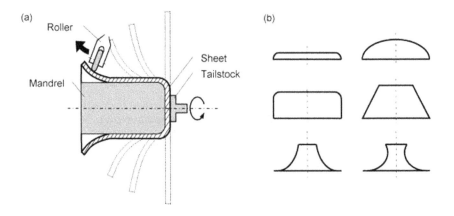

FIGURE 3.3 (a) The conventional spinning process; (b) feasible geometries (Music et al., 2010).

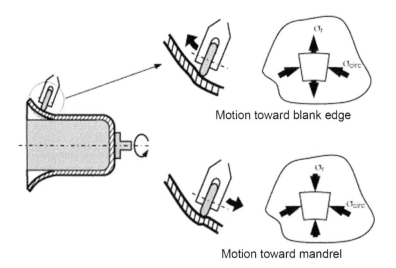

FIGURE 3.4 Stresses in the working zone in conventional spinning (Music et al., 2010).

stresses, they would cost thinning in the final component; if the stresses are balanced, then uniform thickness is obtained in the final part. In the second step, as the roller tool travels away from the circular blank edge and moves toward the tailstock or mandrel edge, the material develops compressive radial and circumferential stress (Runge, 1994; Lange and Pöhlandt, 1985).

3.3.2 SHEAR SPINNING

The significant process advances, such as automation and greater power, in spinning machines eventually led to shear spinning. Shear spinning, in comparison with conventional spinning, reduces the component wall thickness while keeping a constant diameter that is equivalent to the circular blank diameter. While the roller tool follows the profile of the mandrel, at a defined distance, the wall thickness of the circular blank at an initial thickness of t_0 is reduced to a final thickness of t_1, where t_1 represents the wall angle by the known sine law (Figure 3.5 (a)). Figure 3.5 (b) exhibits common shear spinning product geometries. The final component is created in a single pass, often with a single roller, using either a preformed tube or a flat circular blank. Two rollers positioned opposite to one another are occasionally used to make parts with high strength and thickness (Music et al., 2010).

3.3.3 FLEXIBLE OR MANDREL-FREE SPINNING

In flexible spinning, a revolving sheet is clamped around its edges, and a cylindrical tool gradually forms it into desired shape (Figure 3.6). The sheet is then inverted, and the tool is moved along the sheet's outer surface in accordance with its curvature. The concept of inverting the sheet is to use its stiffness to make sharp edges and intricate shapes. Since the thickness variation follows the sine law, this process is identical to shear spinning (Kitazawa, 1994; Spur and Stöferle, 1979).

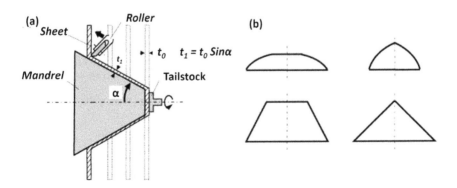

FIGURE 3.5   (a) The shear spinning process; b) feasible geometries (Music et al., 2010).

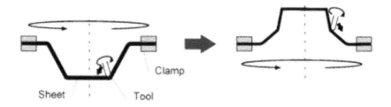

FIGURE 3.6   The mandrel-free spinning process (Music et al., 2010).

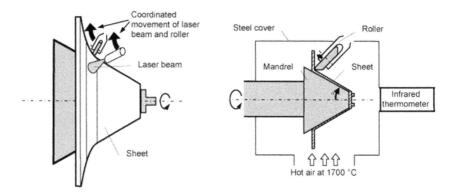

FIGURE 3.7   (a) Spinning with laser beam; (b) spinning with hot air (Music et al., 2010).

3.3.4   HOT SPINNING

Spinning procedures are generally conducted at room temperature; however, heating is often preferred for reducing the forming forces for thick sections and high-strength materials. The focus point of a laser beam or hot air is passed over the sheet in this setup (Figure 3.7 (a and b)), and heating is done just before the contact roller tool. Klocke and Brummer (2014) describe the effective forming of difficult materials such as stainless steel, nickel, and titanium using this method. Stainless steel was successfully formed with improved formability of 15–25% and a decrease in forming forces of up to 40%, as well as good surface quality.

3.4   SPINNING PROCESS PARAMETERS

Conventional spinning and shear spinning differ from each other in terms of the deformation characteristics, but the process parameters that control the conventional spinning also affect the properties of shear-formed and flow-formed products.

3.4.1 FEED RATIO

The feed ratio is defined as the ratio between the feed rate of the roller and that of the spindle speed. As long as the feed ratio remains constant, the roller feed and spindle speed can be changed without compromising the quality of the final product. Maintaining the correct feed ratio is essential, as excessive feed ratios can induce larger forces, which may lead to fractures. In contrast, a low feed ratio results in excess metal flow in the outward direction, which lowers the formability and reduces the wall thickness prematurely (Runge, 1994). Increased spindle speed resulted in two phenomena: First, the spinning force was greater due to the higher deformation rate; second, the deformation energy needed per revolution decreased because the feed rate (mm/rev) was inversely proportional to spindle speed (Wang et al., 1989).

3.4.2 ROLLER PATH

The design of the tool path has a significant impact on the final product quality, which is especially critical in conventional spinning. Several tool paths, including linear, concave, convex, and involute of the circle, have been employed (Liu et al., 2002). The appropriate tool path design results in a high-quality, wrinkle, and crack-free product. In a study of roller paths, researchers evaluated linear, quadratic, and involute tool paths (Figure 3.8) and concluded that an involute tool path produced the greatest outcome in the metal spinning process (Liu et al., 2002).

3.4.3 ROLLER DESIGN

Metal spinning can employ several roller tool designs, as shown in Figure 3.9. The selection of a particular roller tool design depends on the type of starting material used, the geometry or profile made, the spinning process used, and the amount of flatness needed on the final spun component, and the details are given in Table 3.1.

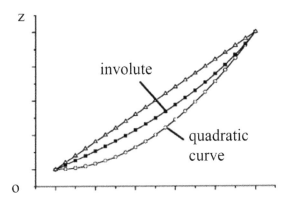

FIGURE 3.8　Various roller tool paths used in the spinning process (Liu et al., 2002).

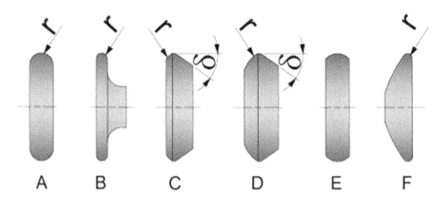

FIGURE 3.9   Various types of roller tools used in the spinning process.

TABLE 3.1

Selection of Suitable Roller Tool Design for Spinning Processes

Type of roller tool	Metal spinning component	Spinning process
A & F	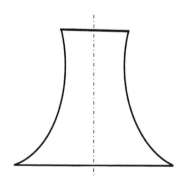	Tools A & F are used to manufacture parts from a flat blank, a finished part, and a circular blank of the same thickness. It is well known that increasing the tool's round-off radius (r) improves the spinning ratio β (Avitzur, 1983).
B		Tool B is used for forming necks or reducing the diameter of the part.

TABLE 3.1 (*Continued*)
Selection of Suitable Roller Tool Design for Spinning Processes

Type of roller tool	Metal spinning component	Spinning process
C	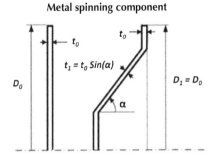	Tool C is used for shear spinning parts and blanks of the same diameter, but the blank thickness is reduced from the initial thickness.
D	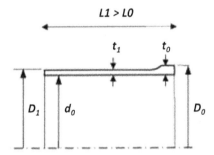	Tool D can be utilized to process shear spinning and flow-forming combinations. Tube-spun preformed and formed part inner dimeters remain constant, but the wall thickness of the formed part is reduced from the initial thickness.
E	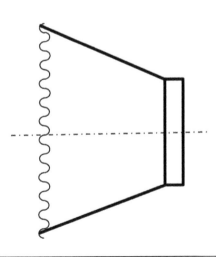	A flat roller, such as type E, is required to flatten or reduce the unevenness of the surface of a component.

3.4.4 Spinning Ratio (β)

The spinning ratio is the ratio of the diameter of the blank to the diameter of the mandrel. The spinning process becomes more complex as the spinning ratio increases. If the ratio of rotation is too high, the remaining cross section of the material will be incapable of supporting the high radial tensile stress produced in the wall. This will cause the sheet to circumferentially separate along the flange-to-wall transition. The spinning ratio, on the other hand, reaches its maximum when the wrinkling in the flange grows so extensive that successive tool passes cannot remove it (Avitzur, 1983).

3.5 ADVANTAGES AND LIMITATIONS OF SPINNING

3.5.1 Advantages

Metal spinning has numerous significant advantages over other sheet metal forming processes due to its incremental forming process. The advantages include flexibility in the process, non-dedicated tooling, lower forming load, and high quality of surface finish (Xu et al., 2001).

- **Process Flexibility**—One of the most significant benefits of metal spinning is its outstanding process flexibility. Metal spinning, unlike other sheet metal forming technologies, allows for the manufacture of a wide variety of sizes and shapes. This flexibility allows manufacturers to easily build complicated and elaborate designs, making it a popular choice in a variety of sectors including automotive, aerospace, and lighting.
- **Non-Dedicated Tooling**—Metal spinning differs from other sheet metal forming methods in that it does not require specialized tools. Manufacturers may easily make different components using a single machine with interchangeable toolings, eliminating the need for costly and time-consuming tooling changes. This adaptability not only reduces manufacturing costs but also improves the overall output.
- **Low-Forming Load**—The minimal forming load required throughout the operation is another key advantage of metal spinning. Metal spinning, in contrast with typical stamping or deep drawing procedures, applies minimum force to the material that is formed. This characteristic is very useful when working with small-gauge materials because it reduces the possibility of material stretching or damage. Furthermore, the low forming load allows for the use of lightweight materials, which helps to reduce weight and increase fuel economy in industries such as automotive and aerospace.
- **Good Surface Finish**—Metal spinning provides excellent surface finish quality, making it a popular choice for applications requiring visually appealing and polished final results. The spinning technique produces smooth and homogeneous surfaces with the absence of defects such as wrinkles, ripples, or tool marks. As a result, metal-spun parts are ideal

for a wide range of applications, including decorative components, lighting fixtures, and architectural features.

3.5.2 LIMITATIONS

- **Size and Shape Limitations**—Because the metal spinning process is circular in nature, its products are available only in concentric and axial symmetry. Examples of easily spinnable shapes are hemispheres, cones, flanged covers, funnel, dished heads, parabolas, and stepped parts. The size of the metal spinning equipment also limits the diameter of the components that can be produced.
- **Production Volume Constraints**—Metal spinning, whether by hand or by machine, needs tooling that is specific to the part being made. That means the manufacturer has to buy or invent spinning tools that are just right for the geometry they want to make. From an economic point of view, it's not practical for a manufacturer to make small amounts of a part with a special geometry.
- **Small Allowance for Errors**—For all practical purposes, cracks and dents in the component during manufacturing are irreparable. Blanks or unfinished components with minor defects must be scrapped because repairing is not a cost-effective option. The metal component may also experience undesired hardening during spinning. It would then need to be heat treated before the next pass of spinning is begun.

3.6   FINITE ELEMENT SIMULATION OF SPINNING

Quigley and Monaghan (2002a) attempted to achieve maximum similarity between the FE model of the spinning process and the experimental work. They used Abaqus 6.4 software in their work. In terms of model geometry, spinning generally employs axisymmetric geometry. Since the roller tool is always in contact with a small part of the workpiece, even small moments of the tool can have significant impact on the process. As a result, the spinning process is modeled in full 3D.

Multiple authors have run FE spinning models (Quigley and Monaghan, 2002b; Grass et al., 2006). FE simulation is the only way to find out the transient stresses and strain that may occur during the spinning process. During the simulation, the mandrel and roller tools are referred to as rigid analysis surfaces that do not need meshing, and the blank material is referred to as an independent deformable body. There are two types of analysis possible in Abaqus/Implicit and Explicit (Manual, 2003).

In Abaqus, implicit analysis can be used to solve a wide range of linear and simple nonlinear problems; explicit analysis is used for dynamic event problems that change quickly, like impacts and explosions, as well as for highly nonlinear problems with variable contact conditions, like forming simulations. If the mesh includes small elements, the stress distribution in the material is quite high, and the increment in time during the explicit dynamic analysis must be kept minimal.

Abaqus/Explicit analysis also has a mass scaling option that permits the mass matrix to be partially modified. This causes the speed of stress waves to slow down and determines the optimum increase in the minimum time increment (Manual, 2003).

The mass scaling should be selected in such a way that the model's overall mass doesn't change much. If it does, then the results won't be accurate. For an almost frictionless surface, the coefficient of friction between the roller tool and the workpiece was reported to be 0.02. The workpiece is divided into two sections with the central cylindrical cell with the radius and is constrained by a mandrel, allowing the tailstock to be excluded from the simulation (Razavi et al., 2005).

Liu (2007) modeled a three-roller spinning setup. In FE simulation, the tail-stock applied a constant load of 100 kN to hold the sheet against the mandrel. Additionally, a small displacement of 0.05 mm was initially imparted. The three rollers rotated simultaneously at the same speed, and the holder maintained the same load until the end of the process. The sheet thickness was assumed to be constant, the roller moved and began to rotate under contact conditions, and the rigid tool was no longer be considered.

Using Coulomb's law of friction (μ), Liu (2007) described the contact and sliding boundary conditions between the sheet and the roller. The author applied surface-to-surface contact with $\mu = 0.2$ between the mandrel and sheet and $\mu = 0.05$ between the roller and sheet (Liu, 2007). In addition, the author selected hourglass control as the default for the sheet metal and implemented adaptive meshing control. All simulations used an optimum mandrel speed of 200 rpm and mass scaling to target time increment of 1×10^{-5} s (Liu, 2007).

The center partition was meshed with the hexahedral element shape and sweep method, while the outer partition of the blank was meshed with the hexahedral element shape and structured technique. The solution time was proportional to the number of elements. The mesh was small enough to permit multiple elements to contact the tool at the same time, thereby increasing contact continuity and accuracy (Liu, 2007). Quigley and Monaghan (2002b) investigated the positive effect of the number of increments and concluded that three to four increments for elements entering the contact area provided a reasonable compromise.

Figure 3.10 shows a typical circular blank on which uniform meshing was used across the entire thickness of the sheet. Along 75.5 mm from the outer edge

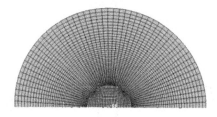

FIGURE 3.10   Circular blank meshing.

of the central partition to the outer edge of the blank, the circumference was divided into 42 segments, and the radius was divided into 18 segments. For better accuracy, the elements of the blank region between 26.5 and 35 mm in radius were divided by two in the radial direction. Simulations were executed on Intel 3.0 GHz single-processor and multi-processor PCs with 512 MB RAM. The simulation of the spinning process took a long time (the solution time radius was between 90 to 120 hours), and this time was dependent on the number of elements used (Razavi et al., 2005).

A circular blank course mesh is known as S1, and it has larger elements than tool mesh. S2 mesh is approximately the same size as tool mesh. S3, the fine mesh, was selected for the workpiece; as shown in Figure 3.11, the element was smaller than the tool mesh. The tool was modeled as a solid material, including

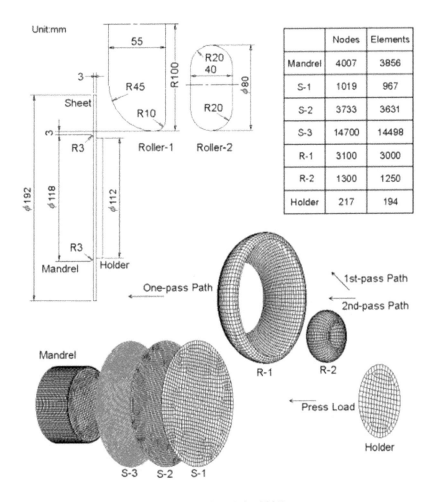

	Nodes	Elements
Mandrel	4007	3856
S-1	1019	967
S-2	3733	3631
S-3	14700	14498
R-1	3100	3000
R-2	1300	1250
Holder	217	194

FIGURE 3.11   FE model and part dimensions (Liu, 2007).

rollers R1 and R2. In Figure 3.11, R2 was designed for multipass spinning (Liu, 2007), and the remaining S-1 to S-3 and R-1 were designed for single-pass spinning. Condition S-1 is denoted as Case A, S-2 is denoted as Case B, and so on until condition R-2 mentioned in the table in Figure 3.11.

3.7 EXPERIMENTAL VALIDATION OF THE MODEL

Spinning experiments were carried out, and the changes in the axial and radial forces of the roller tool between the FE simulation and experiments were compared.

3.7.1 SPINNING OF CYLINDRICAL COMPONENT USING CIRCULAR BLANK

Liu (2007) investigated six examples with various sheet types, roller types, and feed rates and selected case B as the adaptive control strategy as the roller stroke started to deform; the blank meshing type, S-2, was comparable with S-3, as in case C. The two scenarios had identical maximum axial force. Because of the coarse meshing in case B, there was less interface between the blank and the mandrel. Case D was the proposed model for comparison with the experiment because it was easy to analyze and the predefined meshing of the sheet was smaller than the meshing of the tools. Such an approach will result in more consistent simulation interface conditions.

Determining how to build an optimal model is challenging. For instance, in case D, the whole history of strain states appears, as shown in Figure 3.12, and the sheet is formed in four stages. A free-bending-like state occurs before the stroke of 20 mm; then, between the strokes of 20 and 40 mm, deep-drawing-like plastic deformation occurs. During a 40 mm stroke, progressively greater deformations become compressed, bending with oscillation (Liu, 2007).

The axial force of the roller during the spinning process is illustrated in Figure 3.13 for cases A, B, and C. Plastic bending accompanied the maximum forces, and once the bending was completed, the load decreased. As seen in case A, a coarse mesh produced less axial force.

Through modeling and experiments, Figure 3.14 (a) indicates the positional axial force exerted by the roller in single-pass conventional spinning. The difference between case D and the experiment is minor in nature, but there is a discernible difference between case C and the experiment. The main issue is that in the simulation of case C, poor blank meshing resulted in rough and unrealistic interactions. As already mentioned, a fine-mesh model will improve the interface condition in order to acquire the correct loading history. Figures 3.14 (a & b) show good agreement between the FE simulation and experimental force values.

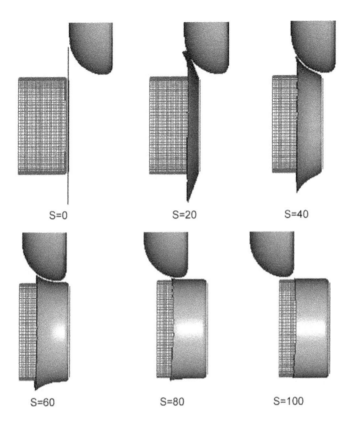

FIGURE 3.12 Stage-by-stage single-pass spinning (Liu, 2007).

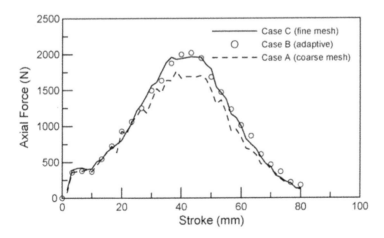

FIGURE 3.13 Various cases of axial load vs stroke length of simulations (Liu, 2007).

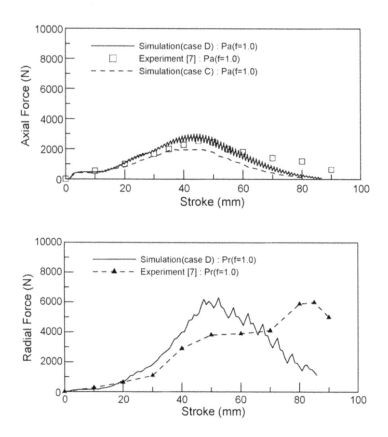

FIGURE 3.14   (a) Axial load validation and (b) radial load validation between FE simulation and experiments (Liu, 2007).

3.8   APPLICATIONS AND FUTURE SCOPE

Metal spinning is a forming method that allows for the creation of hollow, mostly axisymmetric components. These components are widely used in the aviation, aerospace, weapons, and automobile industries, to make components such as jet motor supporting cones, turbo shapes, funnels, gas cylinders, rocket engine nose cones, oil pockets in end plates, and automobile connecting rods (Wong et al., 2003; Runge, 1994; Music et al., 2010; Xia et al., 2014), as shown in Figure 3.15. In recent years, there has been an increase in demand for structural components that are high in strength, light in weight, and corrosion resistant in order to meet the criteria of faster flying speeds, lifetimes, and longer voyages, and researchers have developed alloys of light metals such as magnesium, titanium, and aluminum through spinning (Mei et al., 2015).

Recent metal spinning research is mostly focused on macro formability and forming component quality, but future possibilities should include the material

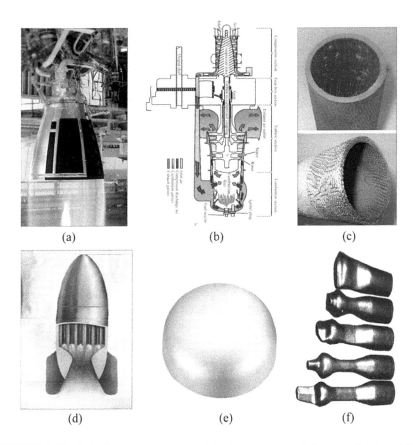

FIGURE 3.15 Spinning components used in the aerospace, defense, and automobile industries: a) jet motor nozzle, b) turboshaft, c) tailing spout, d) rocket engine nose cone, e) end plate of oil pocket, f) automobile connecting rods (Mei et al., 2015).

aspects also, that is, establish a way to regulate the microstructure evolution during the spinning process and optimize the physical properties of spun parts. For instance, hot power spinning of difficult-to-deform metal is a complicated thermal-mechanical coupling method in which sizes are out of tolerance and spinning defects such as surface bulge, surface fracture, surface pileup, and peeling occur often if the processing parameters are excessive. The future scope of hot power spinning research should therefore include improving the forming quality of spun parts by using finite element modeling to optimize the parameters of the metal spinning process.

3.9 SUMMARY

Spinning is an incremental forming process used to make axisymmetric components in a range of materials with different surface finishes. Even though the process is most useful in metal forming industries, it is very expensive, time-consuming, and highly challenging to conduct extensive trials to understand the process and material behaviors that affect the final product quality. Hence, finite element approaches are necessary for fully understanding the process and material behavior, thereby reducing the number of experiments and process development time, and obtaining defect-free products. Researchers have used mass scaling, adaptive techniques, hourglass control, and contact control to perform FE simulations of the single-pass spinning process.

Sheet meshing should be smaller than tool meshing for accurate contact judgment. One mass scaling approach is the constant time increment, which was used to save computing time. Hourglass control was used to prevent zero-energy modes occurring throughout the simulation. Adaptive control reduced the predicted time by simplifying the original mesh data. Researchers found that simulation results were comparable with experimental results, which demonstrates an optimistic trend in understanding the process. These successful FE studies portray the possibility of predicting deformation using dynamic explicit finite element analysis during metal spinning, and FE simulation can also be used to study the effects of process parameters such as roller feed rate, mandrel speed, and material thickness and reduce the defects during spinning. Although the efficiency is impressive, the simulation flaws should be closely monitored for better results.

REFERENCES

Avitzur, B., 1983. Handbook on metal forming processes, Willey Intersc. PubL.

Grass, H., Krempaszky, C. and Werner, E., 2006. 3-D FEM-simulation of hot forming processes for the production of a connecting rod. *Computational Materials Science*, *36*(4), pp. 480–489.

Kitazawa, K., 1994. A CNC incremental sheet metal forming method for producing the shell components having sharp corners. *Journals of the Japan Society for the Technology of Plasticity*, *35*(406), p. 1348.

Klocke, F. and Brummer, C.M., 2014. Laser-assisted metal spinning of challenging materials. *Procedia Engineering*, *81*, pp. 2385–2390.

Lange, K. and Pöhlandt, K., 1985. Handbook of metal forming, McGraw-Hill Companies.

Liu, C.H., 2007. The simulation of the multi-pass and die-less spinning process. *Journal of Materials Processing Technology*, *192*, pp. 518–524.

Liu, J.H., Yang, H. and Li, Y.Q., 2002. A study of the stress and strain distributions of first-pass conventional spinning under different roller-traces. *Journal of Materials Processing Technology*, *129*(1–3), pp. 326–329.

Manual, A.S.U.S., 2003. Abaqus 6.4. *http://130.149*, *89*(2080), p. v6.

Mei, Z., He, Y., Jing, G. and Wang, X.X., 2015. Review on hot spinning for difficult-to-deform lightweight metals. *Transactions of Nonferrous Metals Society of China*, *25*(6), pp. 1732–1743.

Music, O., Allwood, J.M. and Kawai, K., 2010. A review of the mechanics of metal spinning. *Journal of Materials Processing Technology*, *210*(1), pp. 3–23.

Quigley, E. and Monaghan, J., 2002a. Enhanced finite element models of metal spinning. *Journal of Materials Processing Technology*, *121*(1), pp. 43–49.

Quigley, E. and Monaghan, J., 2002b. The finite element modelling of conventional spinning using multi-domain models. *Journal of Materials Processing Technology*, *124*(3), pp. 360–365.

Razavi, H., Biglari, F.R. and Torabkhani, A., 2005, December. Study of strains distribution in spinning process using FE simulation and experimental work. In *Proceedings of the Tehran International Congress on Manufacturing Engineering, Tehran, Iran*.

Runge, M., 1994. Spinning and flow forming. *Leifeld GmbH Werkzeugmaschinenbau/ Verlag Moderne Industrie AG, D-86895, Landsberg/Lech*.

Spur, G. and Stöferle, T., 1979, Handbuch der Fertigungstechnik, vol. 1, Carl Hanser Vlg.

Wang, Q., Wang, T. and Wang, Z.R., 1989, October. A study of the working force in conventional spinning. In *Proceedings of the Fourth International Conference of Rotary Forming, Beijing, China* (pp. 103–108).

Wong, C.C., Dean, T.A. and Lin, J., 2003. A review of spinning, shear forming and flow forming processes. *International Journal of Machine Tools and Manufacture*, *43*(14), pp. 1419–1435.

Xia, Q., Xiao, G., Long, H., Cheng, X. and Sheng, X., 2014. A review of process advancement of novel metal spinning. *International Journal of Machine Tools and Manufacture*, *85*, pp. 100–121.

Xu, Y., Zhang, S.H., Li, P., Yang, K., Shan, D.B. and Lu, Y., 2001. 3D rigid—Plastic FEM numerical simulation on tube spinning. *Journal of Materials Processing Technology*, *113*(1–3), pp. 710–713.

4 High-Energy Rate Forming

Jagadeesha T.
National Institute of Technology, Calicut, India

4.1 INTRODUCTION

High-energy rate forming (HERF) refers to processes that form parts at very high velocities and extremely high pressures. A more accurate term for this process might be "high-velocity forming" because the key feature of these processes is the rate of energy release and the high forming velocity, not the amount of energy used. The rapid application of force is often advantageous for shaping large metal components, as it makes many metals deform more readily. This property proves beneficial for forming sizable parts from a wide range of metals, including those that are typically challenging to work with. Deformation velocities for a few processes are given in Table 4.1. The essence of these techniques is that they involve straining or deformation velocities for the metal that are so high that its inertia plays a very minor role in the deformation process.

The commonly used energy sources for high-energy-rate forming include chemical explosives, high-voltage discharges, electromagnetic fields, and the kinetic energy of fast-moving masses.

The reasons for using high rate forming process are as follows:

- To form very large parts where conventional presses cannot be used.
- For the low-volume production of large, often complex parts from tough metals.
- To minimize production costs by avoiding the need to manufacture dies.
- To deform many difficult-to-form materials like titanium and tungsten alloys under high strain rates.
- To form re-entrant shapes.
- In situations where the use of heavy machinery is difficult, for example, welding in remote areas.
- To reduce the time required for forming. Rough forming followed by finish spinning can be completed quickly.

The energy levels used in HERF vary depending on the material and size of the component. The energy requirement varies from 10 Joules to 109 Joules [1, 15, 22]. HERF employs short bursts of energy that are transmitted through a medium like air or water, the resulting shockwaves act on the part, forcing it to take the shape of the die, and die costs are lower in this process. The springback effect in

DOI: 10.1201/9781003441755-4

TABLE 4.1
Rates of Energy Release

Sr no.	Process	Velocity (m/s)
1	Hydraulic press	0.03
2	Brake press	0.03
3	Mechanical press	0.03–0.07
4	Drop hammer	0.24–4.2
5	Gas-actuated ram	2.4–8.2
6	Explosive forming	9–228
7	Magnetic forming	27–228
8	Electrohydraulic forming	27–228

HERF is almost negligible. Additionally, HERF improves material properties by introducing additional uniform strain. Three commonly used HERF processes are electromagnetic forming, explosive forming, and electro-hydraulic forming, each of which is discussed in this book chapter.

4.2 ELECTROMAGNETIC FORMING/MAGNETIC PULSE FORMING

The first electromagnetic forming (EMF) concept was demonstrated in 1958 by Harvey and Browner (US patent no: US2976907A). The unit they developed had a storage capacity of 6100 N-m and was capable of producing 10 pulses per minute. Figure 4.1 shows the schematic representation of the EMF process.

EMF is applicable to electrically conductive materials. The workpiece is placed near the pulsed magnetic field. When current flows through the coil, a magnetic field is set up around the coil. If the conductor (workpiece) is placed in the magnetic field, eddy currents will be produced in the conductor. The eddy current created in the conductor opposes the magnetic field in the coil. This resulting force between the magnetic field and the eddy currents deforms the part, rapidly pushing it into the die cavity. Forces of approximately 400 MPa can be achieved using this process [1]. Basic electrical elements like capacitors, switches, coils, and power supplies are used in this process.

When current flows through the coil, it generates a high-intensity magnetic field between the coil and the workpiece. As the workpiece intersects with this magnetic field, it induces eddy currents within the work material. The eddy current in the workpiece opposes the magnetic field in the coil, following Lenz's law, and restricts the magnetic field to the surface of the workpiece. The interaction of the magnetic field and eddy current creates an inward force that is used to form the part. There is no physical contact between the tool and workpiece. As a consequence of this high force, the part is accelerated to a high velocity of up to several hundred meters per second, creating a very high strain rate of 10,000 per second [2]. The forming operation is completed within a few seconds.

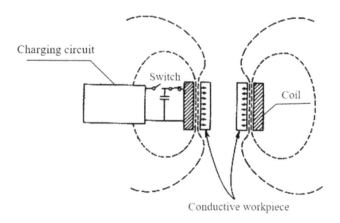

FIGURE 4.1 Schematic representation of an Electromagnetic Processes

4.2.1 Design Consideration for Coils

In EMF, the first step involves applying voltage to charge a large bank of capacitors. These capacitors store a significant amount of energy depending on their capacitance and charging voltage. When the capacitors are fully charged, a switch is used to release this energy across the coil, creating a very high magnetic field. The magnetic forces deform the material, and as a result, equal and opposite forces are applied to the coil. Therefore, both the coil and insulation must be rigid and strong. Pressure of 400 MPa can be generated in just a few microseconds with a 130,000-ampere pulse.

There are two types of coils used: permanent and expendable. Permanent coils are the most commonly used and are designed to run for many production cycles. However, the fabrication cost of these coils is high [3]. Expendable coils are used for batch production applications and are designed for a short lifespan depending on the number of parts to be fabricated. There are three coil configurations used in industries: radial compression, radial expansion, and flat. Radial compression coils are used to form cylindrical surfaces and swage tubes. Radial position coils are placed outside the cylindrical workpiece, with the die fitting inside; when energy is applied, the workpiece is forced to conform to the die's shape. Radial expansion coils are placed inside the part, and when energized, the part expands against the die, creating the desired profile. Flat coils are used for embossing flat workpieces, with parts placed on the coil and then propelled upward into the die [4].

4.2.2 Design Consideration for Dies

Dies are usually made of nonmetallic materials. If a stronger die is required, then dies are made of materials with very poor conductivity. If they are conductors, countercurrents may be generated in the die that can repel the workpiece. The workpiece material should have electrical resistivity of at least 15

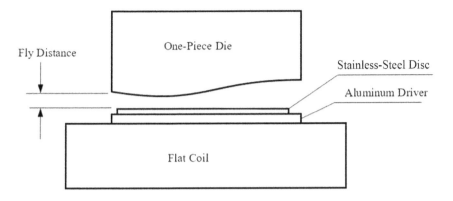

FIGURE 4.2 Schematic representation of the use of a driver plate in EMF.

micro-ohm/cm [5]. If the workpiece is a poor conductor of electrical current, a conducting coating is provided between the coil and the workpiece. To increase the formability of the workpiece, a driver plate is placed between the die and the coil. The driver plate is made of highly conductive material and repels when a magnetic field is applied. The repelling action of the driver plate forces the workpiece into the die. One driver plate is required for one component. A schematic diagram illustrating the use of a driver plate in EMF processes is shown in Figure 4.2.

4.2.3 PROCESS PARAMETERS

Optimizing process parameters is crucial for achieving the desired final product shape and tolerances in high-rate energy forming processes. Understanding the key parameters affecting these processes is essential. The following are some of the important EMF process parameters: capacitance and initial voltage of the capacitor bank, total inductance and resistance of the primary circuit, mutual inductance between the coil and workpiece, coil current, the equivalent induced current in the workpiece, the workpiece equivalent inductance and resistance, the fly and standoff distances, and impact velocity.

4.2.4 EMF MATERIALS

Metal-to-metal seals are useful for fabricating pressure vessels, joining tubes and pipes, and assembling items such as coffee pots, toothpaste tubes, turbine shafts, and missile nose cones. Other successful applications of metal-to-metal seals include forming vacuum-tight and pressure-tight seals between stainless-steel plugs and sealing very-fine-mesh stainless-steel filter screens in stainless-steel end pieces [6]. The cleanliness, speed, and uniform circumferential pressures are particular advantages of this process. It is also used for attaching a turbine wheel to a shaft using a fastening piece with annular grooves.

Joining tubes. EMF is used to join aluminum or copper tubes using an insert. The insert is made of sharp-edged grooves and is sufficiently strong to resist permanent deformation.

Banding. Metal bands made of polished aluminum, brass, and copper are attached to household articles such as furniture, lamps, and utensils using EMF.

Reverse forming. A tube of aluminum can be made to wrap itself around a steel retaining ring.

Electrical connection. Heavy-duty electrical connections are joined by magnetic pulse forming (MPF, another term for EMF). To ensure a good electrical connection with 100% mechanical strength, a copper insert is used to compact fine wires before attaching the terminal [7].

Ceramic assemblies. Stainless-steel, copper, aluminum, and brass caps are assembled to a ceramic insulator to provide a pressure-tight seal by capturing a rubber gasket during the forming of the flange.

Sizing. Sizing a part of a mandrel onto another component of an assembly is a logical use of MPF. Figure 4.2 shows that sizing the finned aluminum sheath to a rod improves thermal conductivity.

Formation of cups. MPF can be used to produce shallow cups. The dynamic action of metal by MPF permits the forming of cups from flat blanks.

Embossing. Ornaments are made using shallow relief dies and a flat coil. These ornaments are embossed in stages. Initially, low-energy forming is performed, followed by high-energy forming to add the details. Ashtrays are economically produced using MPF.

4.2.5 ADVANTAGES OF EMF

- With the help of a single die, several components can be produced, which increases productivity.
- Part lubrication is not required as in conventional metal forming.
- The process can be easily automated, reducing the need for manual labor.
- It is useful for high-volume production.
- Superior metal-to-metal joints can be obtained using this process.
- It is faster than conventional spinning in a lathe machine.

4.2.6 DISADVANTAGES OF EMF

- Coil placement in the workpiece limits certain engineering applications.
- This process involves a high capital cost.
- The size and thickness of the workpiece are limited.
- There is a potential safety hazard.

4.3 ELECTRO-HYDRAULIC FORMING

The basic concept in electro-hydraulic forming (Ehf) is simple: Discharging stored electrical energy into a fluid produces shock waves of very high intensity [8]. The part is produced by constraining these shock waves between a die and a sheet metal plate. Basic research on underwater condenser discharges was conducted by German engineers in 1950, and this process was commercialized in the USA in 1960.

4.3.1 PROCESSES AND EQUIPMENT USED IN EHF

Ehf is also known as electro shape or electro spark forming. Figure 4.3 shows a schematic diagram of the electro-hydraulic process.

The two most common methods of converting electrical energy into mechanical energy are capacitor discharge through a gap and capacitor discharge through a wire. A capacitor bank stores electric charge energy until the switch closes a

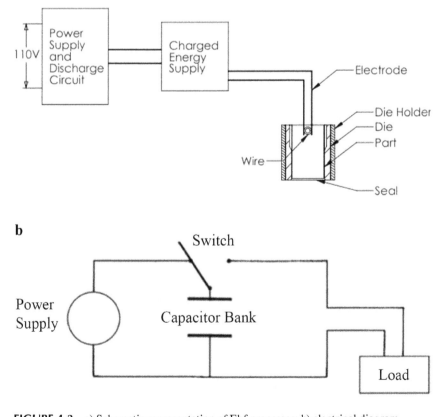

FIGURE 4.3 a) Schematic representation of Ehf processes; b) electrical diagram.

circuit to a water-filled forming tank [9]. When an electrical current is applied, aluminum or magnesium bridge wires positioned between two conductors are heated to the point of vaporization. This vaporization creates a plasma channel within the surrounding water, enabling a spark to pass through it. This spark generates a shockwave that spreads outward in all directions, exerting pressure on the object and pushing it into the dies. The entire system comprises a power supply, storage systems, and a switching unit [10].

Power supply: The power supply requirements depend on the voltage rating of the capacitor bank, the voltage of the power source, and the desired charging time. The major components of the power supply are a step-up transformer and the rectifier. The voltage rating of the power supply must be selected based on the operating voltage rating of the capacitors. The current rating must be sufficient to allow for charging the capacitor bank within the desired time limit.

Storage system: The storage system consists of a suitable bank of capacitors connected with heavy conductors or bus bars to avoid creating detrimental resistances. One side of the capacitor bank is connected to a triggering device for the gap. The stored energy must be delivered by bus bars and rapidly discharged by the triggering device into the work gap. The size of the capacitor bank depends on the amount of energy required to shape a component.

Switching unit: When transmitting a large amount of electrical energy to a wire or gap in a very short time, a large transient current occurs. Peak currents of over 30,000 amperes have been observed when discharging 12,000 joules with their unit. Ionization switches, air gaps, double air gaps, and solenoid-operated air gaps have been used as switching units. The basic requirement for good switching is fast operation and high current-carrying capability.

4.3.2 TYPES OF DISCHARGES USED IN EHF

Capacitor discharge through a gap: A 50,000 volt potential difference can cause an electrical discharge to occur over an approximate 25 mm gap in air. The discharge of a capacitor in a liquid medium is influenced by the thermal and electrical conductivity properties of the fluid. In forming industries, voltage levels typically range from 10,000 to 50,000 volts [11].

Capacitor discharge through a wire: Determining the discharge path within a liquid medium gap can be challenging. However, using a wire offers the advantage of optimizing the discharge path and minimizing power consumption [12]. Nevertheless, a significant drawback of wire usage is the frequent need for replacement, along with the added cost of loading and unloading wire spools. This disadvantage is partially mitigated by the wire's higher efficiency in converting energy into useful work and the potential to operate at lower voltage levels.

4.3.3 DESIGN CONSIDERATION FOR DIES

The design principles for Ehf closely resemble those of traditional die design procedures. Special attention must be given to ensuring effective air venting during the liquid discharge process. Dies designed for creating drawn shapes should incorporate mechanisms to isolate the liquid from the die cavity and facilitate the removal of trapped air behind the workpiece. Unlike traditional forming methods, Ehf does not require the use of a punch [13]. This technique is particularly valuable for shaping hollow body structures and can be employed to perform various operations, including beading, bulging, piercing, and trimming.

4.3.4 EHF PROCESS PARAMETERS

Optimal process parameters are crucial for achieving the desired product shape, form, and acceptable tolerances in high-rate energy forming processes. To optimize these parameters effectively, it is essential to have a clear understanding of the key factors that impact these processes [14–15], including the following:

- Capacitance of the capacitor bank.
- Energy storage capacity of capacitance.
- Impedance of the initiating wire.
- Voltage of the power supply.
- Charging time.
- Thickness, size, and type of material being formed.
- Gap between wires.
- Discharge voltage.
- Circuit inductance.
- Wire material.
- Wire size.
- Difficult to form materials with low critical impact velocity (CIV).

4.3.5 EHF APPLICATIONS

- It is used for bulging, forming, beading, drawing, blanking, and piercing operations.
- It is employed in the fabrication of stainless-steel nozzles that are difficult to create using other processes.
- It is utilized to create irregularly shaped holes in tubes.
- It can be used to form materials with poor conductivity.
- It can be used to produce neat, net-shaped automobile panels.
- It is used to insert insulator mandrels in metallic tubes.

4.3.6 ADVANTAGES OF EHF

- With the help of a single die, several components can be produced, resulting in high productivity.

- Hollow parts can be produced economically.
- The process can be easily automated, reducing the requirement for manpower.
- It is useful for producing small to medium-sized components.
- Superior metal-to-metal joints can be obtained.
- External dimensions with accuracy of ± 0.05 mm can be produced.

4.3.7 DISADVANTAGES OF EHF

- Materials with a CIV below 28–30 m/s are generally unsuitable for the process due to the requirement for a high CIV.
- High capital cost.
- The size and thickness of the workpiece are limited by the maximum energy of the processes.

4.4 EXPLOSIVE FORMING

The explosive forming process leverages the explosive force generated by a chemical reaction to shape metal parts. Although the concept of harnessing energy for forming is not novel, one of its earliest practical applications was pioneered by C.E. Munroe in 1888. This technique utilizes the energy released in the form of high-velocity shockwaves to rapidly reshape a workpiece, achieving deformation rates of several hundred meters per second [16]. Notably, explosive forming boasts the highest forming pressure among all HERF processes, reaching several thousand mega pascals. This substantial energy availability enables the shaping of workpieces as large as 6 meters in diameter with relative ease.

4.4.1 DESCRIPTION OF PROCESS

Explosive forming processes leverage the power of explosives to shape metal components. Explosives are inherently unstable compounds, and when they are disrupted, they generate detonation waves. These detonation reactions occur within mere microseconds, producing a substantial mass of gaseous products at extremely high temperatures and pressures, reaching up to 30,000 megapascals. These detonation or shock waves travel at remarkably high speeds, ranging from 3,000 to 10,000 meters per second. For a brief moment in time, the temperature at the wave front surges to an astonishing 10,000 °C [17, 22]. This tremendous energy can be harnessed directly for shaping a component or for pressurizing another medium, such as water, to facilitate the forming process. Figure 4.4 illustrates the various stages of explosive forming processes.

Facility requirements for explosive forming are as follows:

- Operating areas: These can include natural or river environments, artificial lakes, or ground-based tanks.
- Water supply: Includes water supply pumps, vacuum force pumps, and pressure gauges.

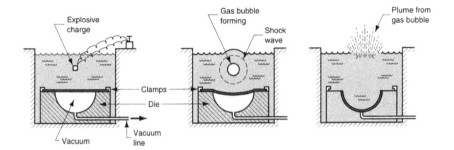

FIGURE 4.4   (a) Charging, (b) explosion, (c) formation of parts.

- Portable power supply: Includes gasoline generators.
- Explosive handling facility: Comprising storage magazines, shelters, and control barricades.
- Material handling facility: Equipped with cranes and equipment for transporting sheets and dies.
- Inspection areas.
- Die fabrication and maintenance area.

4.4.2   TYPES OF EXPLOSIVES

There are two types of explosives used in explosive forming: Low and High.

Gases from low explosives and gun cartridges are commonly used for light-duty applications. The low-explosive forming process is typically performed in a confining die that completely encloses the workpiece. The confinement maximizes the explosive energy, which is transferred to the workpieces. Low explosives are used to directly form parts or to pressurize media water or oil to then form parts. Low explosives also operate pistons, leveraging their kinetic energy to apply pressure to another medium.

High-explosive forming generates extreme pressures, reaching several thousand megapascals, within a few microseconds. When dealing with large workpieces, additional high-power explosives are often necessary. High explosives are known for their exceptional energy efficiency; for instance, a single kilogram of dynamite, which costs just a few hundred rupees, can release over 5 million joules of energy. The shock waves produced by high explosives are incredibly potent and do not require confinement, unlike low explosives [18].

4.4.3   EXPLOSIVE FORMING TECHNIQUES

There are two techniques of high explosive forming: standoff and contact. In the standoff technique, an explosive charge is placed at a stand-off distance from the workpiece, as shown in Figure 4.5. When explosives detonate, shock waves are generated, and a large number of gas bubbles are formed. When the bubbles suddenly collapse, high-pressure energy waves travel toward the workpiece, and the metal is plastically deformed against the die [19, 12].

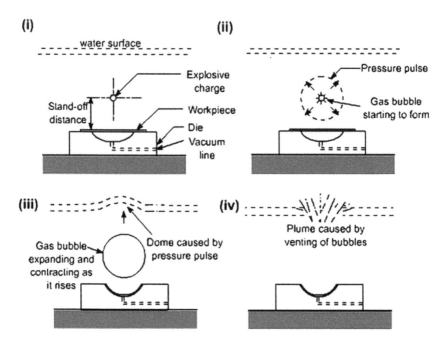

FIGURE 4.5   (a) Charging, (b) pressure development, (c) explosion, (d) formation of parts.

The high explosive is detonated, and its energy is transmitted through a fluid medium, commonly water; the workpiece is typically held on a female die. The space behind the workpiece is evacuated to avoid adiabatic compression and heating of the entrapped air. If not controlled, this could result in burning of the rear face of the workpiece and the die. The energy transfer medium need not always be water. Hot forming calls for molten salts, molten metals, sand, oil, or other media that can sustain the desired forming temperature. Air can also be used as an energy-transferring medium. Many other media are possible, but for the cold fabrication of large workpieces, cost and convenience considerations leave air and water as the most suitable choices. Water has reduced compressibility, and it has shock impedance matching characteristics.

The underwater detonation of a high explosive yields a shock wave and a bubble of hot gaseous detonation products. The shock wave is normally the major energy source in this process, although secondary bubble phenomena can become important under certain conditions. Generally, these explosions occur at relatively shallow depths, sometimes so shallow that the bubble breaks through the surface of the water. In explosive forming, the workpiece is deformed in about one or two milliseconds. Although the average strain rate is of the order of 10 to 100 per second, local strain rates in the vicinity of the die can be as high as 1000 per second [20].

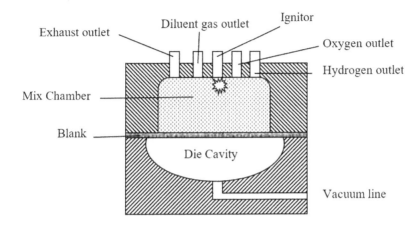

FIGURE 4.6   Contact type explosive forming (gas mixture type).

The gas mixture process is one of the direct contact techniques. A schematic diagram of the gas mixture process is shown in Figure 4.6. High-velocity forming with combustible gas mixtures releases energy through the chemical reaction of a fuel with oxygen. The energy released by the reaction is directly applied to the workpiece. The direct contact process can form a part with little springback. The effects of shock wave reflections from the water surface and the tank wall somewhat modify the simple analysis [21].

In the case of a water–air interface, shock reflections cause an inward-moving rarefaction, which reduces the impulse delivered to the workpiece if the water surface is relatively close to the charge. In general, any homogeneous explosive of reasonable sensitivity is suitable as an energy source. Dynamites and liquid explosives have been used. There appear to be no major advantages from explosives with high detonation velocity.

Gas mixtures have a number of documented characteristics. They can be placed in a ready-to-fire position without any requirement for rigging a charge. They assume the shape of their containers and therefore approach the optimum shape regardless of the quantity used. This removes the variable of charge shape and standoff distance, and shaping is important for providing uniform force. Gas mixtures can be easily adjusted within flammability limits by varying the amount of fuel and oxygen. They can be detonated like high explosives, although they are normally used in the adiabatic combustion range.

When a mixture of combustible gases is confined within a closed vessel and ignited, the reaction can proceed in one of three phases: adiabatic combustion, Unstable or transient detonation, or stable detonation. Under adiabatic combustion, the velocity of the flame front is subsonic, and the changes in pressure, temperature, and volume follow the laws of an adiabatic reaction. Extremely turbulent reactions are accompanied by severe shocks and overpressure, although this is a transient phase and is usually followed by stable detonation. Under stable detonation, the reaction occurs at a fixed supersonic velocity within a thin segment of the flame front, with the unburned gas ahead of the shock front at rest. These changes help to clarify the relationships between the components of the descriptions.

4.4.4 EXPLOSIVE FORMING PROCESS PARAMETERS

Optimizing process parameters is crucial for achieving the desired shape and tolerances in the final product in explosive forming. To do so, it is essential to understand which factors have a significant impact on the process outcome. The following list provides important process parameters for explosive forming processes:

- Fly distance/standoff distance.
- Impact velocity.
- Peak pressure.
- Energy flux density.
- Density of explosives.
- Maximum bubble radius.
- Charge weight.
- Pressure induced in the metal.
- Detonation pressure of the explosive.
- Initial density of the metal.
- Initial density of the explosives.
- Sound velocity of the metal.
- Detonation velocity of the explosives.
- Normal and oblique incidence of pressure wave.
- Combustion pressure in the case of gas mixtures.
- Vacuum in the die.

4.4.5 EXPLOSIVE FORMING

- It is used to make missile nose skin assemblies: 1000 mm in diameter, 3500 mm long, and 12.5 mm thick. The setback was only 0.025 mm for 25 mm.
- It is used to make the nozzle of the space shuttle.
- It is used to make jet engine corrugated panels, V-rocket panels, big domes, titanium alloy helicopter nozzles, and door panels of fighter aircraft.
- Small-scale forging can be done using explosive forming.
- Parts are used in the 3D metal forming of aircraft fuselage skins.
- Used in explosion welding operations.

4.4.6 ADVANTAGES OF EXPLOSIVE FORMING

- With the help of a single die, several components can be produced, which results in high productivity.

- There is very little springback in this process.
- Variation from part to part is minimal.
- It provides good dimensional tolerances.
- Parts that cannot be manufactured by conventional methods can be produced using this method.
- It is a very fast process compared with conventional methods.
- Exotic materials that are difficult to form can be processed using this method.
- This process is suitable for restriking operations.
- Less scrap is produced in this process, as there is no metal removal.
- Shock waves are effectively transmitted using a medium such as water.
- It produces less noise.
- Damage to workpieces is minimal.

4.4.7 Disadvantages of Explosive Forming

- An optimum stand-off distance is required.
- A vacuum is to be maintained in the die.
- Very large and thick dies are required.
- These processes are not suitable for small and thin parts.
- Storage, handling, and maintenance of explosives are challenging.
- Nonavailability of explosives due to government regulations.

REFERENCES

[1] D.J. Mynors and B. Zhang, Application and capabilities of explosive forming, *Journal of Materials Processing Technology*, vol. 1, no. 25, pp. 125–126, 2002.

[2] S.A. Tobias, The state of art of high energy forming processes, *Journal of Mechanical Working Technology*, vol. 9, no. 3, pp. 237–277, 1984.

[3] D. Banabi, *Sheet Metal Forming Processes: Constitutive Modelling and Numerical Simulation*. New York, Heidelberg, Dordrecht, London: Springer, 2010.

[4] M. Tisza and T. Fulop, A general overview of tribology of sheet metal forming a general overview of tribology of sheet, *Journal for Technology of Plasticity*, vol. 6, no. 2, pp. 11–25, 2001.

[5] B.H. Lee, Y.T. Keum and R.H. Wagoner, Modeling of the friction caused by lubrication and surface roughness in sheet metal forming, *Journal of Materials Processing Technology*, vol. 130, no. 31, pp. 60–63, 2002.

[6] D. Gayakwad, M.K. Dargara, P.K. Sharma, R. Purohit and R.S. Rana, A review on electromagnetic forming process, *Procedia Materials Science*, vol. 6, pp. 520–527, 2014.

[7] M.D. Kamal, A uniform pressure electromagnetic actuator for forming flat sheets, *Journal of Manufacturing Science and Engineering*, vol. 129, pp. 369–379, 2007.

[8] J.D. Thomas, M. Seth, G.S. Daehn, J.R. Bradley and N. Triantafyllidis, Forming limits for electromagnetically expanded aluminum alloy tubes: Theory and experiment, *Acta Materialia*, vol. 55, pp. 2863–2873, 2007.

[9] A. Mamalis, D. Manolakos, A. Kladas and A. Koumoutsos, Electromagnetic forming and power processing: Trend and developments, *Applied Mechanics Review (AMR)*, vol. 57, no. 4, pp. 299–324, 2004.

[10] V. Psyk, D. Risch, B.L. Kinsey, A.E. Tekkaya and M. Kleiner, Electromagnetic forming-a review, *Journal of Materials Processing Technology*, vol. 211, pp. 787–829, 2011.

[11] J.P.M. Correia, M.A. Siddiqui, S. Ahzi, S. Belouettar and R. Davies, A simple model to simulate electromagnetic sheet free bulging process. *International Journal of Mechanical Sciences*, vol. 50, no. 10, pp. 1466–1475, 2008.

[12] A. Bhaduri, High energy rate forming, *Mechanical Properties and Working of Metals and Alloys*. Singapore: Springer, 2018.

[13] T. Jimma, Y. Kasuga, N. Iwaki, O. Miyazawa, E. Mori, K. Ito and H. Hatano, An application of ultrasonic vibration to the deep-drawing process, *Journal of Materials Processing Technology*, vol. 80, pp. 406–412, 1998.

[14] S.F. Golovashchenko, Material formability and coil design in electromagnetic forming. *Journal of Materials Engineering and Performance*, vol. 16, no. 3, pp. 314–320, 2007.

[15] V. Psyk and D. Risc, High velocity forming, *Sheet Metal Forming—Processes and Applications*. ASM International, Ohio, USA. pp. 227–245, 2012.

[16] T. Aizawa, Magnetic pressure seam welding method for Aluminum sheets. *Journal of Light Metal Welding and Construction*, vol. 41, no. 3, pp. 20–25, 2003.

[17] N. Nariman-Zadeh, A. Darvizeh, A. Jamali and A. Moeini, Evolutionary design of generalized polynomial neural networks for modelling and prediction of explosive forming process, *Journal of Materials Processing Technology*, vol. 164–165, pp. 1561–1571, 2005.

[18] R. Kumar, K.S. Matwa, K. Punaisiya and D.S. Kore, Numerical and experimental analysis of electrohydraulic forming of Al 5052 and Al 6061 using conical die, *The International Journal of Advanced Manufacturing Technology*, vol. 122, no. 1, pp. 1–23, 2022.

[19] R.M. Miranda, B. Tomás, T.G. Santos and N. Fernandes, Magnetic pulse welding on the cutting edge of industrial applications, *Soldagem & Inspeção*, vol. 19, no. 1, 2014.

[20] S.D. Kore and P.P Kulkarni, Effect of process parameters on electromagnetic welding of Aluminum sheets, *International Journal of Impact Engineering*, vol. 34, pp. 1327–1341, 2006.

[21] Y. Zhang, S.S. Babu, C. Prothe, M. Blakely, J. Kwasegroch, M. LaHa and G.S. Daehn, Application of high velocity impact welding at varied different length scales, *Journal of Materials Processing Technology*, vol. 21, pp. 944–952, 2011. doi: 10.1016/j.jmatprotec.2010.01.001.

[22] M. Groover, *Fundamentals of Modern Manufacturing*, Hoboken, NJ: Wiley, 2007.

5 The Formability of Aluminium and Its Alloy Sheet Metals
A Review

S. Vigneshwaran, C. Sathiya Narayanan,
R. Narayanasamy, and K. Sivaprasad

5.1 INTRODUCTION

Sheet metal forming involves a group of manufacturing processes in which the given flat sheet metal is shaped to the required size without changing the mass, volume and material composition. However, the forming operation changes the surface and mechanical properties of the metals. The required shape and size of sheet metals are obtained by the application of mechanical forces. These forces may be tensile, compressive, a combination of tensile and compression, bending or shear. Forming is usually carried out with the help of machines such as hydraulic or mechanical press and bending machines. Generally, sheet metal forming is carried out at room temperature or sometimes at warm temperatures. Sheet metal is used in automobile bodies, airplane fuselages and wings, cans and buildings, apart from other applications.

Formability is the capacity of sheet metals to undergo plastic deformation to a given shape without any defect; two extreme processes of forming are biaxial stretching and deep drawing, which each impose certain limits on formability. In biaxial stretching, both principal stresses in the plane of the sheet are tensile and cause thinning of the sheet. Material behaviour in biaxial stretching is measured by a bulge test or a stretch test using a hemisphere-shaped punch, whereas deep drawing involves both tensile and compressive stresses that causes relatively little change in thickness; the swift cup test is used to assess deep drawability.

In this cup test, the drawability of sheet metal is limited by some constraints. The most commonly observed failure modes in this drawing process are metal fracture, wrinkling or buckling. The forming limit diagram (FLD) is an effective tool for evaluating the formability of sheet metals under various strain or stress conditions. Formability information is important for both manufacturers and end users of sheet metals.

DOI: 10.1201/9781003441755-5

5.2 FORMING LIMIT DIAGRAM

FLD is a limiting criterion of failure for sheet metal forming operations. It predicts the onset of necking in linear straining paths. Figure 5.1 illustrates a schematic diagram of FLD. FLD is generally used to predict necking and fracture in products that are formed but is also used to redesign the forming process. The two sub-branches of the forming limit curve are the right side for positive minor strain that is stretching and the left side for negative minor strain that is drawing.

FLD is generally used for conventional sheet metal forming processes. As explained by Paul [1], circular grid patterns are turned to ellipses during or after the plastic deformation as explained in the Figure 5.1. The major strain and minor strain can be determined from the major and minor axes of ellipse, as explained by Narayanasamy and Sathiya Narayanan [2]. Since industrial sheet metal forming operations follow non-linear strain paths, a safety margin of 10–20 percent is generally used by offsetting the forming limit curve, as explained by Drotleft and Liewald [3]. This safety margin also controls the statistical scatter of experimental data points.

5.3 NECKING FAILURE

At the time of the uniaxial tensile test, the elongation or strain (which corresponds to the maximum load point or ultimate tensile load point) is followed by diffuse and localized necking of the sheet metal (see Figures 5.2 and 5.3). The rise in stress because of the reduction in cross section of the tensile specimen is

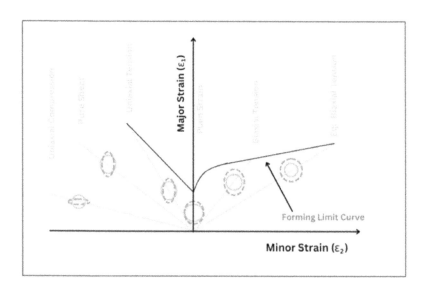

FIGURE 5.1   Schematic representation of a forming limit diagram (redrawn from [1]).

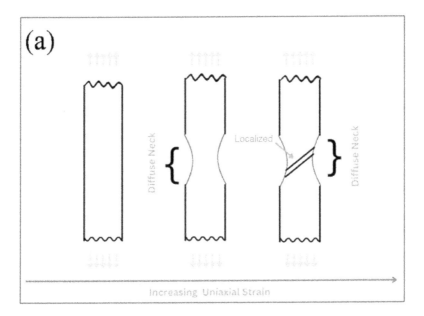

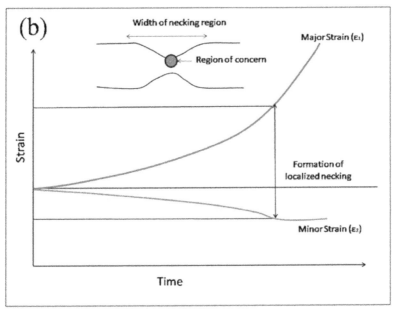

FIGURE 5.2   (a) Represents diffuse necking and localized necking during the uniaxial tension test; (b) variations in major strain and minor strain over time (redrawn from [1]).

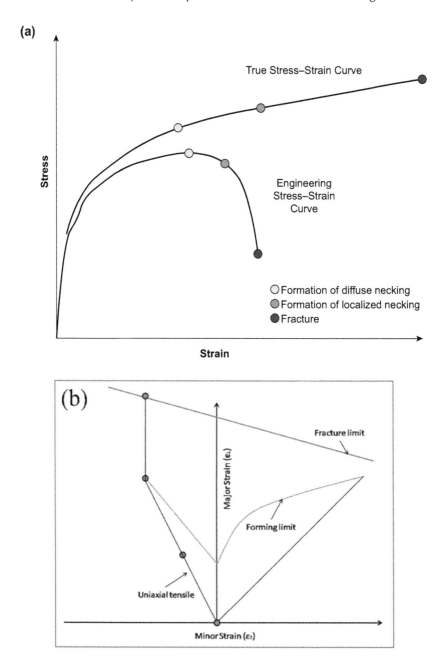

FIGURE 5.3 (a) The difference between the engineering and true stress–strain curve; (b) forming and fracture limit diagram (redrawn from [1]).

greater than the load-bearing capacity of the sheet metal owing to the effect of work hardening at ambient temperature. The true strain, which corresponds to the maximum engineering stress, is nothing but the uniform true strain, which is proportional to strain-hardening exponent (n) of the Ludwik material law.

Two types of necking beyond the maximum engineering stress or ultimate tensile stress, namely, diffuse necking and localized necking, represent instability in plastic flow during either uniaxial or biaxial tensile loading. Once maximum tensile load or maximum engineering stress is reached, diffuse necking immediately follows during the plastic deformation. Diffuse necking may end up with fracture; however, it is followed by localized necking (second instability process) during the tensile loading of sheet metals, as explained by Dieter [4].

In localized necking, the neck is a narrow band having a width more or less equal to the thickness of the sheet metal that is inclined to the specimen axis. Moreover, during localized necking, there is no change in the width of a tensile sheet specimen that corresponds to plane-strain deformation. With increasing deformation, the thickness drastically decreases within the band, finally fracture takes place.

Figure 5.3(b) presents two sheet metal failure curves described by Paul [1], namely, the forming limit curve and the fracture limit curve for failures under various loading conditions. The region between the forming and fracture limit curves is the necking region, in which diffuse and localized necking dominate during the plastic deformation of sheet metals. This process describes the plastic behaviour of sheet metals which are ductile in nature.

FLD can help engineers and researchers in material selection, process design and failure studies, as explained by Embury and Duncan [5]. Both diffuse and localized necking are influenced by the yield-criteria constant (m, 1.6 for Al and its alloys), the strain-hardening exponent (n), the relative density (R_d), and the normal anisotropy coefficient (R). With increasing m or n, as explained by Ponalagusamy and Narayanasamy [6], the major strain of the fracture limit curve increases for any minor strain. As R_d and R increase, both forming limit and fracture limit curve lines shift upwards.

5.4 DRAWABILITY

The drawability of sheet metal is predicted by the limiting draw ratio (LDR), which is the ratio of the maximum blank diameter drawn out from the die (without tearing) to the diameter of the punch. The LDR that can be obtained for Al and its alloys varies from 1.6 to 1.8, compared with that ratio ranges from 2.1 to 2.3 obtained for conventional deep drawing grade steels, as indicated by Narayanasamy Et al. [7]. Redrawing operations are being carried out in industries to increase current LDRs.

Certain specific parameters affect LDR, namely, the radius of the die, punch, clearance gap between punch and die, hold-down pressure and lubrication. Controlling the crystallographic texture can improve the drawability. The correct orientation of the texture (for Al and its alloys {111} planes and <110> directions) improves the drawability, as explained by Narayanasamy [8]. The plastic strain ratio (R) is the ratio of the strain that occurs along the width to the strain that occurs along the thickness during the uniaxial tensile test. This R indicates the sheet metals' resistance along the thickness direction during forming.

R is proportional to the LDR: Increasing R increases in LDR as the sheet metal offers more resistance to thinning. Moreover, when R increases, the major axis of the yield locus ellipse increases and the minor axis decreases, Figure 5.4. This makes the flange deformation (drawing [tensile] stress; hoop [compressive] stress) much easier, which enhances the drawability. As R increases, the major axis of the ellipse increases, which enhances the resistance to thinning (in the thickness direction), thereby improving drawability, as shown in Figure 5.5.

As illustrated in Figure 5.6, the blank hold-down force affects the LDR; specifically, when the blank hold-down force is very low, the drawability is limited by wall wrinkling, whereas for very high force, the drawability is limited by the bottom fracture of the cup during the drawing operation. As explained by Narayanasamy et al. [9], the LDR can be improved with a lower frictional coefficient, a higher strain-hardening exponent (n), lower yield stress, higher normal anisotropy (R) and higher strain rate sensitivity (m) values, as illustrated in Figure 5.7 (a)–(d). For Al and its alloys, R varies from 0.50 to 0.80, as provided by Narayanasamy [7].

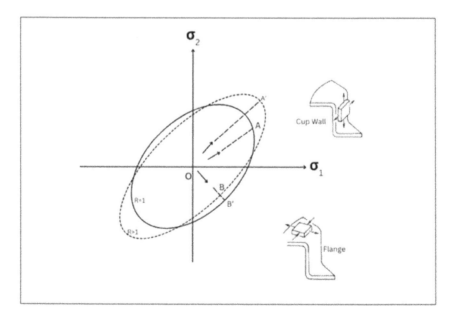

FIGURE 5.4   Variations in the yield locus with respect to R (redrawn from [8]).

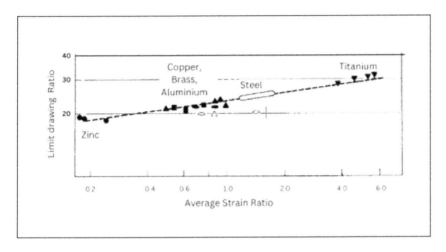

FIGURE 5.5 Variations in LDR with respect to average plastic strain ratio (redrawn from [8]).

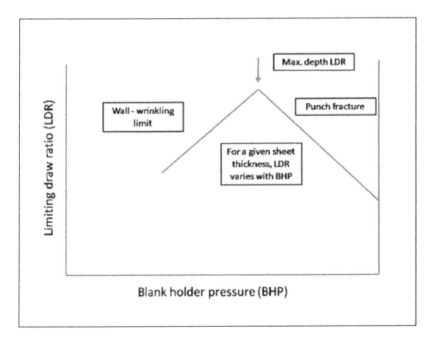

FIGURE 5.6 Changes in LDR with respect to blank hold-down pressure.

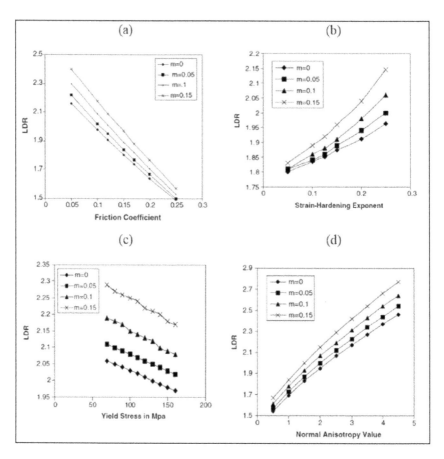

FIGURE 5.7 Variations in LDR with respect to (a) friction coefficient, (b) strain-hardening exponent, (c) yield stress and (d) normal anisotropy R [9].

5.5 WRINKLING

During deep draw operation, the sheet metal experiences negative hoop stress along the circumference. As the hoop stress attains a critical value, the onset of wrinkling takes place. When the hoop strain is plotted against the radial strain, the slope of the variation in the hoop stress can be measured. The slope suddenly changes during the onset of wrinkling.

As explained by Loganathan and Narayanasamy [10], aluminium with higher R, high n, low yield stress and high normalized hardening rate show better resistance against wrinkling during drawing through conical and tractrix dies in which no blank hold-down pressure can be applied. When drawing with the tractrix die, the redundant work can be drastically reduced [11], which improves the drawability compared with drawing with the conical die. Further, the tractrix die can accept high compressive hoop stress before the start of wrinkling when compared to conical die.

Dry frictional condition and flat bottom punch improve the resistance against wrinkling, thereby increasing the LDR or drawability. As provided in [12], higher annealing temperatures of aluminium (Al) alloy grade 5086 improve n and R, thereby increasing the resistance against wrinkling. Higher annealing temperature lowers the yield stress, which makes the Al alloy easy to form during drawing operation. Soft drinks and beer cans are manufactured by drawing Al alloys through tractrix die, followed by ironing as described by Narayanasamy [11]. Controlling the crystallographic texture can reduce the earing (wavy edges) that forms on the Al alloys or make the edges uniform.

As Narayanasamy et al. [12] established, there is a definite limit curve for the onset of wrinkling which can be represented in terms of the strain increment ratio and the stress ratio. In the strain increment ratio diagram, the safe region window enhances for Al 5086 annealed at 300 °C. For this temperature of annealing, the above alloy exhibits high n, high R and high ratio of ultimate tensile stress to yield stress. Further, the plasticity theory [12] results are in good agreement with the experimental values.

5.6 FORMING, FRACTURE AND WRINKLING LIMIT

As explained by Narayanasamy et al. [13], the forming, fracture and wrinkling limit diagrams for different annealing temperatures are illustrated in Figures 5.8–5.10, and the forming, fracture and wrinkling limits for a constant minor strain are shown in Figure 5.11. In biaxial tension (i.e. right side FLD, tension–tension [T–T] region), there are only two limiting lines, and no wrinkling occurs. The central portion of the major strain is called plane strain (PS). The left side of the diagram indicates the deep draw, the tension–compression (T–C) region. This T–C region is essential as the hoop strain that causes wrinkling is negative.

When both the principal and minor strains remain below the forming limit, plastic deformation is considered safe. Once the deformation strain reaches the fracture limit line, it leads to tearing or fracture. Wrinkling occurs when both major and minor strain fall below the wrinkling limit line or curve. A larger gap between the wrinkling limit and forming limit lines, given a fixed minor strain, indicates a greater suitability for forming in the T–C region, specifically in deep drawing conditions.

It is evident that Al 5086 annealed at 300 °C is exceptionally suitable for deep drawing. This is because it exhibits a substantial gap between the forming limit line and wrinkling limit line for a fixed minor strain. This gap for the fixed minor strain was notably larger (0–20) with the annealing at 300 °C than at the other two annealing temperatures, namely, 250 °C (with a gap of 0–16) and 200 °C (with a gap of 0.12). Therefore, it is established that as the annealing temperature increases, the safe working region for this aluminium alloy expands. Similarly, the forming limit and fracture limit lines or curves shift towards higher major strain for any given fixed minor strain as the annealing temperature increases.

As observed earlier, at 300 °C, n (strain-hardening exponent), R (strength coefficient) and UTS/σy (ultimate tensile strength to yield strength ratio) are elevated,

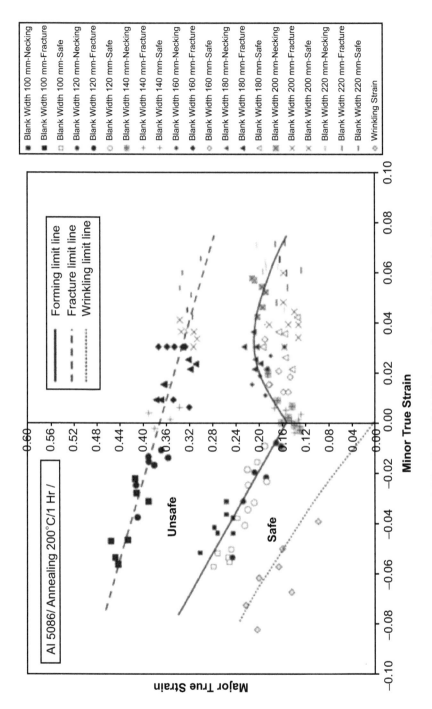

FIGURE 5.8 Forming, fracture and wrinkling limits of Al 5086 annealed at 200 °C for 1 hr [13].

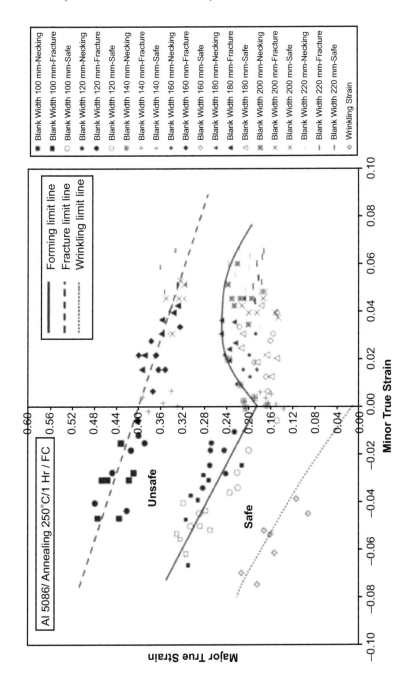

FIGURE 5.9 Forming, fracture and wrinkling limits of Al 5086 annealed at 250 °C for 1 hr [13].

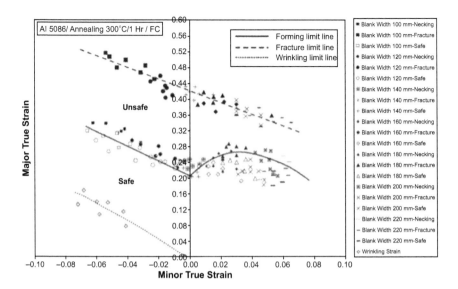

FIGURE 5.10 Forming, fracture and wrinkling limits of Al 5086 annealed at 300 °C for 1 hr [13].

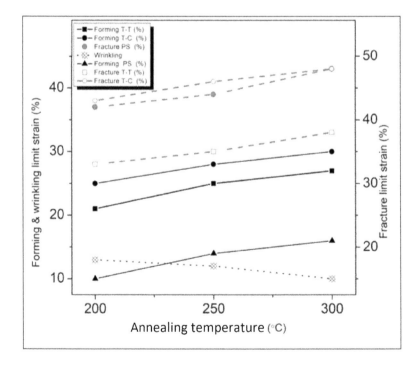

FIGURE 5.11 Forming, fracture and wrinkling limit strain for a constant minor strain of 4 percent.

TABLE 5.1
Average Void Size in Microns for Alloy Al 5086 [13]

S. No.	Blank width (mm)	Average void size in microns for different annealing temperatures		
		200 °C	250 °C	300 °C
1	220	3.7	5.3	5.68
2	200	4.22	5.48	6.36
3	180	4.4	6.82	7.59
4	160	5.21	8.21	11.94
5	140	7.18	9.08	12.08
6	120	7.85	10.56	13.12
7	100	8.55	11.19	14.31

enhancing resistance against wrinkling and improving the formability of Al 5086. These findings align with the average void size measurements observed for Al 5086 annealed at different temperatures, as indicated in Table 5.1. Since the higher annealing temperature allows for more extensive plastic deformation, the average void size at 300 °C is larger than it is at the other two temperatures, as shown in Table 5.1. It can be concluded that aluminium alloys characterized by high n values, high R values and a high UTS/σy ratio enhance overall formability during drawing and stretching processes.

5.7 EFFECTS OF ANNEALING TEMPERATURE ON VOID PROPERTIES

As explained by Velmanirajan et al. [14], the void size and void area fraction increase with an increase in the annealing temperature for Al 1145 as shown in Figures 5.12–5.14. In the T–C region, the plastic deformation is greater due to the combination of tensile and compressive forces than it is in the other two regions, T–T and PS. Therefore, the void size and fraction obtained in the T–C region are always greater than those obtained in other two regions. The ligament thickness (gap between nearby voids) is higher at higher annealing temperature (350 °C) than at annealing temperatures because higher temperatures accommodate more plastic deformation before failure. The length to width (L/W) ratio of the void is lower in T–C than in T–T and at higher annealing temperatures is lower in T–C than in PS and T–T as the main strain (due to hydrostatic stress) increases the void size and void area fraction decreases (Figure 5.15) in general.

When the strain triaxiality is low, the Al alloy accommodates more plastic deformation at higher annealing temperature due to the development of higher

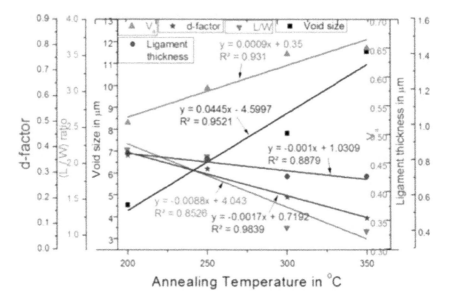

FIGURE 5.12   Changes in void properties with respect to annealing temperature in the T–C region [14].

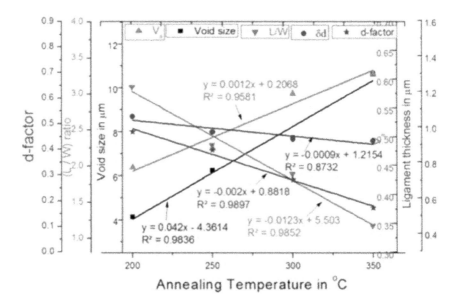

FIGURE 5.13   Changes in void properties with respect to annealing temperature in the PS region [14].

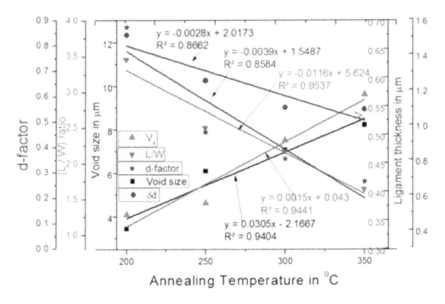

FIGURE 5.14 Changes in void properties with respect to annealing temperature in the T–T region [14].

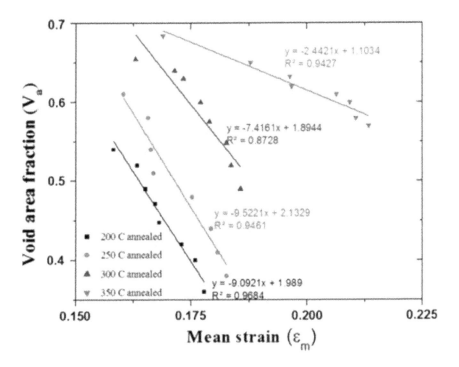

FIGURE 5.15 Void area fraction v/s mean strain at various annealing temperatures [14].

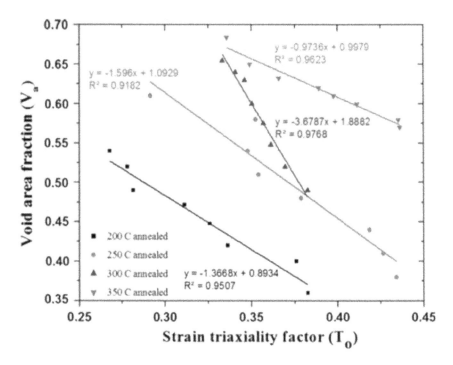

FIGURE 5.16   Strain triaxiality v/s void area fraction at various annealing temperatures [14].

hydrostatic stress as shown in Figure 5.16. The strain triaxiality is just the ratio of mean strain to effective strain.

5.8   EFFECTS OF CRYO-ROLLING ON FORMABILITY

As explained by Sekhar et al. [15, 16], very high strength with moderate formability can be obtained in the case of Al alloys when subjected to cryo-rolling followed by short annealing. High strength with fairly good formability is a requirement for automobile and aerospace applications. Since cryo-rolling is a severe plastic deformation process, very high strength can be obtained because of higher dislocation density developed in Al and its alloys.

Here, the higher strength obtained is due to the development of an ultra-fine-grain microstructure with few nano grains. The ultra-fine-grain structure enhances the strength of Al alloys, and the little coarse grain microstructure accommodates more or moderate plastic deformation during forming operations. As shown in Tables 5.2 and 5.3, the strain-hardening exponent is almost equal to zero. Therefore, Al alloys subject to cryo-rolling demonstrate ideal plastic deformation in which very high necking strain can be obtained during forming operations, in contrast with conventional Al alloys.

TABLE 5.2
Mechanical Properties of Room-Temperature-Rolled AA 5052 Alloy Sheet [15]

Orientation	Yield stress (0.2% strain) (MPa)	Ultimate tensile stress (MPa)	Elongation to failure (%)	Exponent of strain hardening (n)	Strength coefficient (K) in MPa	Plastic strain ratio (R)
0°	76.2	177.4	4.32	0.038	213 MPa	0.172
45°	124.5	190.6	5.18	0.043	228 MPa	0.305
90°	105	178.8	4.80	0.042	216 MPa	0.498
Average	107.5	184.35	4.87	0.041	221 MPa	0.32

TABLE 5.3
Mechanical Properties of Cryo-Rolled AA5052 Alloy Sheet [15]

Orientation	Yield stress (0.2% strain), (MPa)	Ultimate tensile stress (MPa)	Elongation to failure (%)	Exponent of strain hardening (n)	Strength coefficient (K) in MPa	Plastic strain ratio (R)
0°	58	220	4.7%	0.043	261	0.475
45°	202	218	5.0%	0.042	265	0.5
90°	150	223	4.9%	0.041	263	0.335
Average	153	220	4.91%	0.042	263	0.452

The percentage strain is much greater with diffuse necking than with localized necking. The necking strain percentage obtained for cryo-rolled Al alloys (compared with conventional forming of Al and its alloys) is greater in the PS region than in T–T and T–C as shown in Figure 5.17. Experiments with cryo-rolled Al and its alloys followed by short annealing are being carried out in laboratories, but these alloys have not yet been developed for mass production. Many scientists are carrying out research in the area of ultra-fine structured Al alloys in order to improve the formability of sheet metals. This has applications in automotive and aerospace industries where low cost and fuel economy are critical with respect to minimizing negative environmental effects.

The fracture surface morphology of rolled Al-3Mg-0.25Sc alloys [17] after forming under various strain conditions are given in Figure 5.18. The fracture surface observed for room-temperature-rolled sample R-50 (Figure 5.18 (a-c)) and cryo-rolled samples C-50 (Figure 5.18 (d-e)), R-75 (Figure 5.18 (f-h)) and C-75 (Figure 5.18 (i-k)) show microvoids indicating ductile failure.

It is observed that in R-50 (Figure 5.18 (a-c)), the void size (VS) decreases from T–C to T–T strain condition due to the formation of higher shear strain at the time of the forming experiment. This high shear strain increased the VS in

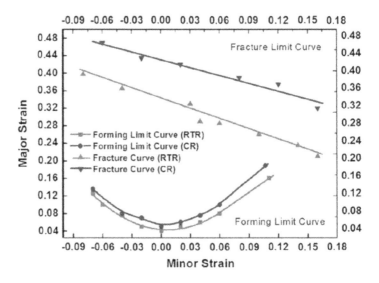

FIGURE 5.17 Comparison of forming limit diagrams and fracture limit curves for room and cryo-rolled Al 5052 [15].

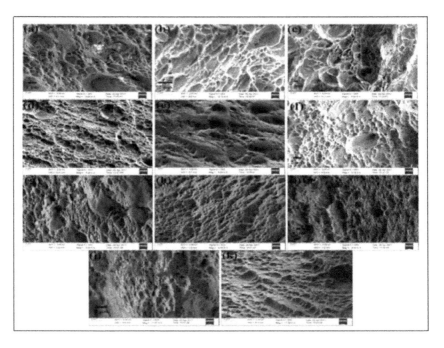

FIGURE 5.18 Fracture surface morphology of room-temperature-rolled (R) and cryo-rolled (C) Al-3Mg-0.25Sc alloy after formability experiments: (a) R-50 (T–T), (b) R-50 (PS), (c) R-50 (T–C), (d) C-50 (T–T), (e) C-50 (T–C), (f) R-75 (T–T), (g) R-75(PS), (h) R-75 (T–C), (i) C-75 (T–T), (j) C-75 (PS), (k) C-75 (T–C) [17].

T–C compared with PS and T–T strain conditions. Additionally, VS decreases from T–C to T–T in all the other conditions as well: C-50, R-75 and C-75. This is because the combined tensile and compressive principal stresses yield maximum shear stress and result in better formability in the T–C region than in PS and T–T.

In line with this, the fracture surface morphology of rolled samples show bigger VS in the T–C strain condition (refer to Figure 5.18 (a), (d), (f) and (i)). On the other hand, the VA increases for lower strain triaxiality and decreases for higher strain triaxiality for the rolled samples, irrespective of the process conditions. The surface morphology also shows more microvoids in cryo-rolled samples because of the formation of more ultra-fine grains (UFGs) that resulted in lower VA for the cryo-rolled samples than those rolled at room temperature. The presence of UFGs is evidenced in the microstructure of the cryo-rolled samples (C-50 and C-75), whereas the room-temperature-rolled samples show micron-sized grains

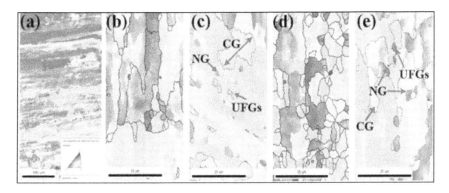

FIGURE 5.19 Electron backscattered diffraction orientation maps of Al-3Mg-0.25Sc alloy: (a) base material, (b) R-50, (c) C-50, (d) R-75, (e) C-75 [18].

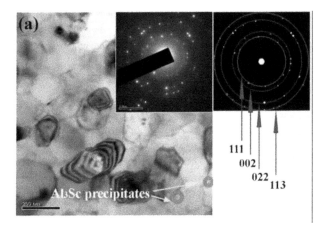

FIGURE 5.20 Transmission electron microscopy images of Al-3Mg-0.25Sc alloy: (a) base material, (b) R-50, (c) C-50, (d) R-75, € C-75 [17].

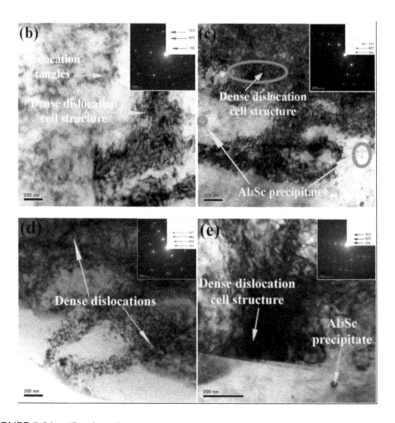

FIGURE 5.20 (Continued)

as depicted in the electron backscattered diffraction orientation images provided in Figure 5.19 (a–e). Apart from the presence of UFGs, the formation of dense dislocation cell structures of cryo-rolled samples (refer Figure 5.20) is responsible for the enhanced strength and fracture resistance of cryo-rolled Al alloys.

5.9 CONCLUSIONS

This chapter presents the influence of limiting strain determination, die geometry, microstructures and tensile properties on forming limit curves. The following conclusions can be derived based on the present study.

- Uniaxial tensile properties, namely, strain-hardening exponent, normal anisotropy, strain rate sensitivity, yield stress, the ratio of ultimate tensile strength to yield stress, normalized hardening rate, yield function constant and relative density have correlations with wrinkling limit, forming limit and forming fracture limit curves.

- Drawability is affected by the coefficient of friction, yield stress, strain-hardening exponent, normal anisotropy and strain rate sensitivity.
- Higher annealing temperature of Al and its alloys makes the sheet metal accommodate more plastic deformation. Therefore, larger void size can be obtained during fracture.
- Die and punch profile affect the drawability of Al alloy sheet metals. A tractrix die gives better drawability than a conical die.
- The void coalescence parameters can be compared with the formability of sheet metals.

5.10 ACKNOWLEDGMENTS

The authors thank Mr. Arun Prasad M. and Mr. Vignesh G., doctoral degree scholars, for their assistance in typing the manuscript.

REFERENCES

[1]. Paul, S.K., 2021. Controlling factors of forming limit curve: A review. *Advances in Industrial and Manufacturing Engineering*, 2, p. 100033.

[2]. Narayanasamy, R. and Narayanan, C.S., 2006. Some aspects on fracture limit diagram developed for different steel sheets. *Materials Science and Engineering: A*, *417*(1–2), pp. 197–224.

[3]. Drotleff, K. and Liewald, M., 2018, September. Application of an advanced necking criterion for nonlinear strain paths to a complex sheet metal forming component. In *IOP Conference Series: Materials Science and Engineering* (Vol. 418, No. 1, p. 012041), Toronto, ON.

[4]. Dieter, G.E. and Bacon, D., 1976. *Mechanical Metallurgy* (Vol. 3, pp. 43–53). New York: McGraw-hill.

[5]. Embury, J.D. and Duncan, J.L., 1981. Formability maps. *Annual Review of Materials Science*, *11*(1), pp. 505–521.

[6]. Ponalagusamy, R., Narayanasamy, R. and Subramanian, K.R., 2007. Prediction of limit strains in sheet metals by using new generalized yield criteria. *Materials & Design*, *28*(3), pp. 913–920.

[7]. Narayanaswamy, R., Doraivelu, S.M., Gopinathan, V. and Venkatesh, V.C., 1982. A comparative study of deep drawing with conventional, isostatic, and hydrostatic pressure. *Journal of Mechanical Working Technology*, *6*(2–3), pp. 227–234.

[8]. Narayanasamy, R., 2000. Metal forming technology. *Ahuja Book Company*, Second edition 2000, ISBN 81-7619-002-0, New Delhi—110002, India.

[9]. Narayanasamy, R., Ponalagusamy, R. and Raghuraman, S., 2008. The effect of strain rate sensitivity on theoretical prediction of limiting draw ratio for cylindrical cup drawing process. *Materials & Design*, *29*(4), pp. 884–890.

[10]. Loganathan, C. and Narayanasamy, R., 2005. Effect of mechanical properties on the wrinkling behaviour of three different commercially pure aluminium grades when

drawn through conical and tractrix dies. *Materials Science and Engineering: A*, *406*(1–2), pp. 229–253.

[11]. Narayanasamy, R., 1992. *Drawability of sheet metals through conical and tractrix dies*, Ph.D. thesis, Regional Engineering College, Tiruchirappalli—620015, TN, India.

[12]. Narayanasamy, R., Satheesh, J. and Loganathan, C., 2008. Effect of mechanical properties on wrinkling limit diagrams for Aluminum 5086 alloy annealed at different temperature. *Journal of Materials Science*, *43*, pp. 43–54.

[13]. Narayanasamy, R., Satheesh, J. and Sathiya Narayanan, C., 2008. Effect of annealing on combined forming, fracture and wrinkling limit diagram of Aluminium 5086 alloy sheets. *International Journal of Mechanics and Materials in Design*, *4*, pp. 31–43.

[14]. Velmanirajan, K., Thaheer, A.S.A., Narayanasamy, R., Madhavan, R. and Suwas, S., 2013. Effect of annealing temperature in Al 1145 alloy sheets on formability, void coalescence, and texture analysis. *Journal of Materials Engineering and Performance*, *22*, pp. 1091–1107.

[15]. Sekhar, K.C., Narayanasamy, R. and Velmanirajan, K., 2014. Experimental investigations on microstructure and formability of cryorolled AA 5052 sheets. *Materials & Design*, *53*, pp. 1064–1070.

[16]. Sekhar, K.C., Narayanasamy, R. and Venkateswarlu, K., 2014. Formability, fracture and void coalescence analysis of a cryorolled Al—Mg—Si alloy. *Materials & Design*, *57*, pp. 351–359.

[17]. Vigneshwaran, S., Sivaprasad, K., Narayanasamy, R. and Venkateswarlu, K., 2018. Formability and fracture behaviour of cryorolled Al-3 Mg-0.25 Sc alloy. *Materials Science and Engineering: A*, *721*, pp. 14–21.

[18]. Vigneshwaran, S., Sivaprasad, K., Narayanasamy, R. and Venkateswarlu, K., 2019. Microstructure and mechanical properties of Al—3Mg—0.25 Sc alloy sheets produced by cryorolling. *Materials Science and Engineering: A*, *740*, pp. 49–62.

6 Analysis of Deep Drawing Quality Steel Using Incremental Hole Flanging with Different Pre-Cut Hole Diameters

G. Praveen Kumar, Din Bandhu,
Ravi Kumar Mandava, Balaji Krushna
Potnuru, and S. Abdul Azeez

6.1 INTRODUCTION

The process of flanging involves deforming the edges of a metal sheet to produce a curved and reinforced edge, resulting in enhanced structural integrity. The process can be partitioned into multiple sub-operations, such as hole flanging, shrink flanging, tube flanging, and stretch flanging [1]. Among these techniques, conventional hole flanging (CHF) has found widespread use in a variety of automotive and aerospace industries [2].

The production of short vertical or conical flanges involves the plastic deformation of a blank that has a pre-cut hole, utilizing a conical or dome-shaped punch. The aforementioned flanges ultimately establish a connection with the primary components of the mainframe. The hole expansion ratio (HER) is a metric used to measure the degree of flanging of a hole. It is used to quantify the deformation of the flanged portion. This ratio is dependent on the inner diameter of the final flange (d_f) and the original pre-cut hole diameter (d_i), as demonstrated in Eq. 6.1 [3]:

$$HER = \frac{d_f}{d_i} \tag{6.1}$$

The limiting forming ratio (LFR) is a common metric used to characterize a material's formability CHF. The LFR is equivalent to the maximal HER at which the material will not fail. According to Eq. 6.2 [4], it is the ratio between the

DOI: 10.1201/9781003441755-6

smallest diameter of the pre-cut hole in the blank and the largest diameter of the final flange:

$$LFR = HER_{max} = \frac{(d_f)_{max}}{(d_i)_{min}}$$ (6.2)

The LFR pertains to the correlation between the circumference of the punch and the minimum circumference of the elliptical pre-cut hole that can be drawn by the punch without experiencing failure. This phenomenon is observed in elliptical flanges. The aperture of the elongated orifice experiences the greatest reduction in thickness towards the terminus of its principal axis. According to the report cited in reference [3], the anticipated results exhibited a typical bell-shaped distribution for the variation of punch load with punch stroke.

Several reports have focused on the relationship between CHF process variables, deformation techniques, and failure. Several significant variables, such as strain-hardening exponent, initial yield stress, sheet thickness, plastic anisotropy, lubrication condition, edge quality of the pre-cut hole, punch shapes, clearance between tools, and blank holding force [5–9], have been demonstrated to impact the process mechanics in CHF. The authors of [5] investigated the effects of varying punch semi-cone angles, ranging from 15° to 90°, on the punch load, formability, flange shape, and thickness of the steel sheets. The results showed that the flange form was strongly impacted by the cone semi-angles. Large cone angles cause springback, while lower cone semi-angles cause spring forward. The LFR does not depend on the punch's semi-cone angle. However, a semi-cone angle dramatically raises the punch's maximal force. The maximum force also decreases linearly as the diameter of the drilled hole grows larger.

A noteworthy observation is that the wall thickness exhibits a linear decrease as the depth increases, and the thickness of the fracture resulting from hole flanging corresponds to the thickness of the fracture observed during a simple tension test. Finite element models [8] were used to examine the impact of punch shape using four punch forms, flat, ellipsoid, conical, and spherical frustum, in CHF. While the strain route was shown to be form independent, the results indicated that maximum punch load was affected by punch shape. The elliptical punch had the lightest load, while the flat punch had the heaviest. The punch shape was observed to alter the hole expansion limit in a similar investigation on DDQ steels using cylindrical, conical, and spherical punches [10]. When flanging, the hole expanded the most when the punch was conical and the least when it was cylindrical (flat-bottomed). No matter the punch geometry, HER will rise as the strain-hardening exponent and plastic anisotropy increase.

Experimental and computational modeling were used to examine the impact of edge quality on pre-cut holes produced with conventional punching (CP) and fine blanking (FB) [6]. For the present study, we employed finite element analysis (FEA) in DEFORM software, utilizing the Ayada fracture framework, to simulate the two blanking techniques. The results of the simulations indicate that CP caused more damage to the hole than FB. This resulted in the effective flange height being lowered since crack propagation began in the central region of the CP hole flanging process. In contrast, we observed that the flange lip remained intact throughout the FB flanging procedure, resulting in an elevation of the flange's height.

Researchers have investigated the formability, lip shape, and fracture attributes of high-strength steel sheets, including transformation-induced plasticity (TRIP) steels and ferrite–bainite steels, through hole-flanging investigations [7]. Researchers have conducted hole flanging using various methods, including a stationary blank holder, no blank holder, a consistent blank-holding force, and a gradually increasing blank-holding force [9]. Since the maximum punch load was found to be unaffected by the blank-holding situation, this meant that only the kinematics of the workpiece's outside edge, which eventually determined the final product's form, were significantly impacted.

Tensile cracks propagate outward from the hole's rim in CHF. Therefore, CHF fracture is predicted using the fracture thickness strain measured during a standard tensile test [11]. Fracture during the hole expansion process has been predicted using a variety of ductile damage models [12–14], including Gurson–Tvergaard–Needleman (GTN), Oyane's, Lemaitre, and Cockroft and Latham. Oyane's ductile damage model constants can be determined through the utilization of basic uniaxial tensile and plane-strain examinations [12]. Pre-cut holes of 10 mm in diameter were drilled into high-strength steel blanks and mild steel, and the created Oyane model was found to be suitable for forecasting fracture start and the critical stroke in hole enlargement. Predictions of oil filter cover fracture in the hole flanging were also made using ductile damage models [13] developed by Cockroft and Latham, Oyane, and Lemaitre. There was a correlation between the macroscopic deformability results and the local strain distributions between phases at the time of failure.

Authors of two literature works [15, 16] on hole flanging demonstrated how material parameters like plastic strain ratio (r_m), total transverse elongation (e_t), and strain-hardening exponent (n) influence the fracture limit. In order to compare the HER of different materials, the researchers suggested many empirical relationships as a result of these factors. The HER was given in terms of the parameters m and et for holes drilled in high-strength low-alloy steel sheets [16]. Other modifications led to the inclusion of additional steel grades, including austenitic stainless steels, ferritic stainless, and nineteen ferritic in the equation [15]. As a function of r_m and e_t, empirical formulae for sheared-hole expansion tests were derived. Experimentation and modeling using electromagnetic forming were used to examine the inhomogeneous deformation characteristics of oblique flanges [17]. It was discovered that the key elements affecting inhomogeneous deformation were the size of the deformation zone, the strength of the electromagnetic force, and the limitations of the die.

CHF is amenable to mass manufacturing and possible using specialized equipment. The die-less nature and straightforward tooling of the recently invented incremental sheet forming (ISF) approach has attracted a lot of interest in the manufacturing industry. Many investigations have been conducted on ISF to learn more about its failure mechanisms, formability, strain distribution, tool path optimization, and form correctness [18–20]. Tool deflection and springback may be estimated with a good understanding of the forming forces. The geometric precision of produced components is enhanced by using these settings for tool path compensation [19].

Closed-loop tool path control was shown to improve the ISF components' form correctness [20]. Funnel, bowl, and two-section cone shapes were formed iteratively to verify the suggested approach. Researchers have suggested a number of tools and methods for enhancing the accuracy of the manufactured component [21], such as partial or complete stiff dies, flexible dies made of foams, a slave tool for local assistance, and peripheral assistance. Experimentation and finite element modeling were used to analyze the L-shaped component's response to the effects of each strategy alone and in combination. Additionally, experimental FLDs were utilized to evaluate the formability of aluminum during ISF. The failure mechanisms were subsequently correlated with microstructural analyses [22].

ISF is a versatile manufacturing method that may be used to create both large and small parts by deforming the material in one or more stages at both higher temperatures and room temperatures. Incremental tube forming, incremental hole flanging (IHF), and two-point incremental forming are all ISF procedures [21, 22]. IHF has demonstrated its feasibility as a substitute for CHF in the production of flanges with diverse geometries while requiring minimal machinery.

IHF involves utilizing a blank holder to securely fasten a blank containing a pre-drilled aperture. Subsequently, the void is subjected to progressive deformation using a hemisphere-headed tool that is operated through computer numerical control. IHF studies have focused on assessing fracture prediction, thickness, stress–strain, and the impact of process factors on formability. Circle grid analysis was used to evaluate the strain mapping, and the plasticity relations were used to expand the strain mapping into stress space [23].

Various ductile damage models, such as Rice–Tracey, Ayada, and Cockcroft and Latham, have been employed to ascertain the ultimate strength of the material. The critical damage variables were calculated based on the maximum effective strain observed at the onset of failure, and when subjected to loading conditions of plane strain, the failure of the flange wall could be attributed to meridional stresses [24]. To provide a consistent thickness throughout an IHF process, several tool path techniques have been developed [25]. Thickness distribution and forming limitations were found to be enhanced with the use of a flanging tool shape comparable to a ball nose milling cutter in the IHF process [26].

Analytical and finite element models have been used to verify the flange tool's performance. With the present study, we investigate the single-stage incremental hole flanging (IHF) process applied to a sheet of AA7075-O aluminum alloy with a thickness of 1.6 mm. We also examine the impacts of different process variables, including spindle speed, pre-cut hole diameter, and tool diameter. We found that CHF experienced sheet failure due to bi-axial stretching, which differs from the conventional uniaxial tension failure observed at the hole edges [27].

Our findings support the literature review results about the CHF process's longevity, but IHF remains in its infancy, and just a few studies have been conducted on it thus far.

In addition, the LFR in the CHF methodology fails to provide sufficient information pertaining to the location of the fracture, mechanism of deformation, type of failure, and thickness. The objective of this investigation was to gain an understanding of the various stages involved in the functioning of the IHF. We investigated the strain distribution of DDQ steel sheets in a multistage incremental hole flanging process with different pre-cut hole sizes, utilizing both empirical and theoretical fracture limits. We examined the impacts of the number of forming steps and the pre-cut hole diameter on forming forces, surface roughness, and thickness through the use of experiments and finite element simulations.

6.2 MATERIALS AND METHODS

6.2.1 MECHANICAL CHARACTERIZATION

The mechanical characterization of the materials was carried out using a universal testing machine. The sheets were sliced at an angle of 0° (rolling), 45° (diagonal), and 90° (transverse) from the sheet's rolling direction. The tensile samples were made according to the requirements of ASTM standard E8/E8M-09 [28, 29]. Ultimate tensile strength (UTS), elongation at break, anisotropy coefficient (r), and yield strength (YS) are only a few of the mechanical parameters shown in Table 6.1. Using Eq. 6.3, it was possible to determine the average r; Table 6.1 lists the mechanical characteristics of the sheet in its many orientations:

$$\bar{r} = \frac{r_0 + 2r_{45} + r_{90}}{4} \qquad (6.3)$$

Holloman's hardening law, as shown in Eq. 6.4, provided a close approximation of the stress–strain curve measured during the tensile test:

$$\sigma = 519\varepsilon^{0.176} \qquad (6.4)$$

TABLE 6.1
Mechanical Characteristics of 1 mm Thick DDQ Steel Sheets

Material	Sheet orientation	YS (MPa)	UTS (MPa)	% Elongation	r	$\bar{r}$
DDQ Steel	0°	204	320.7	30.9	1.22	1.18
	45°	206	330.2	30.9	1.21	
	90°	208	328	30.3	1.08	
Average (P)		206	327.2	30.75		

$P = 0.25(P_0+2P_{45}+P_{90})$, where P = YS, UTS, % elongation $\bar{r} = 0.25\left(r_0 + 2r_{45} + r_{90}\right)$

6.2.2 Multistage IHF Trials

Multistage IHF studies necessitated cutting as-rolled DDQ steel sheets to the requisite diameter of 250 mm. The tests were conducted using a Hardinge Bridgeport three-axis CNC milling machine. Figure 6.1 is a schematic depiction of the IHF operation and the experimental apparatus used to study it. The set-up comprises a blank holder and a backing plate with a hole that measures 110 mm in diameter. The diameters of the holes in the blanks ranged from 45 mm to 70 mm so that their impact on the material's malleability could be measured. After the blanks were formed, we measured the in-plane primary stresses by electrochemically etching 3 mm diameter circular grids onto the surfaces. Using a stereo zoom microscope and image analysis software, we determined the primary and secondary axes of the distorted grid. In addition, we used the following formulae to calculate the primary and secondary principal strains:

$$\varepsilon_1 = \ln\left(\frac{L_{major}}{d_o}\right) \tag{6.5}$$

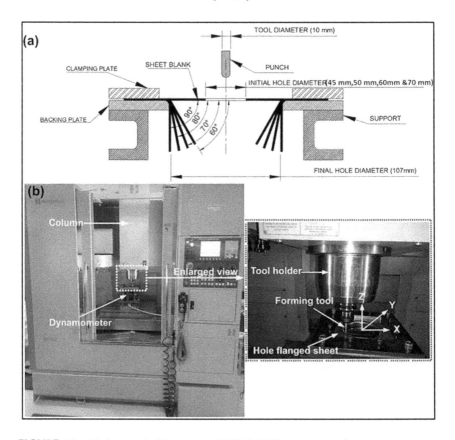

FIGURE 6.1　(a) Schematic illustration of IHF; (b) IHF experimental setup.

$$\varepsilon_2 = \ln\left(\frac{L_{minor}}{d_o}\right) \tag{6.6}$$

where ε_1 and ε_2 characterize the primary and secondary true strains. L_{major}, L_{minor}, and d_0 stand for the original diameter of the grid, the length of the modified primary axis, and the extent of the distorted secondary axis, respectively.

After an initial drawing angle of 60°, the blank was deformed in four phases at 10° intervals to create the vertical flanges. Table 6.2 summarizes the experimental strategy for the IHF's several phases.

We used Mastercam (Tolland, CT, USA) to create the spiral tool path for each level and used a hemisphere-headed tool of 10 mm diameter fabricated from EN26 steel that was heat treated to 60 HRC to shape the blanks. We performed all the tests using grease as the lubricant, with a spindle speed of 250 rpm and a step depth of 0.5 mm. We used a strain-gauge type dynamometer, illustrated in Figure 6.1(b), to determine the IHG forming forces by positioning it between the forming fixture and the machine tool bed. We measured the forces at a sample rate of 10000 s^{-1} on the x, y, and z axes. At last, we used wire-cut EDM (electrical discharge machining) to cut the produced components, and a pointed anvil micrometer accurate to 0.01 mm to determine wall thickness. Additionally, we used Taylor Hubson's Talysurf to assess the surface roughness of produced components at each step.

6.2.3 FINITE ELEMENT ANALYSIS

We ran finite element (FE) models of the multistage (IHF) process using LS-Dyna software. The shell components utilized for the purpose of blank modeling exhibit apertures of 45 mm, 50 mm, 60 mm, and 70 mm in diameter at their center. Figure 6.2 illustrates the FE model of various tools, comprising blank holder, blank, backing plate, and forming tool. All tools were modeled using shell components and were treated as rigid bodies (MAT 20).

The deformable elastoplastic material, which was devoid of any specific identity, was subjected to modeling using a power-law hardening model (MAT 18). Table 6.1 details the blank's material attributes that were integrated.

TABLE 6.2

Experimental Plan for Multistage IHF

Hole diameter	Tool diameter	Plans for multistage IHF (in degrees)			
(Precut, in mm)	(in mm)	1st	2nd	3rd	4th
Ø45	10	60°	70°	80°	90°
Ø50		60°	70°	80°	90°
Ø60		60°	70°	80°	90°
Ø70		60°	70°	80°	90°

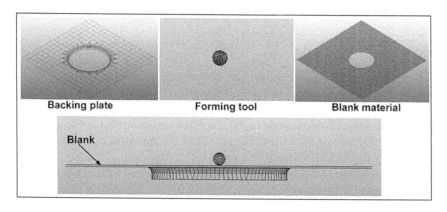

FIGURE 6.2 IHF's FE model.

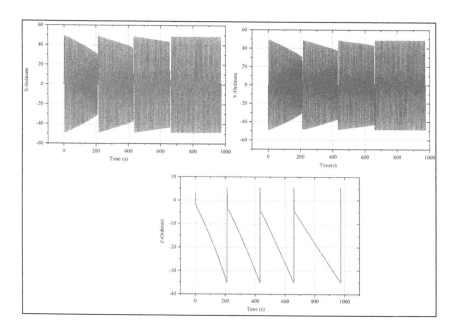

FIGURE 6.3 Plots for the (a) X, (b) Y, and (c) Z ordinates.

We employed the one-way surface-to-surface approach [30] to model the tool-blank interaction and Coulomb's friction law [31] with a coefficient of friction of 0.01. To stop material from being drawn in during deformation, we glued the blank's edges together. As can be seen in Figure 6.3, the simulations were driven by time-position data input from the tool path trajectories created in Mastercam. We used mass scaling and time scaling to reduce the large computing time that longer tool routes entail [32].

6.2.4 MICROTEXTURE AND R_A

To link the mechanical characteristics with the metallurgical attributes, we performed microtexture investigations at varying strain levels on satisfactorily manufactured multistage IHF parts. For a deeper understanding of how the strain route inside a deformed sample shifts in response to an increase in increment angle, we considered the 60 mm starting hole diameter. Cut along RD–TD plane, we mounted the distorted IHF samples with copper-conductive powder to ensure electrical continuity. In addition, the traditional methodology [33] involved using SiC sheets with progressively decreasing granulometry to mechanically polish the mounted samples, succeeded by polishing with diamond paste (3 mm) and colloidal silica (1 mm).

Subsequently, the specimens underwent electrochemical etching for a duration of 24 seconds at a voltage of 35 volts in an A2 electrolytic solution. The A2 electrolyte [34] was composed of 78 milliliters of perchloric acid, 90 milliliters of distilled water, 730 milliliters of ethanol, and 100 milliliters of butoxyethanol. Prior to their examination via field emission scanning electron microscopy, we cleaned the electrochemically etched specimens with ethanol and an ultrasonic cleaner. The utilization of electron backscattered diffraction enabled the acquisition of orientation images and time-versus-position data for the X, Y, and Z coordinates.

We conducted Talysurf surface roughness tests to characterize the IHF parts' quality, evaluating product surface quality using the crucial parameter arithmetic mean of surface roughness (R_a). To prevent any possible mistakes in a conical shape, we cut the samples using EDM machining. We evaluated roughness on the created components' interiors in a direction perpendicular to the tool marks. We took all measurements with a 0.8 mm cut-off and 4 mm sampling lengths. In order to guarantee the reliability of the measurements, we collected three readings from various points on each sample and used the average as the created component's roughness.

6.2.5 ARITHMETICAL STUDIES OF DUCTILE FRACTURE

The anisotropic potential was defined using the constitutive equations related to the Hill48 yield criteria under planar stress circumstances, $\sigma_3 = 0$(Eq. 6.7):

$$\sigma_{Hill} = \bar{\sigma} = \sqrt{\sigma_1{}^2 + \frac{r_0\left(1+r_{90}\right)}{r_{90}\left(1+r_0\right)}\sigma_2{}^2 - \frac{2r_0}{1+r_0}\sigma_1\sigma_2} \qquad (6.7$$

Eq. 6.8 illustrates the link between the strain ratio $\left(\alpha = \varepsilon_2 / \varepsilon_1\right)$ and stress ratio $\left(\beta = \sigma_2 / \sigma_1\right)$ by employing the associative flow rule, $d\bar{\varepsilon} = d\lambda \dfrac{\partial \bar{\sigma}}{\partial \sigma}$, in Eq. 6.7:

$$\alpha = \frac{r_0}{r_{90}}\left[\frac{\beta - r_{90}\left(1-\beta\right)}{1+r_0\left(1-\beta\right)}\right] \qquad (6.8)$$

The term ξ is calculated using the total plastic work per unit volume and plane stress conditions, as shown in Eq. 6.9:

$$\xi = \frac{d\bar{\varepsilon}}{d\varepsilon_1} = \frac{(1+\alpha\beta)\left(\sqrt{r_{90}(1+r_0)}\right)}{\left(\sqrt{r_{90} + r_0\beta^2 + r_0 r_{90}(1-\beta)^2}\right)} \tag{6.9}$$

The Hollomon hardening rule [35] is used to estimate the $(\bar{\sigma} = \sigma_{Hill})$ from knowledge of ξ. In addition, the ratio of the major primary stress (σ_1) to the effective stress, denoted by χ, is calculated using Eq. 6.10:

$$\chi = \frac{\sigma_1}{\bar{v}} = \frac{1}{\left(\sqrt{\dfrac{r_{90} + r_0\beta^2 + r_0 r_{90}(1-\beta)^2}{r_{90}(1+r_0)}}\right)} \tag{6.10}$$

Fracture prediction makes heavy use of stress triaxiality (η), which represents the ratio of hydrostatic stress $(\sigma_m = (\sigma_1 + \sigma_2)/3)$ to effective stress $(\bar{\sigma} = \sigma_{Hill})$, as in Eq. 6.11:

$$\eta = \frac{\sigma_m}{\bar{\sigma}} = \frac{1}{3} \times \frac{(1+\beta)}{\left(\sqrt{\dfrac{r_{90} + r_0\beta^2 + r_0 r_{90}(1-\beta)^2}{r_{90}(1+r_0)}}\right)} \tag{6.11}$$

Table 6.3 displays the most frequently used models for predicting ductile fracture. When a stress function across the effective strain field approaches a critical value, fracture ensues [12, 36–38]. These conditions are straightforward since they only need a single material constant.

TABLE 6.3

Models Employed for Predicting Ductile Fracture

Models for ductile fracture	Mathematical formulation
Cockcroft and Latham (C_{CL}) [39]	$\displaystyle \int_0^{\bar{\varepsilon}_f} \sigma_{max}\, d\bar{\varepsilon} = \int_0^{\bar{\varepsilon}_f} \frac{\sigma_1}{\bar{\sigma}}.\sigma.\frac{d\bar{\varepsilon}}{d\varepsilon_1}.d\varepsilon_1 = \int_0^{\bar{\varepsilon}_f} \chi.\bar{\sigma}.\xi.d\varepsilon_1 = C_{CL}$ (6.12)
Ayada (C_{AY}) [40, 41]	$\displaystyle \int_0^{\bar{\varepsilon}_f} \left(\frac{\sigma_m}{\bar{\sigma}}\right) d\bar{\varepsilon} = \int_0^{\bar{\varepsilon}_f} \left(\frac{\sigma_m}{\sigma_1}.\frac{\sigma_1}{\bar{\sigma}}\right)\frac{d\bar{\varepsilon}}{d\varepsilon_1}.d\varepsilon_1 = \int_0^{\bar{\varepsilon}_f} \left(\frac{1+\beta}{3}\right).\chi.\xi.d\varepsilon_1 = C_{MC}$ (6.13)

TABLE 6.3 (*Continued*)
Models Employed for Predicting Ductile Fracture

Models for ductile fracture	Mathematical formulation

Rice and Tracey (C_{RT}) [42]

$$\int_0^{\bar{\varepsilon}_f} 0.283 \exp\left(\frac{3\sigma_m}{2\bar{\sigma}}\right) d\bar{\varepsilon} = \int_0^{\bar{\varepsilon}_f} 0.283 \exp\left(\frac{3}{2}.\frac{\sigma_m}{\sigma_1}.\frac{\sigma_1}{\bar{\sigma}}\right)\frac{d\bar{\varepsilon}}{d\bar{\varepsilon}_1}.d\bar{\varepsilon}_1$$

$$= \int_0^{\bar{\varepsilon}_f} 0.283 \exp\left(\left(\frac{1+\beta}{2}\right).\chi\right).\xi.d\bar{\varepsilon}_1 = C_{RT} \tag{6.14}$$

Brozzo et al. (C_{BR}) [43]

$$\int_0^{\bar{\varepsilon}_f} \frac{2\sigma_1}{3(\sigma_1 - \sigma_m)} d\bar{\varepsilon} = \int_0^{\bar{\varepsilon}_f} \frac{2}{3(1 - \sigma_m / \sigma_1)}.\frac{d\bar{\varepsilon}}{d\bar{\varepsilon}_1}.d\bar{\varepsilon}_1$$

$$= \int_0^{\bar{\varepsilon}_f} \frac{2}{(2-\beta)}.\xi.d\bar{\varepsilon}_1 = C_{BR} \tag{6.15}$$

Oh et al. (C_{OH}) [44]

$$\int_0^{\bar{\varepsilon}_f}\left(\frac{\sigma_{max}}{\bar{\sigma}}\right) d\bar{\varepsilon} = \int_0^{\bar{\varepsilon}_f}\left(\frac{\sigma_1}{\bar{\sigma}}\right).\frac{d\bar{\varepsilon}}{d\bar{\varepsilon}_1}.d\bar{\varepsilon}_1 = \int_0^{\bar{\varepsilon}_f}(\chi).\xi.d\bar{\varepsilon}_1 = C_{OH} \tag{6.16}$$

Ko et al. (C_{KO}) [36]

$$\int_0^{\bar{\varepsilon}_f} \frac{\sigma_1}{\bar{\sigma}}\left\langle 1+\frac{3\sigma_m}{\bar{\sigma}}\right\rangle d\bar{\varepsilon} = \int_0^{\bar{\varepsilon}_f} \frac{\sigma_1}{\bar{\sigma}}\left\langle 1+3.\frac{\sigma_m}{\sigma_1}.\frac{\sigma_1}{\bar{\sigma}}\right\rangle.\frac{d\bar{\varepsilon}}{d\bar{\varepsilon}_1}.d\bar{\varepsilon}_1$$

$$= \int_0^{\bar{\varepsilon}_f} \chi\left\langle 1+(1+\beta).\chi\right\rangle \xi.d\bar{\varepsilon}_1 = C_{KO} \tag{6.17}$$

In all of these theories, C_{CL}, C_{MC}, C_{RT}, C_{BR}, C_{OH}, and C_{KO}, material constant $\bar{\varepsilon}_f$ is the corresponding plastic strain to fracture. It should be noted that we used the limit of experimental failure as a benchmark against which the correctness of the predictions could be evaluated.

6.3   RESULTS AND DISCUSSION

6.3.1   Limiting Forming Ratio

Mackerle [39] conducted the primary research on hole flanging by press working. The findings demonstrate that in general, stretching and bending together with failure occurring by necking or tearing as a result of high tension at the corners cause the deformation of sheets with a hole inside. Here we introduce the limiting forming ratio (LFR).

Cui and Gao [25] recently identified three key solutions for multistage incremental forming with hole or hole flanging. We conducted the multistage IHF experiments shown in Table 6.2, and Figure 6.4 shows the final efficacious and failed cylindrical flanges at different pre-cut hole diameters. Table 6.4 displays the formability characteristics of the formed components at varying wall angles, including HER, final diameter (d_f), and flange height (h), which are decreased

TABLE 6.4
Findings from Multistage IHF Using Holes of Varying Diameters

Hole dia (Precut)	Intermediate stages	h (mm)	d_f (mm)	h/d_f	HER
φ45	Stage 1 (60°)	40	58.5	0.68	1.30
φ50		38	67.3	0.56	1.30
φ60		25	77	0.36	1.26
φ70		18	82.5	0.22	1.17
φ45	Stage 2 (70°)	44.5	66	0.67	1.46
φ50		40	76.5	0.52	1.53
φ60		29	87.3	0.33	1.41
φ70		20	92	0.21	1.31
φ45	Stage 3 (80°)	–	–	–	–
φ50		43	88.8	0.48	1.78
φ60		32	96.6	0.33	1.58
φ70		21	99	0.21	1.41
φ45	Stage 4 (90°)	–	–	–	–
φ50		50	103.3	0.48	2.06
φ60		34	103.5	0.39	1.73
φ70		22	105	0.20	1.51

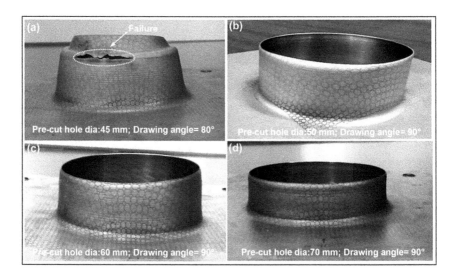

FIGURE 6.4 Unsuccessful and successful parts formed from IHF.

with increasing pre-cut hole diameter. A lower diameter increased the flange height, which gradually thinned the sheet until it fractured prematurely. At hole diameter of 50 mm and 90° wall angle, the LFR is 2.06. At 70 mm and 60°, the LFR is 1.17. Figure 6.4(a) illustrates the commencement of a crack at the edge of the flange while conducting LFR analysis through multistage ISF with a pre-cut hole diameter of 45 mm and a wall angle of 80°.

IHF's cylindrical flange (wall angle 90°) has a potential height of (df − di)/2. Theoretically, a cylindrical flange possessing a pre-cut aperture measuring 70 mm, 60 mm, or 50 mm in diameter is expected to exhibit a corresponding height of 20, 25, or 30 mm. The flange heights reached in the experiments, however, were 22, 34, and 50 mm. That is, the flange stretched notably as the pre-cut hole diameter increased, although at a diameter of 70 mm, the amount of stretching was insignificant. When the pre-cut hole diameter measured 50 mm, the flange stretched 66.66%.

The thickness of the flange underwent a significant reduction as a result of the pronounced stretching that occurred in the meridional direction. Although the pre-cut hole diameter of 50 mm remained intact, it is possible that the flange may not meet the minimum wall thickness requirements necessary for actual manufacturing purposes. Enhancing the flange wall's thickness necessitates exploring novel methodologies. It is noteworthy that when the theoretical flange height was 20 mm or less, there was a negligible occurrence of thinning or stretching. Consequently, it is imperative to conduct experiments on cylindrical flanges that possess a theoretical flange height exceeding 20 mm to evaluate the efficacy of any strategy aimed at enhancing the hole flange thickness.

The preceding discourse elucidates that the thickness of the flange is a crucial criterion in addition to the LFR for evaluating the product's formability or utility. Using h/D_f and HER as nondimensional factors, Eq. 6.18 depicts the average thickness of the produced flange:

$$t_{ave} = \frac{1 - 1/HER^2}{4\left(h/d_f\right)} \times t_0 \tag{6.18}$$

Following Eq. 6.18, the mean flange thicknesses of the DDQ steel blanks featuring pre-cut hole diameters of 70 mm, 60 mm, and 50 mm were determined to be 0.66 mm, 0.47 mm, and 0.39 mm, respectively; that is, flange thickness decreased as LFR increased.

Flanges featuring perpendicular walls were generated exclusively by apertures with a pre-cut diameter of 50 mm or greater. When the diameter of the hole is 45 mm, the flange can be adequately formed for wall angles up to 70°. We observed a circumferential fracture within the 80° wall angle flange that bore a strong resemblance to the circumferential failure that is typically observed in conventional ISF under plane strain conditions. The maximum wall angle achievable for deep-drawing grade steels using conventional ISF is 77°, which is in close proximity to the 80° wall angle observed at the point of failure in incremental hot forming.

This finding suggests that when the hole width is reduced in the IHF process, the plastic flow of material begins to resemble that of the traditional ISF.

6.3.2 Formability Limit

ISF is more formable than conventional deep drawing and stretching thanks to suppressed necking [45]. The forming limit diagram (FLD) obtained from conventional methodologies has limitations in determining material formability because the ISF operation is multistage. As a result, a fracture forming limit diagram (FFLD) is used in the IHF process for the formability assessment. FFLD is depicted in primary strain space as a straight line with a negative slope (−1), as per Eq. 6.19:

$$\varepsilon_{1f} + \varepsilon_{2f} = -\varepsilon_{3f} \tag{6.19}$$

The construction of FFLD is more challenging than that of FLD. To experimentally produce the FFLD, it is imperative to gauge the thickness both before and after fracture alongside the crack, as well as the necking strain under various strain routes. The Nakazima test employs a hemispherical punch to convert rectangular blanks of different widths, thereby providing insights into the necking and fracture thickness strain across different strain paths. Information on the specifics of the tests may be found in [34, 46].

Researchers [38] have shown that from necking to fracture, the material experiences solely plane strain deformation. Under this hypothesis, the minor primary strain at necking (ε_{2n}) is comparable with the second minor principal strain at fracture (ε_{2f}). Circle grid analysis (CGA) of Nakazima test-created blanks can be used to determine the stresses in the necking zone. The third primary fracture strain in the thickness direction is calculated with $\varepsilon_{3f} = \ln\left(\dfrac{t_f}{t_o}\right)$ to measure the fracture thickness (t_f) and the starting thickness of the blank (t_o). The volume consistency relationship is used to get Eq. 6.20, which is the primary principal fracture strain:

$$\varepsilon_{1f} = -\varepsilon_{2n} - \ln\left(\frac{t_f}{t_o}\right) \tag{6.20}$$

The issue at hand pertains to the time-consuming nature of the conventional approach employed for constructing the FFLD. Isik et al. [47] observed that the conventional FFLD forming test bears a striking resemblance to the FFLD derived from the evaluation of fracture stresses in conical and pyramidal frustums produced in ISF, featuring diverse wall angles. Thus, for the present investigation, we acquired the FFLD by utilizing a varying wall angle conical frustum (VWACF) and a varying wall angle pyramidal frustum (VWAPF), both of which featured a wall angle that varied continuously [48]. We used a hemisphere-headed tool with a diameter of 10 mm on a three-axis CNC milling machine to create the VWACF and VWAPF until fracture occurred.

We used the equations we presented to calculate the fracture strains, and the method is discussed in further detail elsewhere [49]. The experimentally deformed VWACF and VWAPF parts are shown in Figure 6.5(a) and (b), and the fracture strains $(\varepsilon_{1f}, \varepsilon_{2f})$ were (0.06, 1.52) and (0.4, 1.15), respectively. Figure 6.5(b) depicts the in-plane fracture stresses derived from VWACF and VWAPF, and the FFLD is generated by fitting a straight line to the data. The resulting FFLD is calculated using Eq. 6.21; to accommodate for experimental ambiguity, a 10% margin of safety is typically used:

$$\varepsilon_{1f} - 1.04\varepsilon_{2f} = 1.58 \qquad (6.21)$$

Eq. 6.21 yields a fracture strain of 1.58 in the plane strain (FLD$_0$) condition. Using Eq. 6.22, it was determined that the greatest wall angle (ψ_{max}) necessary to accommodate this fracture strain is 78° [50]. The experimentally determined

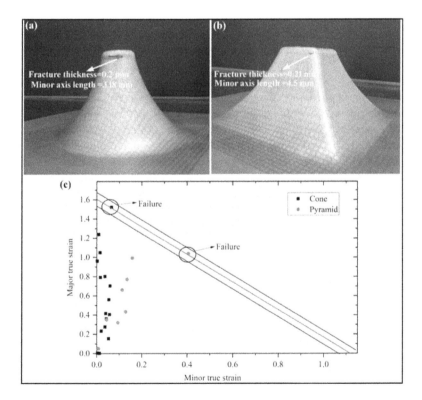

FIGURE 6.5 Deformation characteristics of three distinct geometries: (a) VWACF, (b) VWAPF, and (c) experimental FFLD for cone and pyramid.

maximum wall angle of 77° agreed well with this estimate. In addition, the FFLD
was predicted to have a negative slope, or −1. When the tool radius (r_{tool}) and sheet
thickness (t) were taken into account, the computed slope (m) of FFLD was −1.09
employing Eq. 6.23; the slope prediction from theory was only 3.8% off from the
experimental data:

$$FLD_0 = \ln\left(\frac{\pi}{2} - \psi_{max}\right) \tag{6.22}$$

$$m = \frac{5\left(r_{tool}/t\right) - 2}{3\left(r_{tool}/t\right) + 6} \tag{6.23}$$

6.3.3 EVALUATION AND COMPARISON OF VARIOUS THEORETICAL MODELS UTILIZED TO FORECAST FRACTURE LOCI

The material constants established for several ductile fracture models are discussed in detail in Figure 6.6. These constants are used to predict FFLDs based
on the proportionate loading theory. To evaluate the validity of various ductile
fracture criteria, Figure 6.6 shows the comparison of projected FFLDs with
experimental results. The diagram shows how all models continually underesti-
mate important fracture stresses on the tension–compression side of FFLD.

On the other hand, compared with previous ductile fracture models, the Ayada
fracture criteria proved more reliable in predicting failure limits. As expected by
all ductile fracture models, the principal fracture strains essentially converged at
a single point near the plane strain condition. The projected fracture strain using
experimental data was found to be quite accurate, with variations ranging from
2% to 5%. Ductile fracture regulations, such as the KO, OH, AY, and CL models, tended to overvalue fracture limits when subjected to tension–tension loading conditions, while the RT and BR models underestimated the experimental
results. In general, the AY model outperformed competing models in its potential
to anticipate failure limits. The AY model demonstrated a high level of accuracy
in estimating the fracture limit. This accuracy was observed across the entire
strain ratio domain, ranging from uniaxial stress to balanced biaxial tension, with
minimal variations in trends. Hence, we estimated formability by considering the
influence of vertical forces, thickness =, and strain path development, incorporating both the AY model and the empirically measured FFLD.

6.3.4 STRAIN PATH EVOLUTION

We ran all simulations with the empirically recorded FFLD and the projected AY
model FFLD as failure limits. For the experimental circumstances listed in Table 6.2,
Figure 6.7 depicts the strain path development for varying hole sizes and stages prior
to cutting. All stages with varying pre-cut hole sizes show main primary strains far
below the FFLC curve, with the exception of the stage shown in Figure 6.7(c).

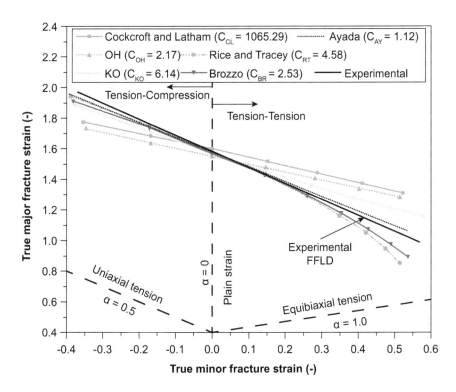

FIGURE 6.6 The calibration of failure limits for different ductile fracture standards incorporating the Hill48 yield model.

The fracture becomes apparent during the third stage, where a pre-cut hole with a diameter of 45 mm is observed. This measurement closely aligns with the experimental data, as depicted in Figure 6.4(a). The high strains along the meridional direction caused the fracture to start at the center of the flange and spread circumferentially; that is, the suggested simulations accurately foretold the fracture in several stages during the IHF process. We employed CGA to evaluate the primary and secondary strains in three specific areas of the generated geometry in order to validate the strain obtained from the simulations. Blanks with a pre-drilled hole are engraved with tiny 3 mm diameter circles for CGA. The initial circular shapes are transformed into ellipses through the hole-flanging procedure, after which the magnitudes of the primary and secondary stresses are determined by employing Eq. 6.5 and Eq. 6.6.

For a 60 mm precut hole, the primary and secondary principal stresses in the flange area are depicted at several phases of formation in Figure 6.8, and Figure 6.9 displays the key strains in the respective FE-simulated cups. As can be seen in Figure 6.8(a), there are three distinct regions within the strain distribution pattern. Zone 1 represents deformation occurring close to the plane strain mode. Major

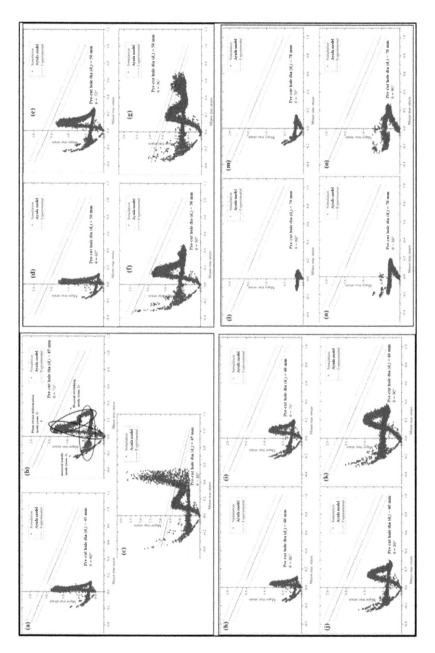

FIGURE 6.7 FE analysis to anticipate the progression of strain paths for various pre-cut hole diameters in the presence of flanging stages: (a–c) 45 mm, (d–g) 50 mm, (h–k) 60 mm, and (l–o) 70 mm.

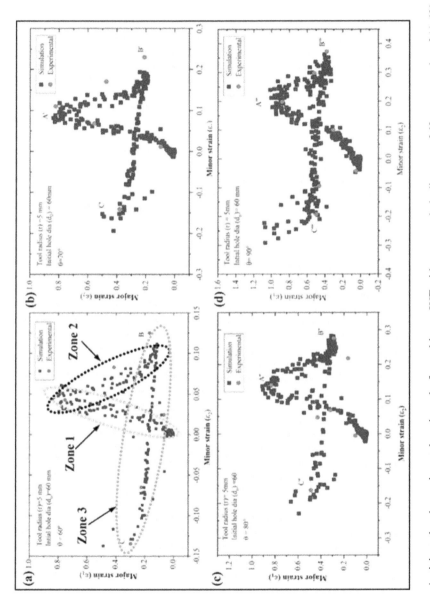

FIGURE 6.8 Anticipated and experimental strain paths for multistage IHF with a pre-cut hole diameter of 60 mm at angles of (a) 60°, (b) 70°, (c) 80°, and (d) 90°.

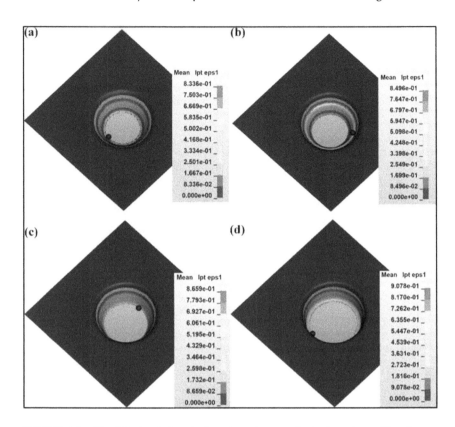

FIGURE 6.9   The FE simulation depicts the primary principal strain profile of a cup with a pre-cut hole diameter of 60 mm at various flanging stages: (a) 60°, (b) 70°, (c) 80°, and (d) 90°.

stresses occurred along the meridional axis, but circumferential strains were minimal. The minor strain changed very slightly to the right, while the major strain grew stronger as one went deeper. Zone 2 is characterized by diminishing major strain and expanding minor strain. In this region, plane strain deformation gave way to bi-axial stretching.

The primary strain grew while the secondary strain decreased in zone 3. In a circumferential direction, the material deformation approached uniaxial tension as it moved near the edge of the drilled hole. We applied circumferential tension to the perimeters of the holes before drilling; as shown in Figure 6.8(b-d), the cumulative stresses shifted toward the right as they were affected by the rise in the number of steps as well as the pre-cut hole's diameter. When the angle increased from 60° to 70°, the minor stress within zone 1 changed from 0.05 to 0.1. The maximum zone 1 minor strains at 80° and 90° were 200% and 300% higher, respectively, than the strain at 60°. The strain distribution along zones 1–3 produced from simulations agreed well with the experimental data under all circumstances.

In addition, Figure 6.7 illustrates that for a given wall angle and initial pre-cut hole width, the strain routes exhibit a linear progression from the starting point to the highest achievable stresses in IHF. This observation implies that localized necking is not expected to occur prior to failure, and instead, the material will experience significant thinning until fracture. In instances of localized necking, the strain path exhibits a predominantly horizontal trajectory until reaching the FLD, at which point it transitions to a predominantly vertical direction until reaching FFLD. The deformation mode known as plane strain, occurring in the material between the necking and fracture regions, results in a vertical strain path. However, the strain path during IHF does not undergo any such transformations.

6.3.5 THE PROFILE OF THE VERTICAL FORCE DURING IHF

A comparison of the vertical forming force at each phase for a variety of pre-cut hole diameters is shown in Figure 6.10 for simulation and experimental conditions. No matter how big the pre-cut hole is, the vertical force is largest in stage 1 since the sheet is stretched and twisted over such a wide angle (0° to 60°). The

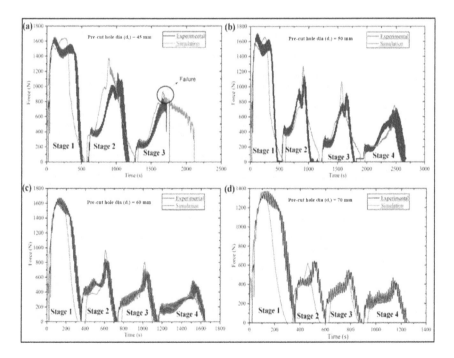

FIGURE 6.10 The vertical forces experienced at various flanging stages during the IHF process for pre-cut hole diameters of (a) 45 mm, (b) 50 mm, (c) 60 mm, and (d) 70 mm.

forming force steadily increases after a given depth to its maximum level. The force necessary to return the sheet to its original shape decreases proportionally to the depth of the hole.

The observed force trend shows a divergence from the conventional force contours typically seen in ISF events. The force in traditional ISF reaches its maximum after the very first bending at the corner radius of the backing plate as long as thinning does not surpass the material's strain hardening. This force remains constant throughout the entire forming procedure. On the other hand, the force experienced in IHF diminishes as the material's resistance to deformation decreases, owing to the presence of a central hole within the blank. The maximum force for stage 1 is 1600 N for pre-cut hole blanks with diameters of 45 mm, 50 mm, and 60 mm, while it is reduced to 1400 N for a pre-cut hole blank with a diameter of 70 mm. The reduction in force can be attributed to the significant influence of the hole on the stiffness of the affected region, as well as the decreased height of the flange.

We found that with 45 and 50 mm diameter pre-cut holes, 1100 N was the highest force achievable in the second stage. With 60 mm and 70 mm hole diameters, the resulting forces were 800 N and 600 N, respectively. Since less deformation energy is needed to further deform from a 60° to a 70° flange, the force needed for the second stage was lower regardless of the pre-cut hole's diameter. Maximum force was exerted towards the hole's edge in stage 1 because the combination of stretching and bending induced by the flanges increased the radial distortion and thickness at the flange's end, but from stage 2 onward, the deformation is driven mostly by stretching.

In the later phases, the forming forces are slightly reduced. The magnitudes of the forming forces in the third stage are 900 N for a pre-cut hole diameter of 50 mm, 700 N for a diameter of 60 mm, and 500 N for a diameter of 70 mm. The fall in forming force observed in the later stages can be attributed to the accelerated hardening of the material. This phenomenon is primarily influenced by the decline in stiffness and thickness of the sheet in the axial direction. Upon reaching the center of the depth during the third stage of the process of developing a 45 mm pre-cut hole, a fracture becomes evident. The simulations demonstrate that once the fracture initiation occurs, there is a decrease in the evolving force pattern during the third stage.

In the concluding stage, the magnitudes of the forming forces are identified as 600 N for pre-cut hole diameters of 50 mm, 500 N for pre-cut hole diameters of 60 mm, and 400 N for pre-cut hole diameters of 70 mm. The same factors that contributed to the third stage's decline in power might be at play here. Leu et al. [51] examined the forces involved in hole flanging on steel sheets with a thickness of 1 mm at different punch geometries. They found that forming forces ranged from 20 kN to 40 kN depending on the punch geometry. Compared with IHF's forming forces, these forces were enormous. The patterns in the forming forces seen in the experimental data are also reflected in the FE simulation findings.

6.3.6 DISTRIBUTION OF THE WALL THICKNESS

The thickness profiles of flanges resulting from computation simulations of a multistage incremental hole flanging process, where the widths of pre-cut holes were varied, are presented in Figure 6.11. Figure 6.12 illustrates the correlations between the flange depth and the significant and insignificant primary stresses across different zones.

The thickness of each sample was initially minimal and subsequently increased, irrespective of the width of the pre-cut hole. The heightened primary stress experienced in zone 1 led to the thinning of the flange up to a specific depth. Consequently, the most significant thinning occurred toward the conclusion of zone 1, which corresponded to the region experiencing plane strain deformation. Subsequently, the material underwent further compression in a downward direction, increasing its overall thickness. Thickness increased in zone 2, characterized by biaxial stretching, and zone 3, characterized by uniaxial tension. The volume consistency necessitates an increase in thickness strain within zone 2 due to the decrease in the main primary strain and a slight increase in the minor principal strain. The growth of thickness strain in zone 3 can be attributed to an augmentation in the most significant principal strain and a reduction in the minor strain.

Table 6.5 summarizes the greatest decrease in thickness that may be achieved at each stage when using holes of varying sizes that have been pre-cut. According

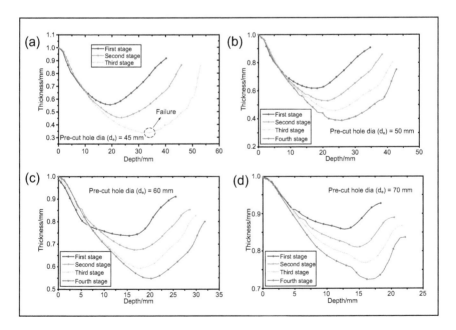

FIGURE 6.11 Distribution of the wall thickness for all deformation settings using the FE method.

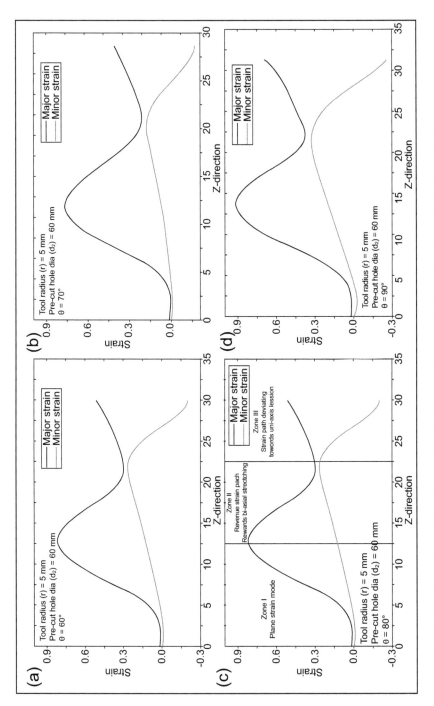

FIGURE 6.12 Distribution outlines of significant and insignificant principal strains in conjunction with the flange depth.

TABLE 6.5

Maximum Percentage Thickness Reductions Observed Across All Aspects of the Experiment

Hole dia. (Precut, in mm)	Stage			
	I	II	III	IV
45	45%	55%	66%	–
50	39%	47%	56%	62%
60	27%	33%	41%	46%
70	14%	19%	24%	28%

to the findings, as the number of phases grows, the thickness decreases, and the thickness decreases more as the pre-cut hole's diameter decreases. For pre-cut holes with sizes of 70 mm, 60 mm, 50 mm, and 45 mm, the greatest drops in thickness are 28%, 46%, 62%, and 66%, respectively. The lowering of stresses along the meridional axis is the primary cause of the reduced thinning in large-diameter holes. As shown in Figure 6.13, the thickness obtained experimentally is compared with the thickness distribution calculated using FE for a hole diameter of 60 mm at various stages.

6.3.7 CHARACTERIZATION OF MICRO-TEXTURE

The purpose of this study was to investigate a specimen that possesses a flanging angle and a 60 mm pre-cut hole diameter. The objective was to gain a deeper comprehension of the presence of three distinct strain routes during a specific incremental flanging step, as well as the subsequent displacement of the strain path. In order to conduct microtexture investigations, it is necessary to obtain a specimen that is extracted from both the base material as well as distinct deformation regions, as depicted in Figure 6.14. Likewise, Figure 6.15 exhibits the inverse pole figure (IPF) at identical coordinates.

The IPF plot in Figure 6.15(a) illustrates the bimodal and coarse distributions of grain sizes of the DDQ grains in the base material. According to estimations, the average grain size was determined to be 69.5 μm. The average grain size exhibited a decrease over time in relation to the initial material as deformation advanced, as depicted in Figure 6.15(b)–(e). Since low-angle grain boundaries (LAGBs) and high-angle grain boundaries (HAGBs) can operate as barriers to fracture propagation and continuous dislocation flow, respectively, they can be used to describe the differences in grain size.

LAGBs are indicated by values less than 69.5 μm, and HAGBs are indicated by values larger than 69.5 μm [52, 53]. The misorientation angles for the base material and the various deformation zones are shown graphically in Figure 6.16(a)–(e). In Figure 6.16(f), we see the averaged proportions of LAGB and HAGB under

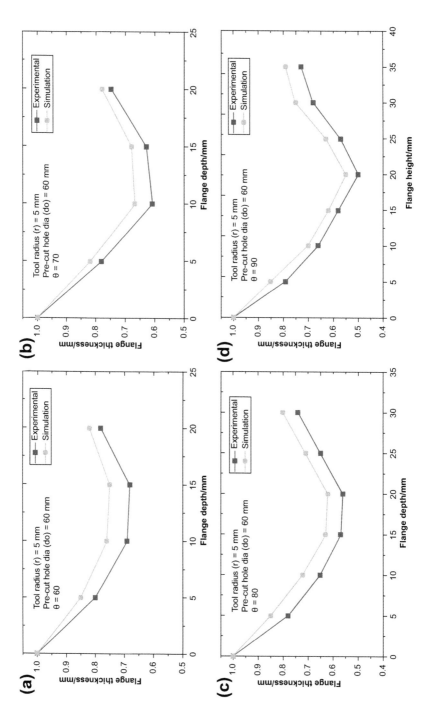

FIGURE 6.13   The simulation and experimental results of the thickness distribution in IHF, specifically focusing on a pre-cut hole diameter of 60 mm.

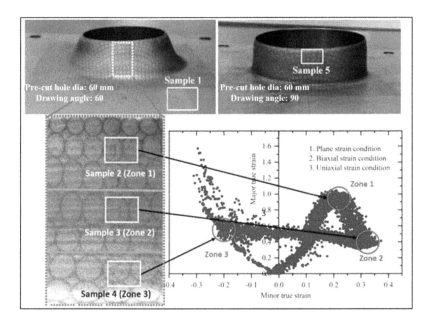

FIGURE 6.14 Samples extracted from various deformation regions and the base material for microtexture evaluations.

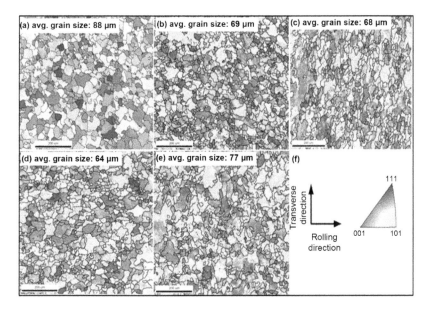

FIGURE 6.15 Obtained IPFs for (a) base material (sample 1); 60° flanging angle at (b) region 1 (sample 2), (c) region 2 (sample 3), (d) region 3 (sample 4); flanging angle of 90° (sample 5); (f) the reference geometries.

various circumstances. The observed decline in grain size during deformation can be attributed to the rising density of LAGBs.

Analyzing the misorientation angle revealed that the volume percentage of LAGBs experienced significant growth from 0.4 in the initial material to 0.8 in the altered specimen, representing an astounding rise of more than 200%. This rise in LAGBs could be attributed to the emergence of deformation bands inside the grains and the activation of dislocations throughout plastic deformation. Prasad et al. [53] reported that the presence of grain interior constraints leads to the subdivision of grain boundaries, thereby facilitating the formation of LAGBs.

Furthermore, we discovered that the grains in the foundational material were virtually evenly scattered along 001. This finding accords well with the Table 6.1 summary of the isotropic nature of the material. Regardless of the deformation zone, as plastic deformation increased, the grains became increasingly aligned with the {111} and {101} planes. Increased fracture resistance due to grain boundaries accompanying the {111} and {110} planes expands the range of plastic deformation possible in the material [54]. Compared with the plane strain condition depicted in Figure 6.15(b), the majority of the grains are noticeably stretched along one axis, as seen in Figure 6.15(c) and Figure 6.15(e). This grain elongation along one axis may have triggered biaxial straining, which in turn would have pushed the plain strain zone closer to the biaxial deformation region.

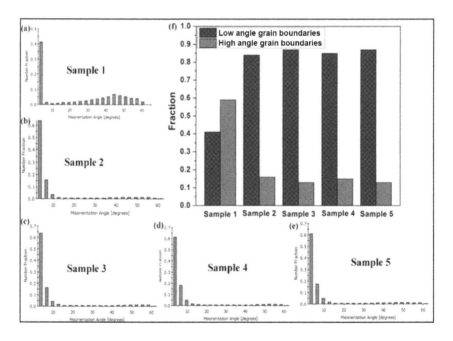

FIGURE 6.16 The misorientation angle determined for various samples, including (a) the base material (sample 1); (b) zone 1 (sample 2); (c) sample 3; (d) sample 4; (e) sample 5; and the consolidated fraction of HAGBs and LAGBs for all situations.

6.3.8 SURFACE ROUGHNESS

Different processes and material characteristics, including forming techniques, lubrication, and sheet material, have noteworthy influence on the surface quality of the produced workpieces. Roughness in the IHF process (R_a) is influenced by the tool diameter (R_t), vertical step size (d_z), and wall angle (α). We employed a Talysurf profilometer to measure the R_a of the fabricated components. Roughness profiles acquired using a tool with a 5 mm radius and a 60 mm diameter pre-cut hole are displayed in Figure 6.17. The undeformed blank had a surface roughness

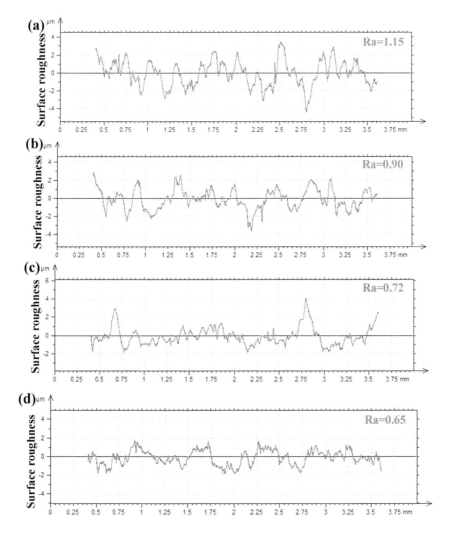

FIGURE 6.17 Investigation of the R_a along the curvilinear distance for a pre-cut hole diameter of 60 mm, considering four different flanging angles: (a) 60°, (b) 70°, (c) 80°, and (d) 90°.

of 1.38 µm. The inner 60° conical surface's R_a dropped from 1.38 µm to 1.15 µm by the conclusion of the first step, a decrease of almost 17%. The observed decrease in R_a was attributable to the stick-slip phenomenon exhibited by the IHF tool during its interaction with the sheet surface.

Wall angles of 70°, 80°, and 90° produced vertical flanges with surface roughnesses of 0.9, 0.72, and 0.65 µm, respectively. There was a 53% reduction in surface roughness between the undeformed base material and the formed surface. Similarly, we distorted pre-cut hole sizes of 45, 50, 60, and 70 mm and used these to measure surface roughness. The findings showed a comparable pattern, with surface roughness decreasing as the number of forming processes increased. It has been hypothesized that less friction on the produced surfaces is accountable for the noticeable intensification in surface quality. By increasing the number of passes, the burnishing action between the forming tool and the sheet is amplified, smoothing down any rough spots on the sheet's surface.

6.4 CONCLUSIONS

In this chapter, we conducted the IHF tests on DDQ steel sheets with pre-cut holes that were 45 mm, 50 mm, 60 mm, and 70 mm in diameter. The cylindrical flanges were created in four phases in increments of 10° from the initial wall angle of 60° to the final angle of 90°. We discovered that the greatest and least LFRs were, respectively, 2.06 and 1.17. The fracture was discovered in the third stage at a depth of 40 mm in a blank with a 45 mm pre-cut hole diameter. According to microstructure study, the strain path in IHF had three deformation modes: plane strain, biaxial, and uniaxial tension.

6.5 CONFLICT OF INTEREST

The authors herein attest that they have no financial or personal stakes in the outcome of this study.

REFERENCES

1. Kalpakjian, S., and Schmid, S.R. (2001). Fundamentals of cutting. In *Manufacturing Engineering and Technology* (4th ed., pp. 534–568). Prentice Hall.
2. Dewang, Y., Purohit, R., and Tenguria, N. (2017). A study on sheet metal hole-flanging process. *Materials Today: Proceedings, 4*(4), 5421–5428.
3. Chen, T.C. (2007). An analysis of forming limit in the elliptic hole-flanging process of sheet metal. *Journal of Materials Processing Technology, 192*, 373–380.
4. Gandla, P.K., Kurra, S., Prasad, K.S., Panda, S.K., and Singh, S.K. (2021). Effect of pre-cut hole diameter on deformation mechanics in multi-stage incremental hole flanging of deep drawing quality steel. *Archives of Civil and Mechanical Engineering, 21*, 1–23.
5. Huang, Y.M., and Chien, K.H. (2002). Influence of cone semi-angle on the formability limitation of the hole-flanging process. *The International Journal of Advanced Manufacturing Technology, 19*, 597–606.

6. Thipprakmas, S., Jin, M., and Murakawa, M. (2007). Study on flanged shapes in fineblanked-hole flanging process (FB-hole flanging process) using Finite Element Method (FEM). *Journal of Materials Processing Technology*, *192*, 128–133.

7. Hyun, D.I., Oak, S.M., Kang, S.S., and Moon, Y.H. (2002). Estimation of hole flangeability for high-strength steel plates. *Journal of Materials Processing Technology*, *130*, 9–13.

8. Tang, S.C. (1981). Large elasto-plastic strain analysis of flanged hole forming. *Computers & Structures*, *13*(1–3), 363–370.

9. Stachowicz, F. (2008). Estimation of hole-flange ability for deep drawing steel sheets. *Archives of Civil and Mechanical Engineering*, *8*(2), 167–172.

10. Stachowicz, F. (2008). Estimation of hole-flange ability for deep drawing steel sheets. *Archives of Civil and Mechanical Engineering*, *8*(2), 167–172.

11. Huang, Y.M., and Chien, K.H. (2001). The formability limitation of the hole-flanging process. *Journal of Materials Processing Technology*, *117*(1–2), 43–51.

12. Takuda, H., Mori, K., Fujimoto, H., and Hatta, N. (1999). Prediction of forming limit in bore-expanding of sheet metals using ductile fracture criterion. *Journal of Materials Processing Technology*, *92*, 433–438.

13. Elbitar, T., and Gemeal, A. (2008). Finite element analysis of deep drawing and hole flanging processing of an oil filter cover. *International Journal of Material Forming*, *1*(Suppl 1), 125–128.

14. Uthaisangsuk, V., Prahl, U., and Bleck, W. (2009). Stretch-flangeability characterisation of multiphase steel using a microstructure based failure modelling. *Computational Materials Science*, *45*(3), 617–623.

15. Comstock, R.J., Scherrer, D.K., and Adamczyk, R.D. (2006). Hole expansion in a variety of sheet steels. *Journal of Materials Engineering and Performance*, *15*, 675–683.

16. Adamczyk, R.D., and Michal, G.M. (1986). Sheared edge extension of high-strength cold-rolled steels. *Journal of Applied Metalworking*, *4*(2), 157–163.

17. Su, H., Huang, L., Li, J., Ma, F., Ma, H., Huang, P., Zhu, H., and Feng, F. (2020). Inhomogeneous deformation behaviors of oblique hole-flanging parts during electromagnetic forming. *Journal of Manufacturing Processes*, *52*, 1–11.

18. Bansal, A., Lingam, R., Yadav, S.K., and Reddy, N.V. (2017). Prediction of forming forces in single point incremental forming. *Journal of Manufacturing Processes*, *28*, 486–493.

19. He, A., Wang, C., Liu, S., and Meehan, P.A. (2020). Switched model predictive path control of incremental sheet forming for parts with varying wall angles. *Journal of Manufacturing Processes*, *53*, 342–355.

20. Raju, C., Haloi, N., and Narayanan, C.S. (2017). Strain distribution and failure mode in Single Point Incremental Forming (SPIF) of multiple commercially pure Aluminum sheets. *Journal of Manufacturing Processes*, *30*, 328–335.

21. Raujol-Veillé, J., Toussaint, F., Tabourot, L., Vautrot, M., and Balland, P. (2015). Experimental and numerical investigation of a short, thin-walled steel tube incremental forming process. *Journal of Manufacturing Processes*, *19*, 59–66.

22. Min, J., Kuhlenkötter, B., Shu, C., Störkle, D., and Thyssen, L. (2018). Experimental and numerical investigation on incremental sheet forming with flexible die-support from metallic foam. *Journal of Manufacturing Processes*, *31*, 605–612.

23. Montanari, L., Cristino, V.A., Silva, M.B., and Martins, P.A.F. (2013). A new approach for deformation history of material elements in hole-flanging produced by single point incremental forming. *The International Journal of Advanced Manufacturing Technology*, *69*, 1175–1183.

24. Cristino, V.A., Montanari, L., Silva, M.B., Atkins, A.G., and Martins, P.A.F. (2014). Fracture in hole-flanging produced by single point incremental forming. *International Journal of Mechanical Sciences, 83*, 146–154.

25. Cui, Z., and Gao, L. (2010). Studies on hole-flanging process using multistage incremental forming. *CIRP Journal of Manufacturing Science and Technology, 2*(2), 124–128.

26. Cao, T., Lu, B., Ou, H., Long, H., and Chen, J. (2016). Investigation on a new hole-flanging approach by incremental sheet forming through a featured tool. *International Journal of Machine Tools and Manufacture, 110*, 1–17.

27. Borrego, M., Morales-Palma, D., Martínez-Donaire, A.J., Centeno, G., and Vallellano, C. (2016). Experimental study of hole-flanging by single-stage incremental sheet forming. *Journal of Materials Processing Technology, 237*, 320–330.

28. E8, ASTM. (2010). ASTM E8/E8M standard test methods for tension testing of metallic materials 1. *Annu B ASTM Standard* (Vol. 4, pp. 1–27). ASTM.

29. Sajun Prasad, K., Panda, S.K., Kar, S.K., Sen, M., Murty, S.N., and Sharma, S.C. (2017). Microstructures, forming limit and failure analyses of Inconel 718 sheets for fabrication of aerospace components. *Journal of Materials Engineering and Performance, 26*, 1513–1530.

30. Prasad, K.S., Panda, S.K., Kar, S.K., Singh, S.K., Murty, S.N., and Sharma, S.C. (2018). Effect of temperature and deformation speed on formability of IN718 sheets: Experimentation and modeling. *IOP Conference Series: Materials Science and Engineering, 418*(1), 012055.

31. Elford, M., Saha, P., Seong, D., Haque, M.Z., and Yoon, J.W. (2013, December). Benchmark 3-incremental sheet forming. In *AIP Conference Proceedings* (Vol. 1567, No. 1, pp. 227–261). American Institute of Physics.

32. Surech, K., and Regalla, S.P. (2014). Effect of time scaling and mass scaling in numerical simulation of incremental forming. *Applied Mechanics and Materials, 612*, 105–110.

33. Panicker, S.S., Prasad, K.S., Sawale, G., Hazra, S., Shollock, B., and Panda, S.K. (2019). Warm redrawing of AA6082 sheets and investigations into the effect of aging heat treatment on cup wall strength. *Materials Science and Engineering: A, 768*, 138445.

34. Basak, S., Prasad, K.S., Mehto, A., Bagchi, J., Ganesh, Y.S., Mohanty, S., Sidpara, A.M., and Panda, S.K. (2020). Parameter optimization and texture evolution in single point incremental sheet forming process. *Proceedings of the Institution of Mechanical Engineers, Part B: Journal of Engineering Manufacture, 234*(1–2), 126–139.

35. Singh, S.K., Limbadri, K., Singh, A.K., Ram, A.M., Ravindran, M., Krishna, M., Reddy, M.C., Suresh, K., Prasad, K.S., and Panda, S.K. (2019). Studies on texture and formability of Zircaloy-4 produced by pilgering route. *Journal of Materials Research and Technology, 8*(2), 2120–2129.

36. Ko, Y.K., Lee, J.S., Huh, H., Kim, H.K., and Park, S.H. (2007). Prediction of fracture in hub-hole expanding process using a new ductile fracture criterion. *Journal of Materials Processing Technology, 187*, 358–362.

37. Habibi, N., Zarei-Hanzaki, A., and Abedi, H.R. (2015). An investigation into the fracture mechanisms of twinning-induced-plasticity steel sheets under various strain paths. *Journal of Materials Processing Technology, 224*, 102–116.

38. Prasad, K.S., Panda, S.K., Kar, S.K., Murty, S.N., and Sharma, S.C. (2018). Prediction of fracture and deep drawing behavior of solution treated Inconel-718

sheets: Numerical modeling and experimental validation. *Materials Science and Engineering: A*, *733*, 393–407.

39. Mackerle, J. (2004) Finite element analyses and simulations of sheet metal forming processes. *Engineering Computations*, *21*(8), 891–940.

40. Cockcroft, M.G. (1968). Ductility and workability of metals. *Journal of Metals*, *96*, 2444.

41. McClintock, F. A. (June 1, 1968). "A Criterion for Ductile Fracture by the Growth of Holes." *ASME. J. Appl. Mech.* June 1968; *35*(2), 363–371. https://doi.org/10.1115/1.3601204

42. Ayada, M. (1987). Central bursting in extrusion of inhomogeneous materials. In *Proc^{ee}dings of 2nd International Conference on Technology for Plasticity, Stuttgart, 1987* (Vol. 1, pp. 553–558).

43. Rice, J.R., and Tracey, D.M. (1969). On the ductile enlargement of voids in triaxial stress fields. *Journal of the Mechanics and Physics of Solids*, *17*(3), 201–217.

44. Brozzo, P., Deluca, B., and Rendina, R. (1972, October). A new method for the prediction of formability limits in metal sheets. In *Proc. 7th Biennal Conference IDDR*.

45. Basak, S., and Panda, S.K. (2023). Use of uncoupled ductile damage models for fracture forming limit prediction during two-stage forming of Aluminum sheet material. *Journal of Manufacturing Processes*, *97*, 185–199.

46. Basak, S., Prasad, K.S., Sidpara, A.M., and Panda, S.K. (2019). Single point incremental forming of AA6061 thin sheet: Calibration of ductile fracture models incorporating anisotropy and post forming analyses. *International Journal of Material Forming*, *12*, 623–642.

47. Isik, K., Silva, M.B., Tekkaya, A.E., and Martins, P.A.F. (2014). Formability limits by fracture in sheet metal forming. *Journal of Materials Processing Technology*, *214*(8), 1557–1565.

48. Kurra, S., and Regalla, S.P. (2014). Experimental and numerical studies on formability of extra-deep drawing steel in incremental sheet metal forming. *Journal of Materials Research and Technology*, *3*(2), 158–171.

49. Cristino, V.A., Silva, M.B., Wong, P.K., and Martins, P.A. (2017). Determining the fracture forming limits in sheet metal forming: A technical note. *The Journal of Strain Analysis for Engineering Design*, *52*(8), 467–471.

50. Silva, M.B., Skjødt, M., Atkins, A.G., Bay, N., and Martins, P.A.F. (2008). Single-point incremental forming and formability—Failure diagrams. *The Journal of Strain Analysis for Engineering Design*, *43*(1), 15–35.

51. Leu, D.K., Chen, T.C., and Huang, Y.M. (1999). Influence of punch shapes on the collar-drawing process of sheet steel. *Journal of Materials Processing Technology*, *88*(1–3), 134–146.

52. Masoumi, M., Silva, C.C., Béreš, M., Ladino, D.H., and Gomes de Abreu, H.F. (2017). Role of crystallographic texture on the improvement of hydrogen-induced crack resistance in API 5L X70 pipeline steel. *International Journal of Hydrogen Energy*, *42*(2), 1318–1326.

53. Prasad, K.S., Panda, S.K., Kar, S.K., Murty, S.N., and Sharma, S.C. (2018). Effect of solution treatment on deep drawability of IN718 sheets: Experimental analysis and metallurgical characterization. *Materials Science and Engineering: A*, *727*, 97–112.

54. Masoumi, M., Silva, C.C. and Gomes de Abreu, H.F. (2016). Effect of crystallographic orientations on the hydrogen-induced cracking resistance improvement of API 5L X70 pipeline steel under various thermomechanical processing. *Corrosion Science*, *111*, 121–131.

7 Applications of Incremental Sheet Forming

Sherwan Mohammed Najm, Valentin Oleksik, and Tomasz Trzepieciński

7.1 INTRODUCTION

Sheet metal forming (SMF) is utilized extensively in diverse industrial sectors for the production of curved components. The process of SMF involves altering the shape of the workpiece without reducing thickness or forming wrinkles. Various SMF techniques are commonly employed, including rolling, stretching, forging, drawing, spinning, bending, and coining.

The processes of sheet forming conventionally require complex and costly punch-and-die tool sets in order to create the components accurately. Because of the complicated structure of the dies, precise equipment and expensive tools are required. However, conventional methods of sheet formation are very suitable for medium/mass production (Schedin, 1992; Arfa, Bahloul, and BelHadjSalah, 2013) due to the cost-sharing nature of punch and die expenses, which helps to minimize tooling costs.

Conventional sheet-forming techniques face challenges in producing small manufacturing batches with precise end components, but due to the pivotal role of SMF in the manufacturing industry, researchers have been actively investigating novel sheet metal forming processes across diverse industrial sectors. For instance, incremental sheet forming (ISF) is a contemporary and inventive way of shaping sheets that deviates from the conventional use of punch and dies (Najm, Paniti, and Viharos, 2020). ISF is widely regarded as a highly effective unconventional technology for prototyping (Filice, Fratini, and Micari, 2002; Yoon and Yang, 2003; Paniti et al., 2020; Trzepieciński, Najm, Oleksik et al., 2022), producing customized products of sheet (Ambrogio et al., 2004), shaping nonsymmetric geometries (Reddy and Lingam, 2018), and fabricating complex parts (Devarajan et al., 2014; Behera et al., 2017; Sisodia et al., 2021).

Authors of a relatively recent publication (Trzepieciński, Oleksik et al., 2021) briefly outlined the latest progress in the ISF field of lightweight materials. Implementing ISF technologies can decrease production time and cost in small-lot or piece manufacturing processes (Park and Kim, 2003). ISF, a punchless and

DOI: 10.1201/9781003441755-7

dieless manufacturing technique, represents a relatively recent approach to sheet shaping. It involves the gradual deformation of a metal sheet through a series of iterative adjustments (Kumar and Kumar, 2015; Rosca *et al.*, 2023).

The device and method patent for incremental dieless forming by ISF can be traced back to Leszak's work (1967). Nevertheless, Berghahn's concept (1967) described in the patent resembles the contemporary technique known as incremental sheet forming. According to Emeens et al. (Emmens, Sebastiani and van den Boogaard, 2010), Berghahn's patents do not indicate the origins of the current advancements in ISF, but the literature review suggests that the roots of ISF can be attributed to Mason (1978). Subsequently, Mason and his colleague Appleton authored a scholarly article elucidating the utilization of surplus tools to fabricate modest quantities of sheet metal (Appleton and Mason, 1984).

Incremental forming involves using an essential tool to induce localized plastic deformation in a sheet based on the principles of layered manufacturing, The component geometry undergoes a transformation whereby it is divided into many layers aligned with the form's direction (Bahloul, Arfa and BelHadjSalah, 2014). The procedure is performed iteratively, with each layer being successively added until the desired final form is achieved. The whole process is under the direction of a numerical control machine. Consequently, when there is a requirement to form a new product with a different shape, the CNC program, which is the path strategy, and the forming tool have to be modified. As a result, unlike traditional procedures, these changes may be easily implemented with little cost.

7.2   INCREMENTAL SHEET FORMING TYPES

Single-point incremental forming (SPIF) and two-point incremental forming (TPIF) are the two primary types of ISF. The first patent for TPIF was obtained by Matsubara (1994). Several subtypes of TPIF have been developed utilizing different support components. Please refer to Figure 7.1 for a hierarchical diagram illustrating the primary ISF approaches.

In ISF, the desired shape is formed by a controlled movement of a basic tool along a different strategy, like inward/outward to the center of the clamped sheet. A tool with a flat end, a tool with a hemispherical/spherical end, or a ball bearing that moves can be used in ISF. From this angle, it is clear that tool names are dependent on the design of the shape of tool's end (see Figures 7.2 and 7.3).

In SPIF, a single tool with a very basic form is used throughout the whole sheet manufacturing process (see Figure 7.4.a); in TPIF, there are two different contact points. The forming tool contacts the sheet metal, and the sheet contacts the supporting item, for instance an auxiliary tool or a partial/full die. To achieve TPIF, one may utilize a counter tool (as depicted in Figure 7.4.b), a

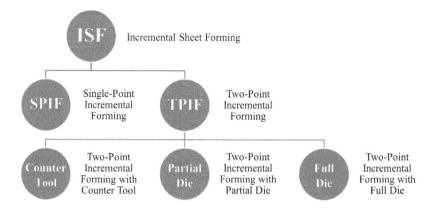

FIGURE 7.1 Principal incremental sheet forming techniques (Trzepieciński, Najm, Pepelnjak et al., 2022).

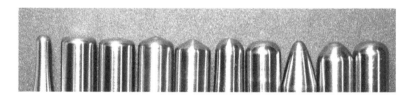

FIGURE 7.2 Some of the shapes suggested for ISF tools (Jeswiet *et al.*, 2015).

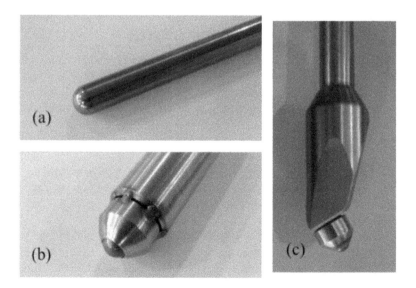

FIGURE 7.3 Suggested ISF tools: (a) rigid tool, (b) vertical roller ball, (c) oblique roller ball (Lu et al., 2014).

partial die (as shown in Figure 7.4.c), or a complete die (as illustrated in Figure 7.4.d). In TPIF utilizing the counter tool, the sheet formation involves two distinct tools, and an additional spindle is positioned in opposition to the primary forming spindle and follows a suitably adjusted trajectory relative to the main tool (see Figure 7.4.b).

In TPIF, an additional assembly movement is employed to ensure that the edges of the formed sheet (blank holder and backing plate) are properly secured, as depicted in Figure 7.4.c and 7.4.d. The motion of the fixture leads to the enhanced geometric accuracy of the produced components, enabling the manipulation of wall thickness. The utilization of TPIF, as opposed to SPIF (as shown in Figure 7.4.a), enhances the geometric accuracy of the produced components.

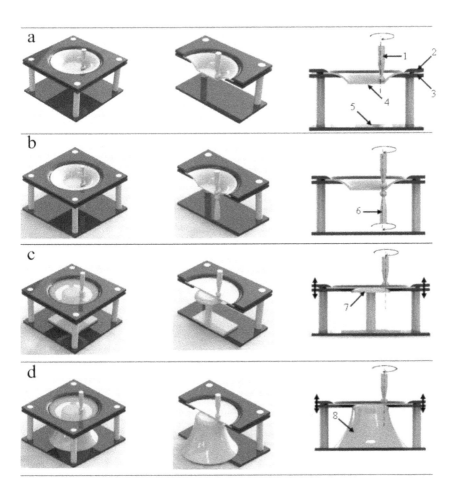

FIGURE 7.4 (a) SPIF, (b) TPIF using the counter tool, (c) TPIF using a partial die, (d) TPIF using a full die: 1–forming tool, 2–blank holder, 3–backing plate, 4–product sheet, 5–rig frame, 6–counter tool, 7–partial die, 8–full die (Trzepieciński, Najm, Pepelnjak et al., 2022).

ISF can be classified as positive or negative contingent on their concavity, which manifests as either concave upward or concave downward (Oleksik et al., 2021). Positive incremental forming, also referred to as PIF with a complete die, is distinguished from other methods, collectively known as negative incremental forming (Trzepieciński, Oleksik et al., 2021).

7.3 EMERGING INCREMENTAL SHEET FORMING METHODS

Over the past few years, there has been significant progress in the development of ISF technologies, which have incorporated novel auxiliary tools to optimize the process. Several researchers have employed a nozzle to apply pressure to water in ISF using a water jet (WJ-ISF). Iseki (2001) was the initial proponent of the adoption of WJ-ISF. WJ-ISF presents several advantages, like reducing friction force, decreasing the roughness, and eliminating contact between the tool and the sheet, thereby preventing the need for lubricant. Duflou et al. (2007) were the first to use laser-assisted ISF, but alternative methods have been developed such as ultrasonic-assisted (Vahdati, Mahdavinejad and Amini, 2017), electromagnetic-assisted (Cui *et al.*, 2014), heat-assisted approaches utilizing cartridge warmers (Saidi et al., 2019), and electric heating (Fan *et al.*, 2008) (see Figure 7.5). In their study, Valoppi et al. (2016) suggested using electrically assisted mixed double-sided incremental forming and accumulative double-sided incremental forming to address the occurrence of spark phenomenon.

Many researchers have investigated these ISF techniques involving using water jets (Jurisevic et al., 2007, 2008; Li et al., 2014; Teymoori, LohMousavi and Etesam, 2016; Lu et al., 2017), using incremental microforming with water jets with supporting dies (Shi et al., 2019), and using ISF for metal bellows

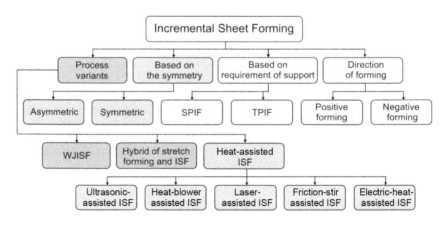

FIGURE 7.5 ISF methods (Trzepieciński, Oleksik et al., 2021).

forming (Wei et al., 2015). Researchers have also studied laser (Mohammadi et al., 2013, 2015, 2016), ultrasonic (Li et al., 2017; Li et al., 2017; Cheng et al., 2019; Sedaghat, Xu and Zhang, 2019; Li et al., 2020), electromagnetic (Cui et al., 2016; Li et al., 2018), and electric heat (Xu et al., 2014; Bao et al., 2015; Grimm and Ragai, 2019; Ao et al., 2020) ISF. In a distinct contribution, Li et al. (2018) introduced a novel approach involving integral electric heating for incremental forming as an alternative to localized heating. In addition to his previous work, Xu (Xu et al., 2016) conducted research on the topic of electrically assisted double-sided incremental shaping. Previous researchers have also investigated alternative methods of heating, including the utilization of halogen lamps (Kim et al., 2007; Okada et al., 2018), induction heating (Al-Obaidi, Kräusel, and Landgrebe, 2016, 2017; Ambrogio et al., 2017; Francesco, Giuseppina, and Luigino, 2017), friction heating (Wang, Li, and Jiang, 2016; Baharudin et al., 2017; Liu, 2017; Grün et al., 2018), and hot air (JI and Park, 2008; Kulkarni, Sreedhara and Mocko, 2016; Leonhardt et al., 2018; Liao et al., 2020).

7.4 MACHINE LEARNING IN INCREMENTAL SHEET FORMING

Artificial neural networks (ANNs) are employed in the development of predictive models for various manufacturing processes, including end-milling, powder metallurgy, and high-speed machining (Ezugwu et al., 2005; Zain, Haron, and Sharif, 2010; Amirjan et al., 2013). Machine learning has emerged as the dominant approach in the development of highly effective predictive models (Lela, Bajić, and Jozić, 2009; Marouani and Aguir, 2012; Kondayya and Gopala Krishna, 2013; Li, 2013; Hussaini, Singh, and Gupta, 2014). Recent years have seen the publication of thousands of studies about neural networks, which have been used by practitioners in many fields such as artists, filmmakers, singers, scientists, and particularly researchers, with the aim of attaining valuable and often groundbreaking results.

Neural networks have extensive scientific applications, although there are variations in the algorithms used for manufacturing optimization. For example, Kurra et al. (2015) conducted a gradual estimation of the surface roughness of extra-deep drawing steel under different forming circumstances. Nasrollahi and Arezoo (2012) employed training data to develop two ANN models and utilized these models to forecast the springback phenomenon in sheet metals containing holes. The authors employed a feed-forward neural network architecture together with the algorithm of backpropagation.

Mekras (2017) utilized an ANN model to address the process setup of sheet metal forming and assessed various factors including the type of aluminum alloy. Kashid and Kumar (2013) conducted an analysis of sixty-three research publications pertaining to the utilization of ANNs in the context of sheet metal production. Interestingly, their investigation revealed a notable absence of any references to the specific application of incremental forming within the examined literature.

ANNs are increasingly employed in contemporary research to model and optimize settings for incremental forming (Najm and Paniti, 2018, 2020). Researchers have attempted to predict surface roughness, formability, pillow effect, and hardness of parts employing models such as ANN, support vector regression, CatBoost, and gradient boosting regressions (Najm and Paniti, 2021a, 2021b). The prediction models have been built specifically for ISF and have been utilized to estimate the process parameters (Najm et al., 2021; Najm and Paniti, 2022).

In their study, Maji and Kumar (2020) observed that utilizing a hybrid approach in conjunction with an adaptive neuro-fuzzy inference system (ANFIS) resulted in improved prediction accuracy. The researchers employed response surface methods and ANFIS to predict the outcomes of the SPIF component and investigated various process parameters and inverse predictions. Oraon and Sharma (2018) employed an ANN model that incorporated feed-forward backpropagation learning for the purpose of forecasting the surface quality of SPIF components, and the ANN simulation achieved 94.744% accuracy with 1.068% mean absolute error. Mulay et al. (2019) employed an ANN model to predict the wall angle and average surface roughness of AA5052-H3 products manufactured through incremental forming. Radu et al. (2013) investigated the utilization of response surface methodology and neural networks to improve and effectively control the accuracy of the SPIF component.

7.5 APPLICATIONS OF INCREMENTAL SHEET FORMING

Several industries, such as the automotive (Scheffler et al., 2019), aerospace (Gupta and Jeswiet, 2019; Gupta, Szekeres and Jeswiet, 2019; Trzepieciński, Najm, et al., 2021), and marine (Gupta, Szekeres and Jeswiet, 2019) sectors, have adopted ISF, and the process exhibits a notable level of adaptability and malleability when employed in medical contexts (Göttmann et al., 2013; Araújo et al., 2014; Potran, Skakun, and Faculty, 2014; Milutinović et al., 2021). Experiments can be conducted under both low- and high-temperature conditions (Sbayti, Bahloul and Belhadjsalah, 2018; Vahdani et al., 2019). Researchers have utilized ISF in the production of intricate shell components (Skjoedt et al., 2008) as well as in the rapid fabrication of prototypes (Lee et al., 2011). Table 7.1 lists a number of current applications of ISF.

The literature review revealed significant efforts to enhance and advance the incremental forming process and fine-tune its parameters, irrespective of the specific product being considered. Researchers have investigated a number of materials, although they notably are not generally studying final products for consumers; rather, they are studying geometric shapes such as cones, pyramids, and spheres simply to improve the process parameters or study the characteristics of ISF forming sheets. The evidence for this statement is that scientists' efforts are not resulting in the sufficient improvement of the incremental forming process, which lacks established criteria and parameters that can be generally applied to diverse technical designs. However, Figures 7.6, 7.7, and 7.8 show several medical and automotive products formed by the ISF process.

TABLE 7.1

Applications for Incremental Forming

Applications	ISF type	Materials	References
Aerospace	SPIF	GLARE	(Kuitert, 2016; Kubit et al., 2020)
	Hybrid ISF	Aluminum Alloy 2A12-O	(Zhang et al., 2017)
Automotive	Different methods	Different materials	(Amino et al., 2014)
	TPIF	Aluminum alloys EN AW-1050 (Al99,5) H14-H24 Novelis Advanz 6F e170	(Scheffler et al., 2019)
	ISF with full die	Low carbon steel Aluminum alloy Pure titanium Ti6Al4V alloy	(Peter et al., 2019)
	SPIF	Different materials Aluminum Alloy 6061	(Kleiner, Geiger, and Klaus, 2003) (Jiménez et al., 2017)
Medical	ISF with full die	Titanium	(Fiorentino, Marzi, and Ceretti, 2012)
		08Al steel	(Han et al., 2010)
	SPIF and TPIF	Biocompatible polymer (ultrahigh molecular weight polyethylene)	(Bagudanch et al., 2018)
	SPIF	Grade 2 titanium	(Vanhove et al., 2017) (Verbert et al., 2008) (AL-Obaidi et al., 2019)
		Low carbon steel EN DC04 and stainless steel EN X6Cr17	(Potran, Skakun, and Faculty, 2014)
		Titanium alloy sheet AA 1050 Grade2 titanium AISI 304	(Saidi et al., 2019) (Duflou et al., 2013)
		Pure titanium	(Sbayti et al., 2018) (Salamati et al., 2015) (Kumar, Kumar, and Singh, 2020) (Lu et al., 2016)
		BFRP laminate structure Aluminium sheets (AA1200H4) Grade 2 titanium	(AL-Obaidi et al., 2019) (Eksteen, 2013)
		Deep drawing quality steel	(Ambrogio et al., 2005)
House pieces	SPIF	Stainless steel and polycarbonate Aluminum 1050–0 Aluminum	(Silva and Martins, 2014) (Jeswiet et al., 2005) (Jeswiet and Hagan, 2002)
Architectural	SPIF	Sandwich panels with different materials	(Jackson, Allwood, and Landert, 2008)
	SPIF and TPIF	Self-supporting lightweight structures	(Bailly et al., 2014)

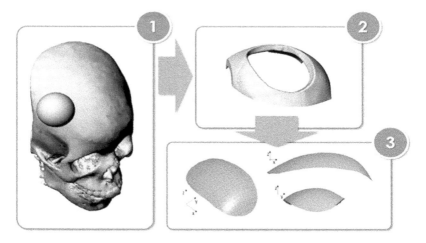

FIGURE 7.6 ISF for medical parts: an ISF-formed skull. 1. Model of a human skull prosthesis in polymeric material, 2. simulating possible accidental damage, 3. the shape of the prosthesis (Palumbo et al., 2020).

FIGURE 7.7 Mounted clavicle implant on a generic synthetic bone (SPIF) (Vanhove et al., 2017).

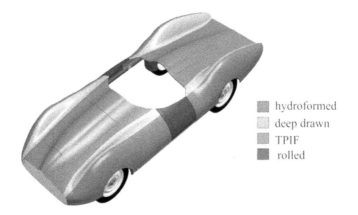

FIGURE 7.8 ISF for automotive products: a. parts of the demonstrator car, b. formed sheet of left and right rear mudguards (Scheffler et al., 2019), c. mounted hood, d. inner and outer form of hood, and e. side panel of Toyota iQ (Amino et al., 2014).

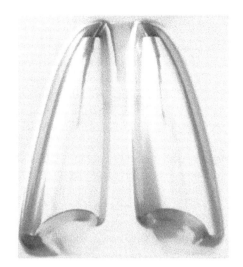

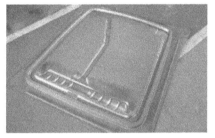

FIGURE 7.8 (Continued)

7.6 CONCLUSIONS

Many industries make curved components using sheet metal forming (SMF). Incremental sheet forming (ISF), a punchless and dieless sheet shaping method, is relatively new. ISF is a manufacturing process that gradually deforms a metal sheet through iterative deformations. Small batches of high-quality finished components challenge traditional and novel sheet-forming procedures. Researchers are interested in inventive sheet metal forming technologies since SMF is the manufacturing industry's backbone.

ISF is best for economic prototypes, different sheet components, complex parts, and nonsymmetric products. ISF methods can save manufacturing time and cost in small-batch production. ISF's creation, types, recent technology, and integration with artificial networks are covered in this chapter; the chapter also covers ISF applications to help researchers learn more about the technology. However, the products researchers have studied are often simple geometric shapes, such as cones, pyramids, and spheres, that are useless for the market and only valuable for scientific investigation. At this point, incremental forming is not standardized and cannot be generalized to different engineering products.

REFERENCES

Al-Obaidi, A., Kräusel, V. and Landgrebe, D. (2016) 'Hot single-point incremental forming assisted by induction heating', *The International Journal of Advanced Manufacturing Technology*, 82(5–8), pp. 1163–1171. doi: 10.1007/s00170-015-7439-x.

Al-Obaidi, A., Kräusel, V. and Landgrebe, D. (2017) 'Induction heating validation of dieless single-point incremental forming of AHSS', *Journal of Manufacturing and Materials Processing*, 1(1), p. 5. doi: 10.3390/jmmp1010005.

Al-Obaidi, A. *et al.* (2019) 'Heat supported single point incremental forming of hybrid laminates for orthopedic applications', *Procedia Manufacturing*, 29, pp. 21–27. doi: 10.1016/j.promfg.2019.02.101.

Ambrogio, G. *et al.* (2004) 'Influence of some relevant process parameters on the dimensional accuracy in incremental forming: A numerical and experimental investigation', *Journal of Materials Processing Technology*, 153–154, pp. 501–507. doi: 10.1016/j.jmatprotec.2004.04.139.

Ambrogio, G. *et al.* (2005) 'Application of incremental forming process for high customised medical product manufacturing', *Journal of Materials Processing Technology*, 162–163, pp. 156–162. doi: 10.1016/j.jmatprotec.2005.02.148.

Ambrogio, G. *et al.* (2017) 'Induction heating and cryogenic cooling in single point incremental forming of Ti-6Al-4V: Process setup and evolution of microstructure and mechanical properties', *The International Journal of Advanced Manufacturing Technology*, 91(1–4), pp. 803–812. doi: 10.1007/s00170-016-9794-7.

Amino, M. *et al.* (2014) 'Current status of "dieless" amino's incremental forming', *Procedia Engineering*, 81, pp. 54–62. doi: 10.1016/j.proeng.2014.09.128.

Amirjan, M. *et al.* (2013) 'Artificial neural network prediction of Cu—Al2O3 composite properties prepared by powder metallurgy method', *Journal of Materials Research and Technology*, 2(4), pp. 351–355. doi: 10.1016/j.jmrt.2013.08.001.

Ao, D. *et al.* (2020) 'Formability and deformation mechanism of Ti-6Al-4V sheet under electropulsing assisted incremental forming', *International Journal of Solids and Structures*, 202, pp. 357–367. doi: 10.1016/j.ijsolstr.2020.06.028.

Appleton, E. and Mason, B. (1984) 'Sheet metal forming for small batches using sacrificial tooling', *Production Engineer*, 63(9), p. 58. doi: 10.1049/tpe.1984.0213.

Araújo, R. *et al.* (2014) 'Single point incremental forming of a facial implant', *Prosthetics and Orthotics International*, 38, pp. 369–378.

Arfa, H., Bahloul, R. and BelHadjSalah, H. (2013) 'Finite element modelling and experimental investigation of single point incremental forming process of aluminum sheets: Influence of process parameters on punch force monitoring and on mechanical and geometrical quality of parts', *International Journal of Material Forming*, 6(4), pp. 483–510. doi: 10.1007/s12289-012-1101-z.

Bagudanch, I. *et al.* (2018) 'Customized cranial implant manufactured by incremental sheet forming using a biocompatible polymer', *Rapid Prototyping Journal*, 24(1), pp. 120–129. doi: 10.1108/RPJ-06-2016-0089.

Baharudin, B. T. H. T. *et al.* (2017) 'Experimental investigation of forming forces in frictional stir incremental forming of Aluminum alloy AA6061-T6', *Metals*, 7(11), p. 484. doi: 10.3390/met7110484.

Bahloul, R., Arfa, H. and BelHadjSalah, H. (2014) 'A study on optimal design of process parameters in single point incremental forming of sheet metal by combining Box—Behnken design of experiments, response surface methods and genetic algorithms', *The International Journal of Advanced Manufacturing Technology*, 74(1–4), pp. 163–185. doi: 10.1007/s00170-014-5975-4.

Bailly, D. *et al.* (2014) 'Analysis into differences between the buckling in single-point and two-point incremental sheet forming of components for self-supporting sheet metal structures', *Key Engineering Materials*, 622–623, pp. 367–374. doi: 10.4028/www.scientific.net/KEM.622-623.367.

Bao, W. *et al.* (2015) 'Experimental investigation on formability and microstructure of AZ31B alloy in electropulse-assisted incremental forming', *Materials & Design*, 87, pp. 632–639. doi: 10.1016/j.matdes.2015.08.072.

Behera, A. K. *et al.* (2017) 'Single point incremental forming: An assessment of the progress and technology trends from 2005 to 2015', *Journal of Manufacturing Processes*, 27, pp. 37–62. doi: 10.1016/j.jmapro.2017.03.014.

Berghahn, W. G. (1967) 'Method of dielessly forming surfaces of revolution', *United States Patent Office*. U.S., Patent No. (3,316,745).

Cheng, R. *et al.* (2019) 'Applying ultrasonic vibration during single-point and two-point incremental sheet forming', *Procedia Manufacturing*, 34, pp. 186–192. doi: 10.1016/j.promfg.2019.06.137.

Cui, X. H. *et al.* (2014) 'Electromagnetic Incremental Forming (EMIF): A novel Aluminum alloy sheet and tube forming technology', *Journal of Materials Processing Technology*, 214(2), pp. 409–427. doi: 10.1016/j.jmatprotec.2013.05.024.

Cui, X. H. *et al.* (2016) 'Incremental Electromagnetic-Assisted Stamping (IEMAS) with radial magnetic pressure: A novel deep drawing method for forming aluminum alloy sheets', *Journal of Materials Processing Technology*, 233, pp. 79–88. doi: 10.1016/j.jmatprotec.2016.02.013.

Devarajan, N. *et al.* (2014) 'Complex incremental sheet forming using back die support on Aluminium 2024, 5083 and 7075 alloys', *Procedia Engineering*, 81, pp. 2298–2304. doi: 10.1016/j.proeng.2014.10.324.

Duflou, J. R. *et al.* (2007) 'Laser assisted incremental forming: Formability and accuracy improvement', *CIRP Annals*, 56(1), pp. 273–276. doi: 10.1016/j.cirp.2007.05.063.

Duflou, J. R. *et al.* (2013) 'Manufacture of accurate Titanium cranio-facial implants with high forming angle using single point incremental forming', *Key Engineering Materials*, 549, pp. 223–230. doi: 10.4028/www.scientific.net/KEM.549.223.

Eksteen, P. D. W. (2013) 'Development of incrementally formed patient-specific titanium knee prosthesis', (March), pp. 1–119.

Emmens, W. C., Sebastiani, G. and van den Boogaard, A. H. (2010) 'The technology of Incremental sheet forming—A brief review of the history', *Journal of Materials Processing Technology*, 210(8), pp. 981–997. doi: 10.1016/j.jmatprotec.2010.02.014.

Ezugwu, E. O. *et al.* (2005) 'Modelling the correlation between cutting and process parameters in high-speed machining of Inconel 718 alloy using an artificial neural network', *International Journal of Machine Tools and Manufacture*, 45(12–13), pp. 1375–1385. doi: 10.1016/j.ijmachtools.2005.02.004.

Fan, G. *et al.* (2008) 'Electric hot incremental forming: A novel technique', *International Journal of Machine Tools and Manufacture*, 48(15), pp. 1688–1692. doi: 10.1016/j.ijmachtools.2008.07.010.

Filice, L., Fratini, L. and Micari, F. (2002) 'Analysis of material formability in incremental forming', *CIRP Annals*, 51(1), pp. 199–202. doi: 10.1016/S0007-8506(07)61499-1.

Fiorentino, A., Marzi, R. and Ceretti, E. (2012) 'Preliminary results on Ti Incremental Sheet Forming (ISF) of biomedical devices: Biocompatibility, surface finishing and treatment', *International Journal of Mechatronics and Manufacturing Systems*, 5(1), pp. 36–45. doi: 10.1504/IJMMS.2012.046146.

Francesco, G., Giuseppina, A. and Luigino, F. (2017) 'Incremental forming with local induction heating on materials with magnetic and non-magnetic properties', *Procedia Engineering*, 183, pp. 143–148. doi: 10.1016/j.proeng.2017.04.037.

Göttmann, A. *et al.* (2013) 'Manufacturing of individualized cranial implants using two point incremental sheet metal forming', in *Future Trends in Production Engineering*. Berlin, Heidelberg: Springer, pp. 287–295. doi: 10.1007/978-3-642-24491-9_28.

Grimm, T. J. and Ragai, I. (2019) 'An investigation of liquid metal lubrication during electrically-assisted incremental forming of Titanium', *Procedia Manufacturing*, 34, pp. 118–124. doi: 10.1016/j.promfg.2019.06.128.

Grün, P. A. *et al.* (2018) 'Formability of titanium alloy sheets by friction stir incremental forming', *The International Journal of Advanced Manufacturing Technology*, 99(5–8), pp. 1993–2003. doi: 10.1007/s00170-018-2541-5.

Gupta, P. and Jeswiet, J. (2019) 'Manufacture of an aerospace component by single point incremental forming', *Procedia Manufacturing*, 29, pp. 112–119. doi: 10.1016/j.promfg.2019.02.113.

Gupta, P., Szekeres, A. and Jeswiet, J. (2019) 'Design and development of an aerospace component with single-point incremental forming', *The International Journal of Advanced Manufacturing Technology*, 103(9–12), pp. 3683–3702. doi: 10.1007/s00170-019-03622-4.

Han, F. *et al.* (2010) 'A digital manufacture technology for skull prosthesis using incremental sheet forming method', *Advanced Materials Research*, 102–104, pp. 348–352. doi: 10.4028/www.scientific.net/AMR.102-104.348.

Hussaini, S. M., Singh, S. K. and Gupta, A. K. (2014) 'Experimental and numerical investigation of formability for austenitic stainless steel 316 at elevated temperatures', *Journal of Materials Research and Technology*, 3(1), pp. 17–24. doi: 10.1016/j.jmrt.2013.10.010.

Iseki, H. (2001) 'Flexible and incremental bulging of sheet metal using high-speed water jet', *JSME International Journal Series C*, 44(2), pp. 486–493. doi: 10.1299/jsmec.44.486.

Jackson, K. P., Allwood, J. M. and Landert, M. (2008) 'Incremental forming of sandwich panels', *Journal of Materials Processing Technology*, 204(1–3), pp. 290–303. doi: 10.1016/j.jmatprotec.2007.11.117.

Jeswiet, J. *et al.* (2005) 'Custom manufacture of a solar cooker—A case study', *Advanced Materials Research*, 6–8, pp. 487–492. doi: 10.4028/www.scientific.net/AMR.6-8.487.

Jeswiet, J. *et al.* (2015) 'Single point and asymmetric incremental forming', *Advances in Manufacturing*, 3(4), pp. 253–262. doi: 10.1007/s40436-015-0126-1.

Jeswiet, J. and Hagan, E. (2002) 'Rapid proto-typing of a headlight with sheet metal', *SME Technical Paper, Society of Manufacturing Engineers*, MF02–205, p. 6.

Ji, Y. H. and Park, J. J. (2008) 'Incremental forming of free surface with magnesium alloy AZ31 sheet at warm temperatures', *Transactions of Nonferrous Metals Society of China*, 18, pp. s165–s169. doi: 10.1016/S1003-6326(10)60195-1.

Jiménez, I. *et al.* (2017) 'Investigation of residual stress distribution in single point incremental forming of Aluminum parts by X-ray diffraction technique', *The International Journal of Advanced Manufacturing Technology*, 91(5–8), pp. 2571–2580. doi: 10.1007/s00170-016-9952-y.

Jurisevic, B. *et al.* (2007) 'Experimental and numerical study of the tool in water jet incremental sheet metal forming', in *Advances in Integrated Design and Manufacturing in Mechanical Engineering II*. Dordrecht, Netherlands: Springer, pp. 79–91. doi: 10.1007/978-1-4020-6761-7_6.

Jurisevic, B. *et al.* (2008) 'Introduction of laminated supporting tools in water jet incremental sheet metal forming', *The International Journal of Advanced Manufacturing Technology*, 37(5–6), pp. 496–503. doi: 10.1007/s00170-007-0994-z.

Kashid, S. and Kumar, S. (2013) 'Applications of artificial neural network to sheet metal work—A review', *American Journal of Intelligent Systems*, 2(7), pp. 168–176. doi: 10.5923/j.ajis.20120207.03.

Kim, S. W. *et al.* (2007) 'Incremental forming of Mg alloy sheet at elevated temperatures', *Journal of Mechanical Science and Technology*, 21(10), pp. 1518–1522. doi: 10.1007/BF03177368.

Kleiner, M., Geiger, M. and Klaus, A. (2003) 'Manufacturing of lightweight components by metal forming', *CIRP Annals*, 52(2), pp. 521–542. doi: 10.1016/S0007-8506(07)60202-9.

Kondayya, D. and Gopala Krishna, A. (2013) 'An integrated evolutionary approach for modelling and optimization of laser beam cutting process', *The International Journal of Advanced Manufacturing Technology*, 65(1–4), pp. 259–274. doi: 10.1007/s00170-012-4165-5.

Kubit, A. *et al.* (2020) 'Strength analysis of a rib-stiffened GLARE-based thin-walled structure', *Materials*, 13(13), p. 2929. doi: 10.3390/ma13132929.

Kuitert, R. (2016) 'Feasibility study on single point incremental forming', *Master of Science thesis*, Delft University of Technology.

Kulkarni, S., Sreedhara, V. S. M. and Mocko, G. (2016) 'Heat assisted single point incremental forming of polymer sheets', in *Volume 4: 21st Design for Manufacturing and the Life Cycle Conference; 10th International Conference on Micro- and Nanosystems*. American Society of Mechanical Engineers. doi: 10.1115/DETC2016-60031.

Kumar, R., Kumar, G. and Singh, A. (2020) 'An assessment of residual stresses and microstructure during single point incremental forming of commercially pure titanium used in biomedical applications', *Materials Today: Proceedings*, 28, pp. 1261–1266. doi: 10.1016/j.matpr.2020.04.147.

Kumar, Y. and Kumar, S. (2015) 'Incremental Sheet Forming (ISF)', in, pp. 29–46. doi: 10.1007/978-81-322-2355-9_2.

Kurra, S. *et al.* (2015) 'Modeling and optimization of surface roughness in single point incremental forming process', *Journal of Materials Research and Technology*, 4(3), pp. 304–313. doi: 10.1016/j.jmrt.2015.01.003.

Lee, S. U. *et al.* (2011) 'Application of incremental sheet metal forming for automotive body-in-white manufacturing', *Transactions of Materials Processing*, 20(4), pp. 279–283. doi: 10.5228/KSTP.2011.20.4.279.

Lela, B., Bajić, D. and Jozić, S. (2009) 'Regression analysis, support vector machines, and Bayesian neural network approaches to modeling surface roughness in face milling', *The International Journal of Advanced Manufacturing Technology*, 42(11–12), pp. 1082–1088. doi: 10.1007/s00170-008-1678-z.

Leonhardt, A. *et al.* (2018) 'Experimental study on incremental sheet forming of Magnesium alloy AZ31 with hot air heating', *Procedia Manufacturing*, 15, pp. 1192–1199. doi: 10.1016/j.promfg.2018.07.369.

Leszak, E. (1967) 'Apparatus and process for incremental dieless forming, [US3342051A]', *United States Patent Office. U.S.*

Li, E. (2013) 'Reduction of springback by intelligent sampling-based LSSVR metamodel-based optimization', *International Journal of Material Forming*, 6(1), pp. 103–114. doi: 10.1007/s12289-011-1076-1.

Li, J. *et al.* (2014) 'Modeling and experimental validation for truncated cone parts forming based on water jet incremental sheet metal forming', *The International Journal of Advanced Manufacturing Technology*, 75(9–12), pp. 1691–1699. doi: 10.1007/s00170-014-6222-8.

Li, J. *et al.* (2018) 'Innovation applications of electromagnetic forming and its fundamental problems', *Procedia Manufacturing*, 15, pp. 14–30. doi: 10.1016/j.promfg.2018.07.165.

Li, P. *et al.* (2017) 'Evaluation of forming forces in ultrasonic incremental sheet metal forming', *Aerospace Science and Technology*, 63, pp. 132–139. doi: 10.1016/j.ast.2016.12.028.

Li, Y. *et al.* (2017) 'Effects of ultrasonic vibration on deformation mechanism of incremental point-forming process', *Procedia Engineering*, 207, pp. 777–782. doi: 10.1016/j.proeng.2017.10.828.

Li, Y. *et al.* (2020) 'Investigation on the material flow and deformation behavior during ultrasonic-assisted incremental forming of straight grooves', *Journal of Materials Research and Technology*, 9(1), pp. 433–454. doi: 10.1016/j.jmrt.2019.10.072.

Li, Z. *et al.* (2018) 'Electric assistance hot incremental sheet forming: An integral heating design', *The International Journal of Advanced Manufacturing Technology*, 96(9–12), pp. 3209–3215. doi: 10.1007/s00170-018-1792-5.

Liao, J. *et al.* (2020) 'Influence of heating mode on orange peel patterns in warm incremental forming of Magnesium alloy', *Procedia Manufacturing*, 50, pp. 5–10. doi: 10.1016/j.promfg.2020.08.002.

Liu, Z. (2017) 'Friction stir incremental forming of AA7075-O sheets: Investigation on process feasibility', *Procedia Engineering*, 207, pp. 783–788. doi: 10.1016/j.proeng.2017.10.829.

Lu, B. *et al.* (2014) 'Mechanism investigation of friction-related effects in single point incremental forming using a developed oblique roller-ball tool', *International*

Journal of Machine Tools and Manufacture, 85, pp. 14–29. doi: 10.1016/j. ijmachtools.2014.04.007.

Lu, B. *et al.* (2016) 'Titanium based cranial reconstruction using incremental sheet forming', *International Journal of Material Forming*, 9(3), pp. 361–370. doi: 10.1007/ s12289-014-1205-8.

Lu, B. *et al.* (2017) 'A study of incremental sheet forming by using water jet', *The International Journal of Advanced Manufacturing Technology*, 91(5–8), pp. 2291–2301. doi: 10.1007/s00170-016-9869-5.

Maji, K. and Kumar, G. (2020) 'Inverse analysis and multi-objective optimization of single-point incremental forming of AA5083 Aluminum alloy sheet', *Soft Computing*, 24(6), pp. 4505–4521. doi: 10.1007/s00500-019-04211-z.

Marouani, H. and Aguir, H. (2012) 'Identification of material parameters of the Gurson—Tvergaard—Needleman damage law by combined experimental, numerical sheet metal blanking techniques and artificial neural networks approach', *International Journal of Material Forming*, 5(2), pp. 147–155. doi: 10.1007/s12289-011-1035-x.

Mason, B. (1978) 'Sheetmetal forming for small batches', *Bachelor thesis*, University of Nottingham.

Matsubara, S. (1994) 'Incremental backward bulge forming of a sheet metal with a hemispherical head tool-a study of a numerical control forming system II-', *Journal of the Japan Society for Technology of Plasticity*, 35(406), pp. 1311–1316. Available at: https://ci.nii.ac.jp/naid/10026457537/en/.

Mekras, N. (2017) 'Using artificial neural networks to model Aluminium based sheet forming processes and tools details', *Journal of Physics: Conference Series*, 896, p. 012090. doi: 10.1088/1742-6596/896/1/012090.

Milutinović, M. *et al.* (2021) 'Characterisation of geometrical and physical properties of a stainless steel denture framework manufactured by single-point incremental forming', *Journal of Materials Research and Technology*, 10, pp. 605–623. doi: 10.1016/j.jmrt.2020.12.014.

Mohammadi, A. *et al.* (2013) 'Influence of laser assisted single point incremental forming on the accuracy of shallow sloped parts', in, pp. 864–867. doi: 10.1063/1.4850107.

Mohammadi, A. *et al.* (2015) 'Formability enhancement in incremental forming for an automotive Aluminium alloy using laser assisted incremental forming', *Key Engineering Materials*, 639, pp. 195–202. doi: 10.4028/www.scientific.net/ KEM.639.195.

Mohammadi, A. *et al.* (2016) 'Towards accuracy improvement in single point incremental forming of shallow parts formed under laser assisted conditions', *International Journal of Material Forming*, 9(3), pp. 339–351. doi: 10.1007/s12289-014-1203-x.

Mulay, A. *et al.* (2019) 'Prediction of average surface roughness and formability in single point incremental forming using artificial neural network', *Archives of Civil and Mechanical Engineering*, 19(4), pp. 1135–1149. doi: 10.1016/j.acme.2019.06.004.

Najm, S. M. and Paniti, I. (2018) 'Experimental investigation on the single point incremental forming of AlMn1Mg1 foils using flat end tools', *IOP Conference Series: Materials Science and Engineering*, 448, p. 012032. doi: 10.1088/1757-899X/448/1/012032.

Najm, S. M. and Paniti, I. (2020) 'Study on effecting parameters of flat and hemispherical end tools in SPIF of Aluminium foils', *Tehnicki vjesnik—Technical Gazette*, 27(6). doi: 10.17559/TV-20190513181910.

Najm, S. M. and Paniti, I. (2021a) 'Artificial neural network for modeling and investigating the effects of forming tool characteristics on the accuracy and formability of thin Aluminum alloy blanks when using SPIF', *The International Journal of*

Advanced Manufacturing Technology, 114(9–10), pp. 2591–2615. doi: 10.1007/s00170-021-06712-4.

Najm, S. M. and Paniti, I. (2021b) 'Predict the effects of forming tool characteristics on surface roughness of Aluminum foil components formed by SPIF using ANN and SVR', *International Journal of Precision Engineering and Manufacturing*, 22(1), pp. 13–26. doi: 10.1007/s12541-020-00434-5.

Najm, S. M. and Paniti, I. (2022) 'Investigation and machine learning-based prediction of parametric effects of single point incremental forming on pillow effect and wall profile of AlMn1Mg1 Aluminum alloy sheets', *Journal of Intelligent Manufacturing*. doi: 10.1007/s10845-022-02026-8.

Najm, S. M., Paniti, I. and Viharos, Z. J. (2020) 'Lubricants and affecting parameters on hardness in SPIF of AA1100 Aluminium', in *17th IMEKO TC 10 and EUROLAB Virtual Conference 'Global Trends in Testing, Diagnostics and Inspection for 2030'*, Dubrovnik, Croatia, pp. 387–392.

Najm, S. M. *et al.* (2021) 'Parametric effects of single point incremental forming on hardness of AA1100 Aluminium alloy sheets', *Materials*, 14(23), p. 7263. doi: 10.3390/ma14237263.

Nasrollahi, V. and Arezoo, B. (2012) 'Prediction of springback in sheet metal components with holes on the bending area, using experiments, finite element and neural networks', *Materials & Design (1980–2015)*, 36, pp. 331–336. doi: 10.1016/j.matdes.2011.11.039.

Okada, M. *et al.* (2018) 'Development of optical-heating-assisted incremental forming method for carbon fiber reinforced thermoplastic sheet—Forming characteristics in simple spot-forming and two-dimensional sheet-fed forming', *Journal of Materials Processing Technology*, 256, pp. 145–153. doi: 10.1016/j.jmatprotec.2018.02.014.

Oleksik, V. *et al.* (2021) 'Single-point incremental forming of Titanium and Titanium alloy sheets', *Materials*, 14(21), p. 6372. doi: 10.3390/ma14216372.

Oraon, M. and Sharma, V. (2018) 'Prediction of surface roughness in single point incremental forming of AA3003-O alloy using artificial neural network', *International Journal of Materials Engineering Innovation*, 9(1), p. 1. doi: 10.1504/IJMATEI.2018.092181.

Palumbo, G. *et al.* (2020) 'Design of custom cranial prostheses combining manufacturing and drop test finite element simulations', *The International Journal of Advanced Manufacturing Technology*, 111(5–6), pp. 1627–1641. doi: 10.1007/s00170-020-06213-w.

Paniti, I. *et al.* (2020) 'Experimental and numerical investigation of the single-point incremental forming of Aluminium alloy foils', *Acta IMEKO*, 9(1), pp. 25–31. doi: 10.21014/acta_imeko.v9i1.750.

Park, J.-J. and Kim, Y.-H. (2003) 'Fundamental studies on the incremental sheet metal forming technique', *Journal of Materials Processing Technology*, 140(1–3), pp. 447–453. doi: 10.1016/S0924-0136(03)00768-4.

Peter, I. *et al.* (2019) 'Incremental sheet forming for prototyping automotive modules', *Procedia Manufacturing*, 32, pp. 50–58. doi: 10.1016/j.promfg.2019.02.182.

Potran, M., Skakun, P. and Faculty, M. (2014) 'Application of single point incremental', *Journal for Technology of Plasticity*, 39(2), pp. 16–23.

Radu, C. *et al.* (2013) 'Improving the accuracy of parts manufactured by single point incremental forming', *Applied Mechanics and Materials*, 332, pp. 443–448. doi: 10.4028/www.scientific.net/AMM.332.443.

Reddy, N. V. and Lingam, R. (2018) 'Double sided incremental forming: Capabilities and challenges', *Journal of Physics: Conference Series*, 1063, p. 012170. doi: 10.1088/1742-6596/1063/1/012170.

Rosca, N. *et al.* (2023) 'Minimizing the main strains and thickness reduction in the single point incremental forming process of polyamide and high-density polyethylene sheets', *Materials*, 16(4), p. 1644. doi: 10.3390/ma16041644.

Saidi, B. *et al.* (2019) 'Hot incremental forming of Titanium human skull prosthesis by using cartridge heaters: A reverse engineering approach', *The International Journal of Advanced Manufacturing Technology*, 101(1–4), pp. 873–880. doi: 10.1007/s00170-018-2975-9.

Salamati, M. *et al.* (2015) 'Creating custom Titanium implants for cranioplasty using incremental sheet forming', *13th International Cold Forming Congress*, (October 2016), pp. 72–79. Available at: www.researchgate.net/publication/309359259.

Sbayti, M., Bahloul, R. and Belhadjsalah, H. (2018) 'Numerical modeling of hot incremental forming process for biomedical application', in, pp. 881–891. doi: 10.1007/978-3-319-66697-6_86.

Sbayti, M. *et al.* (2018) 'Optimization techniques applied to single point incremental forming process for biomedical application', *The International Journal of Advanced Manufacturing Technology*, 95(5–8), pp. 1789–1804. doi: 10.1007/s00170-017-1305-y.

Schedin, E. (1992) 'Sheet metal forming', *Materials & Design*, 13(6), pp. 366–367. doi: 10.1016/0261-3069(92)90017-c.

Scheffler, S. *et al.* (2019) 'Incremental sheet metal forming on the example of car exterior skin parts', *Procedia Manufacturing*, 29, pp. 105–111. doi: 10.1016/j.promfg.2019.02.112.

Sedaghat, H., Xu, W. and Zhang, L. (2019) 'Ultrasonic vibration-assisted metal forming: Constitutive modelling of acoustoplasticity and applications', *Journal of Materials Processing Technology*, 265, pp. 122–129. doi: 10.1016/j.jmatprotec.2018.10.012.

Shi, Y. *et al.* (2019) 'Experimental study of water jet incremental micro-forming with supporting dies', *Journal of Materials Processing Technology*, 268, pp. 117–131. doi: 10.1016/j.jmatprotec.2019.01.012.

Silva, M. B. and Martins, P. A. F. (2014) *Incremental Sheet Forming, Comprehensive Materials Processing*. Elsevier. doi: 10.1016/B978-0-08-096532-1.00301-0.

Sisodia, V. *et al.* (2021) 'Experimental investigation on geometric error in single-point incremental forming with dummy sheet', in, pp. 81–92. doi: 10.1007/978-981-15-6619-6_9.

Skjoedt, M. *et al.* (2008) 'Multi stage strategies for single point incremental forming of a cup', *International Journal of Material Forming*, 1(S1), pp. 1199–1202. doi: 10.1007/s12289-008-0156-3.

Teymoori, F., LohMousavi, M. and Etesam, A. (2016) 'Numerical analysis of fluid structure interaction in water jet incremental sheet forming process using coupled Eulerian—Lagrangian approach', *International Journal on Interactive Design and Manufacturing (IJIDeM)*, 10(2), pp. 203–210. doi: 10.1007/s12008-013-0197-9.

Trzepieciński, T., Oleksik, V., *et al.* (2021) 'Emerging trends in single point incremental sheet forming of lightweight metals', *Metals*, 11(8), p. 1188. doi: 10.3390/met11081188.

Trzepieciński, T., Najm, S. M., *et al.* (2021) 'New advances and future possibilities in forming technology of hybrid metal—Polymer composites used in aerospace applications', *Journal of Composites Science*, 5(8), p. 217. doi: 10.3390/jcs5080217.

Trzepieciński, T., Najm, S. M., Oleksik, V., *et al.* (2022) 'Recent developments and future challenges in incremental sheet forming of Aluminium and Aluminium alloy sheets', *Metals*, 12(1), p. 124. doi: 10.3390/met12010124.

Trzepieciński, T., Najm, S. M., Pepelnjak, T., *et al.* (2022) 'Incremental sheet forming of metal-based composites used in aviation and automotive applications', *Journal of Composites Science*, 6(10), p. 295. doi: 10.3390/jcs6100295.

Vahdati, M., Mahdavinejad, R. and Amini, S. (2017) 'Investigation of the ultrasonic vibration effect in incremental sheet metal forming process', *Proceedings of the Institution of Mechanical Engineers, Part B: Journal of Engineering Manufacture*, 231(6), pp. 971–982. doi: 10.1177/0954405415578579.

Vahdani, M. *et al.* (2019) 'Electric hot incremental sheet forming of Ti-6Al-4V Titanium, AA6061 Aluminum, and DC01 steel sheets', *The International Journal of Advanced Manufacturing Technology*, 103(1–4), pp. 1199–1209. doi: 10.1007/s00170-019-03624-2.

Valoppi, B. *et al.* (2016) 'A hybrid mixed double-sided incremental forming method for forming Ti6Al4V alloy', *CIRP Annals*, 65(1), pp. 309–312. doi: 10.1016/j.cirp.2016.04.135.

Vanhove, H. *et al.* (2017) 'Production of thin shell clavicle implants through single point incremental forming', *Procedia Engineering*, 183, pp. 174–179. doi: 10.1016/j.proeng.2017.04.058.

Verbert, J. *et al.* (2008) 'Multi-step toolpath approach to overcome forming limitations in single point incremental forming', *International Journal of Material Forming*, 1(S1), pp. 1203–1206. doi: 10.1007/s12289-008-0157-2.

Wang, J., Li, L. and Jiang, H. (2016) 'Effects of forming parameters on temperature in frictional stir incremental sheet forming', *Journal of Mechanical Science and Technology*, 30(5), pp. 2163–2169. doi: 10.1007/s12206-016-0423-z.

Wei, S. *et al.* (2015) 'Experimental study on manufacturing metal bellows forming by water jet incremental forming', *The International Journal of Advanced Manufacturing Technology*, 81(1–4), pp. 129–133. doi: 10.1007/s00170-015-7202-3.

Xu, D. K. *et al.* (2014) 'A comparative study on process potentials for frictional stir- and electric hot-assisted incremental sheet forming', *Procedia Engineering*, 81, pp. 2324–2329. doi: 10.1016/j.proeng.2014.10.328.

Xu, D. K. *et al.* (2016) 'Enhancement of process capabilities in electrically-assisted double sided incremental forming', *Materials & Design*, 92, pp. 268–280. doi: 10.1016/j.matdes.2015.12.009.

Yoon, S. J. and Yang, D. Y. (2003) 'Development of a highly flexible incremental roll forming process for the manufacture of a doubly curved sheet metal', *CIRP Annals*, 52(1), pp. 201–204. doi: 10.1016/S0007-8506(07)60565-4.

Zain, A. M., Haron, H. and Sharif, S. (2010) 'Prediction of surface roughness in the end milling machining using artificial neural network', *Expert Systems with Applications*, 37(2), pp. 1755–1768. doi: 10.1016/j.eswa.2009.07.033.

Zhang, H. *et al.* (2017) 'Thickness control in a new flexible hybrid incremental sheet forming process', *Proceedings of the Institution of Mechanical Engineers, Part B: Journal of Engineering Manufacture*, 231(5), pp. 779–791. doi: 10.1177/0954405417694061.

8 Finite Element Analysis of the Incremental Forming Process

Ömer Seçgin, and Vedat Taşdemir

8.1 INTRODUCTION

Incremental forming is a quick and low-cost manufacturing technique used in the production of metal parts. It has a wide range of uses in many different industries (such as automotive, aviation, and white goods). In incremental forming, the geometry of the part to be produced is first created with a computer-aided design (CAD) program. Then, tool paths are created in a computer-aided manufacturing (CAM) program to move the form tool. The sheet is formed by applying local pressure to the sheet with the form tool moving according to this tool path (Martins et al., 2008; Mulay, Ben, et al., 2017a; Seçgin, Ata, and Özsert, 2018; Khan and Pradhan, 2019).

Examining the incremental forming literature reveals a focus on parameters such as the diameter of the form set and spindle revolution and the effects of wall thickness change and formability on those parameters (Malyer, 2013; Azevedo et al., 2015; Sbayti et al., 2016; Mulay et al., 2019). In addition to the previous works, Lu et al. used a water jet instead of a metal forming tool and obtained better wall thickness than with traditional incremental forming (Lu, Cao, and Ou, 2011). Devarajan et al. used two-point contact incremental forming to study the formability of several aluminum sheets (AA2024, 5083, and 7075) (Devarajan et al., 2014). They showed that the wall thickness in the regions with right angles on the piece was much lower than in the other regions. They stated that tearing occurs earlier in AA2024 and AA7075 due to plastic instability.

Seçgin and Özsert investigated the incremental formability of titanium GR 2 sheet with their new method: two-point incremental forming–rolling blank holder (TPIF–RL) (Seçgin and Özsert, 2022). They set parameter levels that optimized forming force, surface roughness, and thickness distribution. They determined that the most important parameters for vertical forming force were increment and forming tool diameter. Then, Şen et al. applied the TPIF–RL method, which makes the wall thickness of the sheet more homogeneous, to HC380LA (Şen, Taşdemir, and Seçgin, 2020).

Li et al. (2016) applied a flexible die system to incremental forming, and Karbowski (2015) demonstrated the applicability of incremental forming in the

DOI: 10.1201/9781003441755-8

biomedical industry. AL-Obaidi et al. applied incremental forming to fiberglass material using hot air (AL-Obaidi, Kunke and Kräusel, 2019). Ao et al. (2020) used electropulsing for the incremental forming of Ti-6Al-4V sheet. Mulay et al. investigated the formability of AA5052-H32 by SPIF (Mulay, Ben, et al., 2017b). They revealed that step depth and sheet thickness are the most important parameters. In this study, we investigated the formability of titanium GR 2 sheets with incremental forming. We performed a series of finite element analyses to understand the effects of wall thickness variation on formability.

8.2 INCREMENTAL FORMING

In incremental forming, first, the geometry of the part to be formed is drawn in the SolidWorks program (Figure 8.1). Then the forming parameters (feed, speed, tool diameter, etc.) are determined. The toolpath is created using the SolidWorks Cam. Local pressure is applied on the sheet metal with a tool with a rounded tip following this tool path. Thus, the desired form is given to the sheet (Mohanty, Prakash, and Daseswara, 2019; Milutinovic et al., 2021).

In traditional single-point incremental forming, the sheet is fixed in a hollow mold. For fastening, the sheet can be tightened directly with bolts, or a ring piece can also be used (Figure 8.2). Then the form tool approaches the sheet and pushes

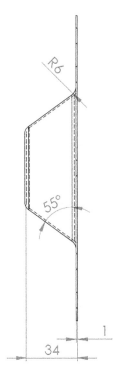

FIGURE 8.1   Dimensions of the part.

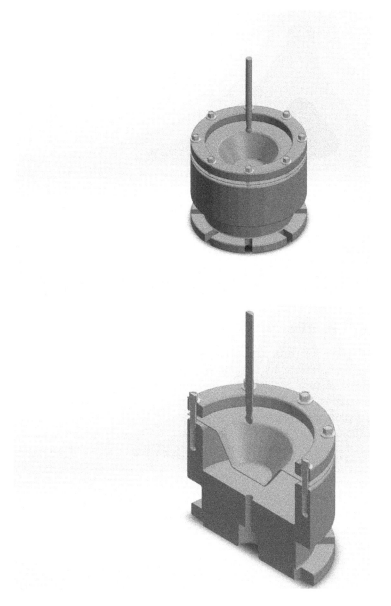

FIGURE 8.2   Single-point incremental forming. A: Full view. B: Section view.

the sheet depending on the tool path (Figure 8.3). As a result of this pressure, deformation occurs in the sheet metal. As the form team continues to move, the piece gains a certain form.

In this method, the basic parameters such as increment amount, form tool diameter, and feed rate significantly affect the forming time, forming force, surface roughness, etc.

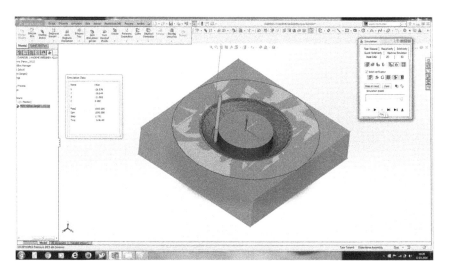

FIGURE 8.3   Creating a toolpath by the CAM program.

8.3   FINITE ELEMENT ANALYSIS

Finite element (FE) analysis is an effective method used in mechanical analysis such as static, flow, dynamic, and modal analysis. Many analyses can be performed successfully in the field of metal forming (deep drawing, bending, incremental forming, etc.). The FE analysis of incremental forming, the subject of this study, is generally performed in the Abaqus program (Peng et al., 2019; Li et al., 2020; Zheng et al., 2020; Seçgin, Nart and Özsert, 2021), including in this study.

First, the data obtained from the tensile tests of the sheet material were entered into the program. The ultimate tensile strength of the material was 116 MPa, and elongation was 0.45%. Then, the parts (sheet metal, mold, ring and form set) drawn in the CAD program were assembled, and their placement was confirmed. In the next step, the step was created, and the interactions between the parts were defined. Boundary conditions and motion properties of the form set obtained from the CAM program were also defined. Then, all other parts except the form set were meshed. Because the form set was defined as analytical rigid, it did not need to be meshed. While the mesh was being thrown, an element definition suitable for the analysis characteristics was also made.

In the FE analysis of incremental forming, the correct definition of the tool motion completely affects the calculation result. The tool path file obtained from the CAM program contains only X-axis codes in some lines and only Y-axis codes in others; some lines also have F and M codes. In FE, only the time data and X, Y, and Z coordinates must be entered regularly. Therefore, this tool path file must be edited and processed in order to be entered into the FE program correctly. For this purpose, three different amplitudes were defined to define the tool movement (Figure 8.4).

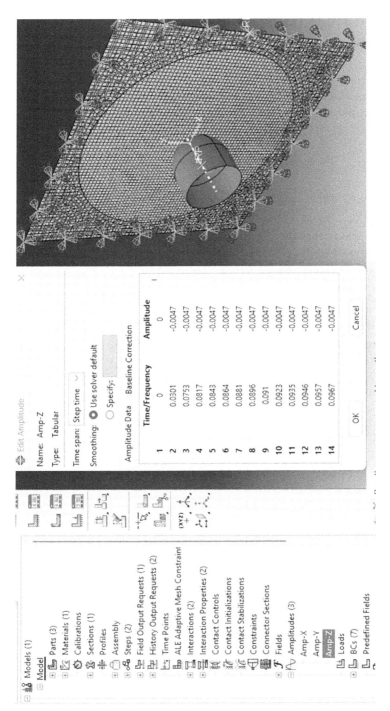

FIGURE 8.4 Amplitude identification phase

8.4   RESULTS AND DISCUSSION

The scope of the study included analyses of sheet metal with different wall thicknesses from 0.5 mm to 2.5 mm, a total of 21 different analyses (Table 8.1). Table 8.1 gives the wall thickness change in each analysis. Thickness deviation was obtained by subtracting the smallest wall thickness formed in the analysis from the initial wall thickness (Equation 1). Thinning rate is obtained by dividing thickness deviation by $t_{initial}$ (Equation 2):

$$Thickness\,Deviation = t_{initial} - t_{minimum} \qquad (8.1)$$

$$Thinning\,Rate = \frac{Thickness\,Deviation}{t_{initial}} \qquad (8.2)$$

Table 8.1 shows that the thickness deviation increased with the increase in wall thickness and the thinning rate did not change much. The sheet thickness

TABLE 8.1

Finite Element Analysis Results

İnitial sheet thickness (mm)	t max (mm)	Thickness deviation (mm)	Thinning rate (%)
0.5	0.141	0.36	72
0.6	0.163	0.44	73
0.7	0.175	0.53	75
0.8	0.219	0.58	73
0.9	0.245	0.65	73
1	0.278	0.72	72
1.1	0.301	0.80	73
1.2	0.324	0.88	73
1.3	0.353	0.95	73
1.4	0.376	1.02	73
1.5	0.419	1.08	72
1.6	0.451	1.15	72
1.7	0.485	1.22	71
1.8	0.524	1.28	71
1.9	0.565	1.34	70
2	0.598	1.40	70
2.1	0.638	1.46	70
2.2	0.671	1.53	70
2.3	0.708	1.59	69
2.4	0.753	1.65	69
2.5	0.792	1.71	68

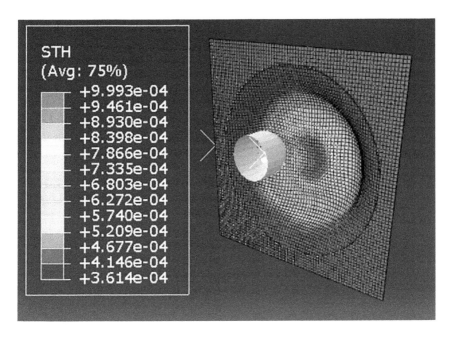

FIGURE 8.5   Sheet thickness distribution (1 mm thick sheet).

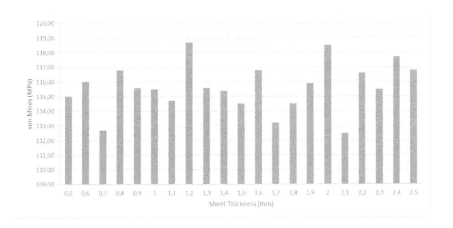

FIGURE 8.6   von Mises stresses.

distribution (1 mm sheet thickness) is given in Figure 8.5. Von Mises stress was also investigated, and the resulting stress varied between approximately 113 MPa and 118 MPa (Figure 8.6). The von Mises stress for a sheet 0.7 mm thick is given in Figure 8.7.

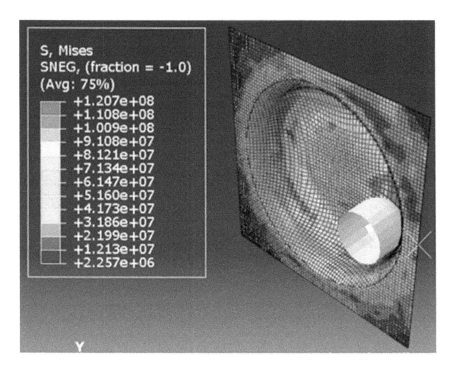

FIGURE 8.7   von Mises stress, 0.7 mm thick sheet.

8.5 CONCLUSIONS

The incremental formability of titanium grade 2 sheet was examined in this work. Within the scope of the study, 21 different thicknesses were investigated. The effects of sheet thickness on the wall thickness variation ratio and stress distribution were investigated. It was observed that the wall thickness change rate was around 70% maximum. Stress distribution, on the other hand, does not show large variability.

REFERENCES

AL-Obaidi, A., Kunke, A. and Kräusel, V. (2019) 'Hot single-point incremental forming of glass-fiber-reinforced polymer (PA6GF47) supported by hot air', *Journal of Manufacturing Processes*, 43(August 2018), pp. 17–25. doi: 10.1016/j.jmapro.2019.04.036.

Ao, D. *et al.* (2020) 'Formability and deformation mechanism of Ti-6Al-4V sheet under electropulsing assisted incremental forming', *International Journal of Solids and Structures*, 202, pp. 357–367. doi: 10.1016/j.ijsolstr.2020.06.028.

Azevedo, N. G. *et al.* (2015) 'Lubrication aspects during single point incremental forming for steel and Aluminum materials', *International Journal of Precision Engineering and Manufacturing*, 16(3), pp. 589–595. doi: 10.1007/s12541-015-0079-0.

Devarajan, N. *et al.* (2014) 'Complex incremental sheet forming using back die support on Aluminium 2024, 5083 and 7075 alloys', *Procedia Engineering*, 81(October), pp. 2298–2304. doi: 10.1016/j.proeng.2014.10.324.

Karbowski, K. (2015) 'Application of incremental sheet forming', *Management and Production Engineering Review*, 6(December), pp. 55–59. doi: 10.1515/mper-2015-0036.

Khan, S. and Pradhan, S. K. (2019) 'Experimentation and FE simulation of single point incremental forming', *Materials Today: Proceedings*, 27, pp. 2334–2339. doi: 10.1016/j.matpr.2019.09.123.

Li, J. *et al.* (2016) 'A new type of incremental bending method for complicated curved sheet metal', *Transactions of the Canadian Society for Mechanical Engineering*, 40(4), pp. 433–443.

Li, Y. *et al.* (2020) 'Investigation on the material flow and deformation behavior during ultrasonic-assisted incremental forming of straight grooves', *Journal of Materials Research and Technology*, 9(1), pp. 433–454. doi: 10.1016/j.jmrt.2019.10.072.

Lu, B., Cao, J. and Ou, H. (2011) 'Theoretical and numerical analysis of incremental sheet forming by using high pressure water jet', *ASME 2011 International Manufacturing Science and Engineering Conference, MSEC 2011*, 1(February 2016). doi: 10.1115/MSEC2011-50235.

Malyer, E. (2013) 'The influence of toolpath strategy on geometric accuracy in incremental forming', *Key Engineering Materials*, 554–557(May), pp. 1351–1361. doi: 10.4028/www.scientific.net/KEM.554-557.1351.

Martins, P. A. F. F. *et al.* (2008) 'Theory of single point incremental forming', *CIRP Annals—Manufacturing Technology*, 57(1), pp. 247–252. doi: 10.1016/j.cirp.2008.03.047.

Milutinovic, M. *et al.* (2021) 'Characterisation of geometrical and physical properties of a stainless steel denture framework manufactured by single-point incremental forming', *Journal of Materials Research and Technology*, 10, pp. 605–623. doi: 10.1016/j.jmrt.2020.12.014.

Mohanty, S., Prakash, S. and Daseswara, R. Y. V. (2019) 'Robot - assisted incremental sheet metal forming under the different forming condition', *Journal of the Brazilian Society of Mechanical Sciences and Engineering*, 6. doi: 10.1007/s40430-019-1581-6.

Mulay, A., Ben, S., *et al.* (2017a) 'Artificial neural network modeling of quality prediction of a single point incremental sheet forming process', *Advanced Science and Technology Letters*, 147, pp. 244–250.

Mulay, A., Ben, B. S., *et al.* (2017b) 'Experimental investigation and modeling of single point incremental forming for AA5052-H32 Aluminum alloy', *Arab J Sci Eng*, (July), pp. 1–12. doi: 10.1007/s13369-017-2746-1.

Mulay, A. *et al.* (2019) 'Prediction of average surface roughness and formability in single point incremental forming using artificial neural network', *Archives of Civil and Mechanical Engineering*, 19(4), pp. 1135–1149. doi: 10.1016/j.acme.2019.06.004.

Peng, W. *et al.* (2019) 'Experimental and finite element investigation of over-bending phenomenon in Double-Sided Incremental Forming (DSIF) of Aluminium sheets', *Procedia Manufacturing*, 29, pp. 59–66. doi: 10.1016/j.promfg.2019.02.106.

Sbayti, M. *et al.* (2016) 'Finite element analysis of hot single point incremental forming of hip prostheses', in *The 12th International Conference on Numerical Methods in Industrial Forming Processes*. Troyes, France. doi: 10.1051/matecconf/20168014006.

Seçgin, Ö., Ata, E. and Özsert, İ. (2018) 'A new approach for tool path definition in finite element analysis of incremental forming method used in the manufacturing industry', in *International Conference on Automotive Technologies (OTEKON 2018), 7–8 Mayıs 2018*. Bursa, pp. 1219–1225.

Seçgin, Ö., Nart, E. and Özsert, İ. (2021) 'Incremental forming of Titanium grade 2 sheet by TPIF-RL method', *International Journal of Automotive Science and Technology*, 5, pp. 386–389. doi: 10.30939/ijastech.999466.

Seçgin, Ö. and Özsert, İ. (2022) 'Optimization of Titanium grade 2 sheet forming by roller blank holder method', *Proceedings of the Institution of Mechanical Engineers, Part E: Journal of Process Mechanical Engineering*. doi: 10.1177/09544089221143349.

Şen, N., Taşdemir, V. and Seçgin, Ö. (2020) 'Investigation of formability of HC380LA material via the TPIF-RL incremental forming method', *Ironmaking and Steelmaking*, 47(10), pp. 1199–1205. doi: 10.1080/03019233.2019.1711351.

Zheng, C. *et al.* (2020) 'Laser shock induced incremental forming of pure copper foil and its deformation behavior', *Optics and Laser Technology*, 121(June 2019), p. 105785. doi: 10.1016/j.optlastec.2019.105785.

9 The Incremental Sheet Forming of Light Alloys

G. Karthikeyan, D. Nagarajan, and B. Ravisankar

9.1 INTRODUCTION

Metal forming involves mechanically shaping metals into desired shapes, and it is essential in automobiles, aircraft, construction, and electronics industries. Metal forming is cost-effective and fast and can create complex structures, but the capital cost and lead time involved in making metal-forming dies is quite high, as the process demands separate dies for every individual component (Mori, Yamamoto, and Osakada, 1996; Araghi et al., 2009). These procedures can only be utilized for mass production since creating toolings is expensive and requires technical expertise (Jeswiet et al., 2005), whereas aerospace and automotive components require small-batch fabrication of complex-shaped parts (Bhandari et al., 2022).

New production methods for small-volume, batch-type, or prototype manufacturing had to be developed to meet user demands. Many techniques evolved in response, and incremental sheet forming (ISF) is one such technique that offers flexibility, cost-effectiveness, and the ability to produce complex shapes with simple tooling and a common setup. In ISF, complex 3D geometries can be formed with a simple CNC milling machine in a short period and at low cost (Matsubara, 2001). The localized deformation of ISF allows for forming difficult-to-deform materials, and it provides an alternative to volume-forming processes like press forming and deep drawing. ISF is better than conventional volume-forming techniques (Allwood, Braun and Music, 2010). The tool path is created using computer-aided manufacturing (CAM) software and are based on the required component shape and size. The basic experimental setup for the ISF process is depicted in Figure 9.1(a).

Three ISF processes are commonly used in practice: single-point incremental forming (SPIF); two-point incremental forming (TPIF); and double-sided incremental forming (DSIF). SPIF is one of the simplest and most commonly used processes (Martins et al., 2008). Using a single-sided hemispherical or cylindrical tool, the sheet/blank metal is stretched out in small increments. The material is shaped by localized forces as the tool moves along the path that has already been set up. SPIF is known for being flexible and able to make complicated shapes.

TPIF uses two tools at the same time to shape the metal: the forming tool that does most of the deforming and the support tool that controls the flow of the

DOI: 10.1201/9781003441755-9

157

FIGURE 9.1   (a) ISF experimental setup; (b) ISF tool (Markanday and Nagarajan, 2018).

material to keep it from becoming too thin or breaking (Martins et al., 2008). TPIF can make materials easier to shape and more stable. Finally, DSIF uses two forming tools to shape the sheet metal from both sides at the same time. The tools work together to stretch and form the blank metal incrementally to the required geometry. DSIF possesses several advantages over SPIF such as improved process control, enhanced formability, and less springback (Moser et al., 2016).

Robot-assisted incremental forming is an advanced forming process that is carried out using industrial automation. The blank metal is shaped by a robot arm with a tool that follows a defined tool path, and the process offers flexibility in terms of tool path generation (Mohanty, Regalla, and Daseswara Rao, 2019). The selection of the appropriate ISF process depends on factors such as the desired part geometry, material properties, production requirements, and available resources.

9.2   PRINCIPLES AND PROCEDURES OF ISF

The principle of ISF is localized plastic deformation; the process is similar to conventional spinning, but there are some fundamental differences. In the spinning process, the workpiece rotates, whereas in ISF, the tool rotates along the periphery of the designed tool path. The differences between the processes are listed in Table 9.1.

TABLE 9.1
Characteristics between ISF and Conventional Spinning

Factor	ISF	Spinning
Clamping part	Fixture	Mandrel
Rotating part	Forming tool	Sheet
Geometries made possible	Complex shapes	Symmetrical only possible
Sheet	Clamped in Fixture	Clamped on the mandrel.
Final thickness of the workpiece	Varies according to the wall angle	Remains constant

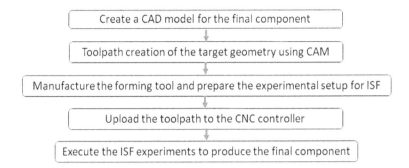

FIGURE 9.2   Flow diagram depicting the steps involved in ISF.

In the spinning process, the roller or tool pushes the sheet, and the thickness remains constant for both the base material and the final component, but in ISF, the tool moves based on the tool path, and the thickness varies between the base material and the final component. ISF is a completely dieless forming process that uses common hemispherical tool runs in the designed tool path to attain the specified final geometry. ISF can be carried out by employing a lab-scale three-axis CNC milling machine as shown in Figure 9.1.

The ISF procedure is to fix the sheet metal in between the clamping plate and the spherical ring. The whole fixture is placed over the machine bed and the coolant in the workspace is filled as shown in Figure 9.1(b). The forming tool is inserted into the spindle of the CNC machine, which pushes the sheet downward according to the designed toolpath. Figure 9.2 explains the steps that are involved in the ISF process.

9.3   ISF PARAMETERS

This dieless technique is competitive and feasible based on component features, quality, and cost, which are determined by the manufacturing parameters. The ISF process parameters control the forming force, formability, surface finish/roughness, strain, dimensional accuracy, and microstructure of produced components. ISF process innovations have shown their usefulness in the medical, automotive, and aerospace sectors.

However, a lack of process parameter information has prevented high-value sectors from completely implementing this unique technology. Significant data explaining the relationships between input parameters and process responses (forming force, formability, surface roughness, strain, dimensional accuracy, the microstructure of formed components, etc.) can improve industrial ISF applicability. The process parameters determining the above-mentioned output parameters are feed rate, spindle speed, pitch size, tool diameter, tool shape, sheet thickness, lubricant, and forming temperature.

9.3.1 FEED RATE

The feed rate is the velocity at which the hemispherical tool runs over the sheet during the process. The rate at which components are fed is another critical process variable. A faster feed rate reduces the amount of time required to form or cycle through components. In ISF, the feed rate is often given in (mm/min) and can be very low (50 mm/min) or very high (5000 mm/min).

The feed rate, or how fast the tool moves along the tool path, affects how the part is deformed and how well it is made. A higher feed rate can increase the productivity, but it can lead to reduced accuracy and surface finish. Balancing the feed rate with other process parameters is crucial for achieving the desired results.

9.3.2 SPINDLE SPEED

The spindle speed is critical in the ISF process since it directly affects material deformation and overall part quality. The rate at which the tool contacts and deforms the sheet metal is determined by the speed at which the spindle rotates; a faster spindle results in faster material deformation, whereas a lower speed provides more accurate control of the forming process. The spindle speed plays an important role in the precision of the part being formed depending on the material properties of the sheet metal: When forming softer materials or when a higher forming rate is required, the higher spindle speeds are frequently used, although extremely high spindle speeds can cause material tearing or surface defects.

When forming the intricate shapes that require finer control or working with harder materials, lower spindle speeds are typically used. Higher spindle speeds can improve the material flow and reduce the forming forces but can generate more heat and affect the surface quality. Balancing the spindle speed with cooling mechanisms and lubrication is necessary to optimize the process.

9.3.3 PITCH SIZE

Pitch is the vertical distance that the making tool is pushed into the sheet. Step-down is specified in a direction perpendicular to the undeformed sheet, whereas step-over is specified in a direction parallel to the undeformed sheet. The step-down/pitch size is the distance that the forming tool travels after each contour. In the same way, the step-over size is how far the forming tool travels in the horizontal direction after every contour in the forming process. Step size is the most vital ISF process parameter because it affects the forming force, formability, dimensional accuracy, surface roughness, and quality of formed components, along with the specifications for the forming machinery and forming tool.

9.3.4 TOOL GEOMETRY

The geometry of the forming tool, including its shape, size, and surface characteristics, plays an important role in ISF. Tool geometry determines the contact area,

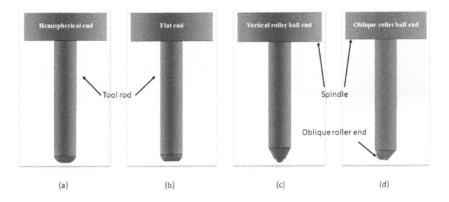

FIGURE 9.3 ISF tool geometries: (a) hemispherical tool, (b) flat-end tool, (c) vertical roller ball tool (Durante, Formisano, and Langella, 2011), and (d) oblique roller ball tool (Li et al., 2014).

localized deformation, and material flow patterns. Tools should be designed and optimized to ensure proper material flow, avoid excessive thinning or tearing, and achieve the desired shape accurately. The tools are of different types as shown in Figure 9.3.

In ISF, hemispherical tools were mostly used, which initiated scratches in the sheet metal due to high friction acting in between the tool and the blank metal. To eliminate these issues, a ball tool was introduced that eliminated the high friction and enhanced the formability but had wall angle limitations. To eliminate the wall angle, the oblique roller ball tool was introduced.

9.3.5 Sheet Thickness

In ISF process, sheet metal thickness, which dictates the amounts of force required, is essential in forming parts successfully. ISF researchers have used sheets of various thicknesses ranging from 0.5 mm to 2.0 mm. Sheet thickness affects formability, dimensional accuracy, and surface roughness. A thicker material requires higher forming forces to deform plastically, but higher thickness also increases material formability. Per the sine law, the sheet thickness decreases during the forming process as the wall angle increases. As a result, at the critical wall angle region, the thinned sheet material cannot resist the applied force and eventually fractures. Thus, severe thinning causes sheet failure.

9.3.6 Lubricant

The lubricant used in ISF serves several purposes, including reducing friction between the sheet metal and the forming tool, improving material flow, and preventing defects on the formed part's surface. Lubricants must resist forming temperatures and not squeeze out throughout the ISF process to keep the tool blank lubricated. The appropriate lubricant can reduce friction at the tool–blank metal interface, lowering forming forces during the process and thereby boosting the heat dissipation and extending the tool life.

Reduced forming forces allow for optimizing the material flow, thereby bringing down the number of stages needed to form the final shape; this process calls for smaller, less expensive for forming. Efficient lubrication can also achieve uniform deformation and reduced interior residual stresses in informed parts. Most lubricants extend tool life by thermally insulating the contact zone.

9.3.7 Forming Temperature

ISF can be carried out using either cold or hot forming. Cold forming takes place at temperatures lower than the material's recrystallization temperature. Unless it is necessary to raise the specimen's temperature to the recrystallization temperature, cold forming is typically carried out at ambient temperature. Using a variety of techniques and tools, the specimen's temperature is kept above the recrystallization temperature during hot forming.

9.3.8 Tool Path

The ISF process's deformation pattern and part shape depend on the toolpath. CNC or CAM software can be used to program the toolpath. The software controls the forming tool movement relative to the sheet metal. A set of coordinates or commands defines the tool path's position and movement at each step. The tool moves incrementally, deforming the part in small steps. This incremental

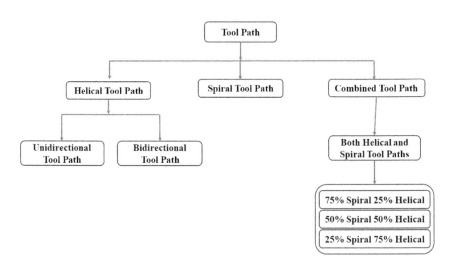

FIGURE 9.4 Classification of tool paths.

approach makes complex geometries easier to shape without requiring specialized dies or molds.

A proper tool path greatly improves the surface finish, formability, and dimensional accuracy of the formed component. The tool paths are classified into three types as shown in Figure 9.4. In the helical path, the z movement (pitch size) is constant for the whole designed geometry, but in the spiral path, the X, Y, and Z coordinates are varied according to the designed geometry; indeed, the spiral path is even called a continuous tool path. However, one of this chapter's coauthors, Dr. Nagarajan, designed a novel combined tool path strategy to improve the formability of the material at ambient temperature by combining the helical and spiral paths.

9.4 ADVANTAGES AND LIMITATIONS OF ISF

9.4.1 ADVANTAGES

- Flexibility and rapid prototyping: ISF allows for faster prototyping and small batch-type production as it eliminates the need for dedicated dies. It provides the flexibility to produce complex and customized parts without significant setup costs or time-consuming die fabrication.
- Material savings and waste reduction: ISF minimizes material waste since it deforms the sheet incrementally without requiring a complete blank. This can result in substantial cost savings, especially for expensive or scarce materials.
- Enhanced formability: ISF enables the forming of materials with high strain-hardening characteristics or low formability, expanding the range of feasible applications. It also allows the manufacturing of parts with nonuniform thickness, which may be challenging with traditional forming methods.
- Lower energy consumption: Compared with conventional forming processes, ISF generally requires less energy due to its localized deformation and reduced reliance on heavy machinery.

9.4.2 LIMITATIONS

- Limited production rate: ISF is not suitable for high-volume production due to its relatively slower forming speed compared with traditional stamping methods.
- Surface finish and accuracy: The surface finish of ISF parts may not match with the quality achieved by traditional forming processes, particularly for complex shapes or fine details. Achieving high accuracy and dimensional tolerances may also be challenging.
- Process monitoring and control: Real-time process monitoring and control in ISF can be complex due to the incremental plastic deformation nature of the process. Ensuring consistent part quality requires careful attention to process parameters and tool path optimization.

9.5   ISF OF LIGHT ALLOYS

Their very low density, better strength-to-weight ratio, high damping resistance, and excellent corrosion resistance made the alloys of light metals such as aluminum, magnesium, and titanium excellent structural engineering materials in many industries. Because of its ability to achieve precise forming with minimal springback, ISF is ideally suited for these materials, allowing for the creation of lightweight and complex components. Most studies reported that pitch size was an important process parameter and that the formability of the material decreased as its value is increased. Only Liu, Li, and Meehan (2013) reported that the formability of AA7075-O ultimately went up as the vertical step-down increased.

Some researchers showed that the best way to improve formability was to use the minimum pitch size, while Khalatbari et al. (2015) suggested that the best way to improve formability was to use the maximum step size. Feed rate and spindle speed determine SPIF part formability because friction profile, heat generation, and sheet metal properties change with increasing feed rate and spindle speed. Most researchers reported that increasing feed rates were unfavorable, whereas increasing spindle speed is beneficial for material formability.

Mr. Nagarajan has also been a coauthor on a study of the formability of commercial pure aluminum by varying the feed rate and spindle speed (Markanday and Nagarajan, 2018). The authors studied the maximum depth of forming, thickness variation in the wall region, and springback of the axisymmetric hemispherical dome created using the helical ISF tool path, and found that increasing the tool speed while maintaining the feed rate resulted in greater forming depth and thinner walls. When the feed rate was raised and the tool speed was kept constant, the depth of forming and thinning were reduced, but the springback was greater because of increased forming forces. Maximum formability was achieved in the combination of the S3000F50 parameter (Markanday and Nagarajan, 2018). Figure 9.5 visually displays the decreased thinning with the studied parameters, and Figure 9.6 displays the decreased springback.

The most common tool paths used in ISF are spiral or contoured, although complex tool path generation has become an essential topic related to ISF. Some researchers advise that running the tool path from the part edge to the center increases the accuracy. Similarly, the tool path developed for a particular geometry should have a higher wall angle than the required geometry for springback correction and enhanced component accuracy. All tool path work to date has utilized feature-based tool paths or modified tool path strategies (ex: multipass, multistep forming) utilizing the predictive approach.

Mr. Nagarajan coauthored a study on the effect of combining helical and spiral tool paths with different proportions (Prasad and Nagarajan, 2018). For the identical process parameter values, the helical tool path had higher formability, less thinning, and more springback than the spiral tool path. The spiral tool path with 75% helical tool path (25S75H) enhanced wall thickness by 10–15%, as shown in Figure 9.7, around the component's maximum wall angle region, in contrast with the complete spiral or other tool paths. Thus, to improve the final product

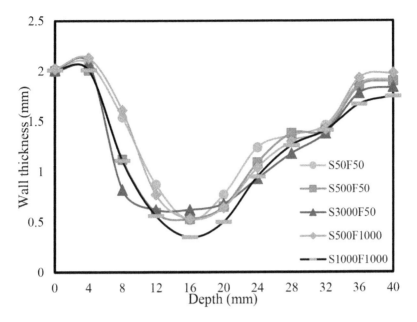

FIGURE 9.5 Thinning of ISF parts with various process parameter combinations (Markanday and Nagarajan, 2018).

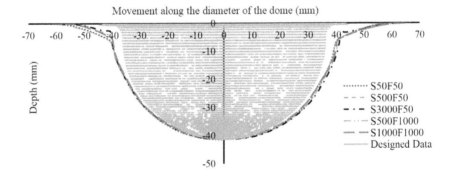

FIGURE 9.6 Springback comparison of formed and designed ISF profiles (Markanday and Nagarajan, 2018).

dimensional accuracy, reduce plastic instability, and delay component failure, the spiral tool path (25S75H) must have enough helical tool path (Figure 9.8).

Due to its light weight, high specific strength, good shock-absorption capacity, and high conductivity, magnesium and its alloys have become popular in automotive and aerospace structural components. However, their wide range of applications is restricted due to the metal's very low ductility at room temperature

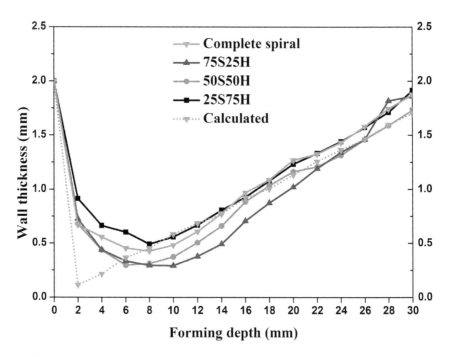

FIGURE 9.7   ISF part springback profiles compared with designed tool path profile (Prasad and Nagarajan, 2018).

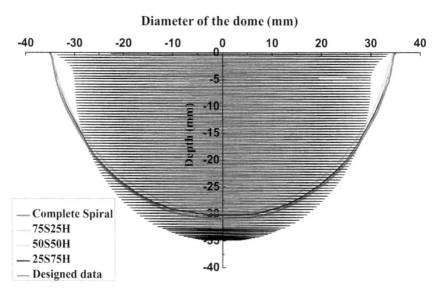

FIGURE 9.8   ISF part thinning profiles using different tool path combinations (Prasad and Nagarajan, 2018).

(Dziubińska et al., 2015; Joost and Krajewski, 2017; Chaudry, Hamad, and Kim, 2019). Park, Hong, and Lee (2013) carried out the cupping test on rolled Mg-3Al-1Zn alloy at room temperature and reported a maximum forming depth of 3.1 mm. Subsequent researchers have attempted to improve the ductility of magnesium alloys (AZ31) either by modifying the metallurgical aspects (Yu et al., 2015; Nagarajan, Ren, and Cáceres, 2017; Nagarajan, 2022) or by increasing the working temperature (Agnew and Duygulu, 2005; Huang et al., 2018). Better formability in the magnesium alloys is achieved between 150 °C and 300 °C (Liu, Li, and Meehan, 2013).

Researchers have conducted some studies on the ISF of magnesium AZ31 alloy to check the formability, but these studies were conducted directly under warm conditions rather than at room temperature (Iwanaga et al., 2004; Ambrogio, Filice, and Gagliardi, 2012; Mohanraj, Elangovan, and Pratheesh Kumar, 2022). The different slip and twinning systems or deformation systems in magnesium are shown in Figure 9.9. The improved formability of AZ31B alloy sheets at high temperature during the ISF process is a result of newer deformation systems being activated at elevated temperatures (Malhotra, Reddy, and Cao, 2010).

Even though the formability of Mg alloys is improved at higher temperatures, it is imperative that the warm working of magnesium alloys minimize the surface quality and dimensional accuracy of structurally formed products, thereby confining the usage of these alloys. Therefore, it is necessary to quantify the formability of magnesium alloys at room temperature during the ISF process to improve their surface properties and expand their use in a variety of structural applications, and the present authors are currently focusing in this direction.

ISF has become a good way to make medical devices and parts that are focused on the patient. The process is good for making medical implants because they

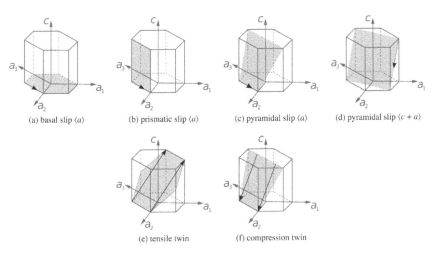

(a) basal slip $\langle a \rangle$ (b) prismatic slip $\langle a \rangle$ (c) pyramidal slip $\langle a \rangle$ (d) pyramidal slip $\langle c + a \rangle$

(e) tensile twin   (f) compression twin

FIGURE 9.9 Schematic view of the slip and twin systems observed in Mg (Chang and Kochmann, 2015).

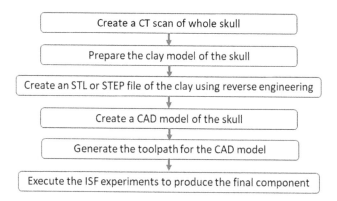

FIGURE 9.10   Flow diagram for making skull implants.

TABLE 9.2

Titanium ISF Applications in the Medical Field

Material	Formed Components	Reference
Ti grade 2	Palate prosthesis	(Fiorentino et al., 2012)
Ti grade 2	Cranial plate	(Fiorentino, Marzi, and Ceretti, 2012)
Ti grade 2 and pure Ti	Knee prosthesis	(Oleksik et al., 2010)
Ti grade 2	Facial implant	(Araújo et al., 2014)

need to be made to fit the shape of a person's body. Titanium is mainly used in medical applications, and ISF of different grades of titanium, such as titanium grade 1 (Behera, Lu, and Ou, 2016) and grade 2 (Duflou et al., 2013) have been tried. Ambrogio et al. (2005) used ISF to make a patient-specific foot orthosis. The procedure to make a skull implant is shown in Figure 9.10. Several research groups have also found out that ISF is the best way to make skull implants (Duflou et al., 2018), and Vanhove et al. (2017) made titanium clavicular implants. Table 9.2 shows the titanium applications in the medical field using ISF.

9.6   ISF APPLICATIONS IN INDUSTRY

ISF shapes sheet metal into complicated 3D geometries. In prototyping, ISF is often utilized for the quick manufacture of low-volume products; its flexibility permits quick design iterations and adjustments without expensive equipment, making it cost-effective for prototyping numerous products and components. ISF is also ideal for making customized or small-batch parts; it allows efficient and cost-effective small-batch manufacture of customized components without specific tooling.

In the automotive industry, ISF is used to make intricate and customized car body panels, fenders, and interior trim. It allows cost-effective prototyping, customization, and limited production runs. In the aircraft industry, the aircraft structure needs complicated, lightweight components with precise shapes. ISF can build complicated designs and meet weight constraints for engine, airframe, and ducting system elements. In medical device manufacturing, ISF can customize and make patient-specific implants, surgical equipment, and prosthetic components. ISF also works with art and sculpture, used to make elaborate metal sculptures, artworks, and decorations. ISF is useful in art since it can generate complicated shapes and very fine details. These applications demonstrate the ISF process's applicability across various industries, providing cost-effective and efficient manufacturing solutions for a wide range of products and components.

9.7 FUTURE SCOPE AND COMMERCIALIZATION

ISF has significant potential in manufacturing. Since ISF is a flexible and innovative forming technology, it has the potential to be employed in many applications. This assessment will analyze ISF's commercialization opportunities and challenges.

Process simulation: Advanced process modeling and simulation technologies are ISF's future. These tools will improve the material behavior prediction, process parameter optimization, and forming process. Simulations will help manufacturers to understand and control the ISF process, thereby improving the part quality and production time.

Robotics integration: Automation and robots can help to commercialize ISF. Complex part geometries and high accuracy are possible with robot-controlled forming tools. Intelligent control systems and machine learning algorithms may improve automation and efficiency. In ISF, robotic systems and modern automation technologies will enhance production rates and decrease manpower costs.

Material advancements: Materials research will enhance the applications and process capabilities of ISF. Applying it to difficult-to-deform lightweight alloys and advanced composites will expand its usage in aerospace, automotive, and consumer goods. Exploring innovative combinations of material and hybrid manufacturing procedures will help to create complex parts.

Surface finish and postprocessing: Many applications require very good surface finish to avoid fracture of the components and increase the service life. Tool path, lubrication, and post-processing improvements will improve ISF surface quality. Laser polishing and chemical etching can improve the surface finish of ISF parts, making them more usable in sectors that value aesthetics and function.

Design standards: Standardizing and developing ISF will help commercialize and implement the process. Manufacturers and end users will have belief in the process and obtain consistent and dependable outcomes with specified testing procedures, material standards, and quality control. These standards and recommendations require business, academia, and regulatory collaboration. ISF's promising future requires addressing a few obstacles to commercializing it. For instance, commercial viability requires cost-effectiveness. To compete with traditional forming techniques, research must optimize production costs, reduce energy usage, and increase production rates.

Characterization and process control: Understanding complicated material behavior during ISF and creating dependable process control strategies are ongoing problems. Quality and dimensional accuracy require accurate material characterisation models and effective monitoring systems.

Adoption in large-scale production: ISF excels at small-batch production and customization, but its scalability is limited. Improving production rates, reducing cycle times, and overcoming equipment restrictions are most important for commercializing the ISF process.

9.8 SUMMARY

ISF is an excellent approach for making complex parts out of sheet metal that can be used in a wide range of applications. The process can become a useful solution to the variety of challenges encountered in the conventional sheet-forming industry. ISF is a versatile forming method that is also suitable for Industry 4.0, but it has yet to make the transition from the laboratory to the shop floor.

Compared with traditional sheet-forming methods, ISF offers the benefits of superior formability, lower required forming force, and less-expensive experimental setups. These advantages make ISF an easy and cost-effective process for the automobile, aerospace, and medical industries. The tool path is an important parameter in deciding the maximum formability of the alloy sheets used, and novel tool paths such as bidirectional and combined paths need to be studied in detail, especially on the difficult-to-deform light alloys at room temperature to enhance their formability.

Similarly, studies on the ISF of magnesium and titanium alloys have not been carried out sufficiently compared with alloys of other metals like steel or aluminum. Such studies on ISF will be helpful for improving the room temperature formability of these materials, thereby enhancing the geometrical accuracy and surface finish of the formed components. This will reduce the shop floor cost, material cost, and eventually overall cost of the components, and ISF can be successfully used as a substitute for the conventional volume-forming techniques.

REFERENCES

Agnew, S. R. and Duygulu, Ö. (2005) 'Plastic anisotropy and the role of non-basal slip in Magnesium alloy AZ31B', *International Journal of Plasticity*, 21(6), pp. 1161–1193. doi: 10.1016/j.ijplas.2004.05.018.

Allwood, J. M., Braun, D. and Music, O. (2010) 'The effect of partially cut-out blanks on geometric accuracy in incremental sheet forming', *Journal of Materials Processing Technology*, 210(11), pp. 1501–1510. doi: 10.1016/j.jmatprotec.2010.04.008.

Ambrogio, G., Filice, L. and Gagliardi, F. (2012) 'Formability of lightweight alloys by hot incremental sheet forming', *Materials and Design*, 34, pp. 501–508. doi: 10.1016/j.matdes.2011.08.024.

Ambrogio, G. *et al.* (2005) 'Application of incremental forming process for high customised medical product manufacturing', *Journal of Materials Processing Technology*, 162–163(SPEC. ISS.), pp. 156–162. doi: 10.1016/j.jmatprotec.2005.02.148.

Araghi, B. T. *et al.* (2009) 'Investigation into a new hybrid forming process: Incremental sheet forming combined with stretch forming', *CIRP Annals—Manufacturing Technology*, 58(1), pp. 225–228. doi: 10.1016/j.cirp.2009.03.101.

Araújo, R. *et al.* (2014) 'Single point incremental forming of a facial implant', *Prosthetics and Orthotics International*, 38(5), pp. 369–378. doi: 10.1177/0309364613502071.

Behera, A. K., Lu, B. and Ou, H. (2016) 'Characterization of shape and dimensional accuracy of incrementally formed Titanium sheet parts with intermediate curvatures between two feature types', *International Journal of Advanced Manufacturing Technology*, 83(5–8), pp. 1099–1111. doi: 10.1007/s00170-015-7649-2.

Bhandari, K. S. *et al.* (2022) 'Formability of Aluminum in incremental sheet forming', *Solid State Phenomena*, 330 SSP, pp. 77–81. doi: 10.4028/p-25oilk.

Chang, Y. and Kochmann, D. M. (2015) 'A variational constitutive model for slip-twinning interactions in hcp metals : Application to single- and polycrystalline Magnesium', *International Journal of Plasticity*, 73, pp. 39–61. doi: 10.1016/j.ijplas.2015.03.008.

Chaudry, U. M., Hamad, K. and Kim, J. G. (2019) 'On the ductility of magnesium based materials: A mini review', *Journal of Alloys and Compounds*, 792, pp. 652–664. doi: 10.1016/j.jallcom.2019.04.031.

Duflou, J. R. *et al.* (2013) 'Manufacture of accurate titanium cranio-facial implants with high forming angle using single point incremental forming', *Key Engineering Materials*, 549, pp. 223–230. doi: 10.4028/www.scientific.net/KEM.549.223.

Duflou, J. R. *et al.* (2018) 'Single point incremental forming: State-of-the-art and prospects', *International Journal of Material Forming*, 11(6), pp. 743–773. doi: 10.1007/s12289-017-1387-y.

Durante, M., Formisano, A. and Langella, A. (2011) 'Observations on the influence of tool-sheet contact conditions on an incremental forming process', *Journal of Materials Engineering and Performance*, 20(6), pp. 941–946. doi: 10.1007/s11665-010-9742-x.

Dziubińska, A. *et al.* (2015) 'The microstructure and mechanical properties of AZ31 Magnesium alloy aircraft brackets produced by a new forging technology', *Procedia Manufacturing*, 2(February), pp. 337–341. doi: 10.1016/j.promfg.2015.07.059.

Fiorentino, A., Marzi, R. and Ceretti, E. (2012) 'Preliminary results on Ti Incremental Sheet Forming (ISF) of biomedical devices: Biocompatibility, surface finishing and treatment', *International Journal of Mechatronics and Manufacturing Systems*, 5(1), pp. 36–45. doi: 10.1504/IJMMS.2012.046146.

Fiorentino, A. *et al.* (2012) 'Rapid prototyping techniques for individualized medical prosthesis manufacturing', *Innovative Developments in Virtual and Physical Prototyping—Proceedings of the 5th International Conference on Advanced Research and Rapid Prototyping*, pp. 589–594. doi: 10.1201/b11341-94.

Huang, Z. *et al.* (2018) 'Observation of non-basal slip in Mg-Y by in situ three-dimensional X-ray diffraction', *Scripta Materialia*, 143, pp. 44–48. doi: 10.1016/j.scriptamat.2017.09.011.

Iwanaga, K. *et al.* (2004) 'Improvement of formability from room temperature to warm temperature in AZ-31 Magnesium alloy', *Journal of Materials Processing Technology*, 155–156(1–3), pp. 1313–1316. doi: 10.1016/j.jmatprotec.2004.04.181.

Jeswiet, J. *et al.* (2005) 'Asymmetric single point incremental forming of sheet metal', *CIRP Annals—Manufacturing Technology*, 54(2), pp. 88–114. doi: 10.1016/s0007-8506(07)60021-3.

Joost, W. J. and Krajewski, P. E. (2017) 'Towards magnesium alloys for high-volume automotive applications', *Scripta Materialia*, 128, pp. 107–112. doi: 10.1016/j.scriptamat.2016.07.035.

Khalatbari, H. *et al.* (2015) 'High-speed incremental forming process: A trade-off between formability and time efficiency', *Materials and Manufacturing Processes*, 30(11), pp. 1354–1363. doi: 10.1080/10426914.2015.1037892.

Li, Y. *et al.* (2014) 'Simulation and experimental observations of effect of different contact interfaces on the incremental sheet forming process', *Materials and Manufacturing Processes*, 29(2), pp. 121–128. doi: 10.1080/10426914.2013.822977.

Liu, Z., Li, Y. and Meehan, P. A. (2013) 'Experimental investigation of mechanical properties, formability and force measurement for AA7075-O Aluminum alloy sheets formed by incremental forming', *International Journal of Precision Engineering and Manufacturing*, 14(11), pp. 1891–1899. doi: 10.1007/s12541-013-0255-z.

Malhotra, R., Reddy, N. V. and Cao, J. (2010) 'Automatic 3D spiral toolpath generation for single point incremental forming', *Journal of Manufacturing Science and Engineering, Transactions of the ASME*, 132(6), pp. 1–10. doi: 10.1115/1.4002544.

Markanday, H. and Nagarajan, D. (2018) 'Formability behavior studies on CP-Al sheets processed through the helical tool path of incremental forming process', *IOP Conference Series: Materials Science and Engineering*, 314(1). doi: 10.1088/1757-899X/314/1/012026.

Martins, P. A. F. *et al.* (2008) 'Theory of single point incremental forming', *CIRP Annals—Manufacturing Technology*, 57(1), pp. 247–252. doi: 10.1016/j.cirp.2008.03.047.

Matsubara, S. (2001) 'A computer numerically controlled dieless incremental forming of a sheet metal', *Proceedings of the Institution of Mechanical Engineers, Part B: Journal of Engineering Manufacture*, 215(7), pp. 959–966. doi: 10.1243/0954405011518863.

Mohanraj, R., Elangovan, S. and Pratheesh Kumar, S. (2022) 'Experimental investigations of warm incremental sheet forming process on Magnesium AZ31 and Aluminium 6061 alloy', *Proceedings of the Institution of Mechanical Engineers, Part L: Journal of Materials: Design and Applications*. doi: 10.1177/14644207221110783.

Mohanty, S., Regalla, S. P. and Daseswara Rao, Y. V. (2019) 'Robot-assisted incremental sheet metal forming under the different forming condition', *Journal of the Brazilian Society of Mechanical Sciences and Engineering*, 41(2), pp. 1–12. doi: 10.1007/s40430-019-1581-6.

Mori, K., Yamamoto, M. and Osakada, K. (1996) 'Determination of hammering sequence in incremental sheet metal forming using genetic algorithm', *Nippon Kikai Gakkai Ronbunshu, A Hen/Transactions of the Japan Society of Mechanical Engineers, Part A*, 62(603), pp. 2456–2461. doi: 10.1299/kikaia.62.2456.

Moser, N. *et al.* (2016) 'Effective forming strategy for double-sided incremental forming considering in-plane curvature and tool direction', *CIRP Annals—Manufacturing Technology*, 65(1), pp. 265–268. doi: 10.1016/j.cirp.2016.04.131.

Nagarajan, D. (2022) 'Quantitative measurements of strain-induced twinning in Mg, Mg-Al and Mg-Zn solid solutions', *Materials Science and Engineering A*, 860(October), p. 144292. doi: 10.1016/j.msea.2022.144292.

Nagarajan, D., Ren, X. and Cáceres, C. H. (2017) 'Anelastic behavior of Mg-Al and Mg-Zn solid solutions', *Materials Science and Engineering A*, 696(April), pp. 387–392. doi: 10.1016/j.msea.2017.04.069.

Oleksik, V. *et al.* (2010) 'The influence of geometrical parameters on the incremental forming process for knee implants analyzed by numerical simulation', *AIP Conference Proceedings*, 1252, pp. 1208–1215. doi: 10.1063/1.3457520.

Park, S. H., Hong, S. G. and Lee, C. S. (2013) 'Enhanced stretch formability of rolled Mg-3Al-1Zn alloy at room temperature by initial {10–12} twins', *Materials Science and Engineering A*, 578, pp. 271–276. doi: 10.1016/j.msea.2013.04.084.

Prasad, M. D. and Nagarajan, D. (2018) 'Optimisation of tool path for improved formability of commercial pure Aluminium sheets during the incremental forming process', *AIP Conference Proceedings*, 1960. doi: 10.1063/1.5035049.

Vanhove, H. *et al.* (2017) 'Production of thin shell clavicle implants through single point incremental forming', *Procedia Engineering*, 183, pp. 174–179. doi: 10.1016/j.proeng.2017.04.058.

Yu, Z. *et al.* (2015) 'High strength and superior ductility of an ultra-fine grained Magnesium-Manganese alloy', *Materials Science and Engineering A*, 648, pp. 202–207. doi: 10.1016/j.msea.2015.09.065.

10 Formability and Surface Integrity in Incremental Sheet Metal Forming

Kuntal Maji, Tushar Banerjee, Gunda Yoganjaneyulu, and Gautam Kumar

10.1 INTRODUCTION

This chapter presents a critical review on formability and surface integrity in incremental sheet metal forming considering different process variants like warm forming, hot forming, and high-speed forming considering various tool–sheet interactions and different influencing factors with characterizations of deformed sheets for surface integrity. Incremental sheet forming (ISF) is a flexible dieless deformation process for low-quantity productions of sheet metal parts with improved formability. However, application of this process is limited by poor surface quality and higher forming force, especially for difficult-to-form sheet materials like titanium and magnesium alloys.

Forming operations are carried out at elevated temperature to reduce the forming load for the ISF of hard-to-form sheet metals. In hot and warm forming, the surface finish of the deformed sheets reduces because of the frictional conditions at the tool–sheet interfaces. The use of suitable lubricant coatings on the sheet material and on the tool surface to be formed could decrease the forming load and enhance the surface quality of the incremental formed sheets at room temperature as well as under warm and hot working conditions.

Incremental forming gives higher formability than traditional forming/ stamping because of smaller deformation zones and less shear deformation in the thickness direction [1]. However, it has some drawbacks like forming speed, difficult-to-form sheets, and surface quality, which researchers have investigated and attempted to overcome through different techniques like forming at elevated temperature and using coatings and lubricants. The properties of ISF-formed components depend on different process variables, namely rpm of the forming tool; tool feed; forming step height; wall angle; and forming conditions like temperature, coating, and lubricants, as described by Kumar [2] as well as depicted in a schematic diagram in Figure 10.1 and the depiction of an actual experimental setup in Figure 10.2.

DOI: 10.1201/9781003441755-10

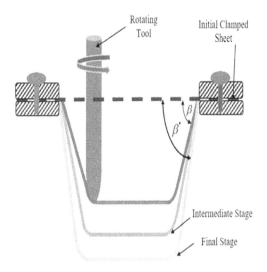

FIGURE 10.1 Schematic of the incremental sheet forming process.

FIGURE 10.2 Experimental ISF setup in a CNC machining center.

Kumar et al. [3] investigated the formability of AA2024-O aluminium alloy sheet in ISF considering forming tool size, shape, and speed; forming angle; step depth; and sheet thickness; among these, forming angle and step height had the most significant effects on formability. Gupta et al. [4] demonstrated the fabrication of a complex flat-bottomed part for aerospace applications using ISF considering different aspects like thickness, geometrical error analysis, and cost; the authors concluded that the most appropriate multistage tool path could be determined through multiple experiments and analysis to fabricate a desired component with reduced cost and geometric accuracy. Riaz et al. [5] showed that heat generated due to high tool rotational speed in ISF of AA2219 alloy sheet affected the microstructure and consequently formability of the sheet.

Wang et al. [6] found that incremental forming of AA2024 alloy sheet could significantly increase the forming limit compared with that for stamping. They also showed that the forming limit could be increased through warm incremental forming, and the forming limit curve improved with increased temperature under different heat treatments. Kumar and Maji [7–9] also reported that the formability with ISF was better than that with stamping and that warm forming could improve the formability for different metallic and nonmetallic sheets.

10.2　HOT INCREMENTAL FORMING

Researchers have determined that single-point incremental formability (SPIF) can be enhanced by conducting the forming process at higher temperatures. However, hot incremental forming can have adverse effects such as deteriorating the surface finish of the formed surface and undesired microstructural changes may take place due to heating and mechanical deformation. Liu [10] described various ISF heating techniques to increase formability and other features for different sheet materials. Bressan et al. [11] developed a novel technique to predict forming limit strains in ISF at elevated temperature and found improved formability compared with that at room temperature.

Shrivastava et al. [12] concluded that preheating sheets resulted in reduced deformation force, deformed sheet thickness along the deformation height or depth, and improved dimensional accuracy in the incremental forming of aluminium alloy AA1050 sheets. Shrivastava and Tandon [13] investigated the effects of microstructural features and texture characteristics on the forming behaviour and results for incremental forming of AA1050. They investigated and characterized surface quality considering the orange peel effects of the deformed sheet surface and obtained improved formability in terms of increased limiting strain values and smaller springback deformation as a result of the very fine microstructural features of the preheated sheets. Kumar [2] has investigated hot incremental forming of AA5083 sheet through hot air using electrical resistance heating as shown in Figure 10.3.

Kumar reported that heating improved the formability of AA5083, expressed as the forming limit curve, in SPIF at higher temperature, as shown in Figure 10.4 [2]. However, the surface quality deteriorates with hot incremental forming, as shown in Figure 10.5 [2].

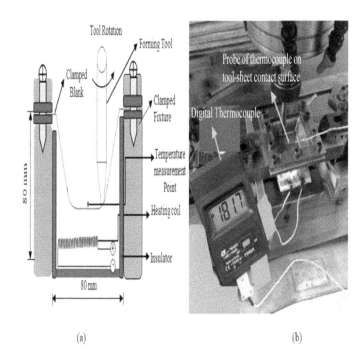

FIGURE 10.3   Hot ISF experiment: (a) line diagram, (b) actual setup.

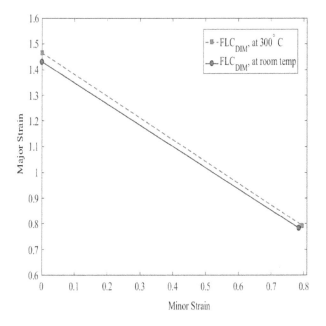

FIGURE 10.4   Improved formability in hot ISF of AA5083 sheet.

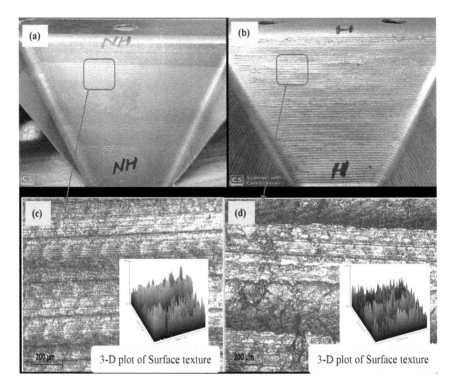

FIGURE 10.5   Comparison of ISF and hot ISF sheet surfaces and their surface textures.

10.3   HIGH-SPEED INCREMENTAL FORMING

To improve the inherent low SPIF productivity, researchers have also investigated incremental forming at high speed considering different sheet materials and their formability and properties of the deformed sheets. Gupta and Jeswiet [14] studied the influence of elevated temperature generated by increased rpm of the forming tool in SPIF. Khalatbari et al. [15] obtained enhanced formability in ISF performed at higher speed for AA3003-H12 measured as a function of wall angle and fabrication time and considering different input factors (forming tool diameter and velocity, step size, and thickness of the sheet to be formed.

Mulay et al. [16] conducted experiments to improve the incremental forming speed of Al-alloy AA5754-H22 sheet and DC04 steel sheets at higher speed and feed. The authors found a significant increase in formability and decrease in fabrication time operating under optimal processing conditions and forming parameters. Shrivastava and Tandon [17] studied the increased formability and mechanics of deformation in the incremental forming of AA1050 sheet through microstructural and textural study. They identified bending as one of the significant modes of controlling deformation at the beginning of the forming operation. In case of higher wall depth or height, the deformation modes consist of both bending and shearing that generates dislocations leading to instability in deformation causing fracture. Kumar [2] also found enhanced formability at higher speeds, as shown in Figure 10.6; the effects of the tool feed are shown in Figure 10.7.

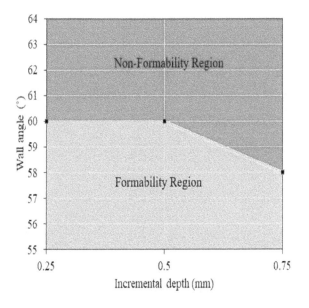

FIGURE 10.6   Parametric study of formability in high-speed incremental forming.

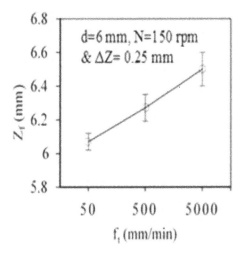

FIGURE 10.7   Effect of tool feed rate on formability in ISF: (a) fracture depth, (b) forming limit curve.

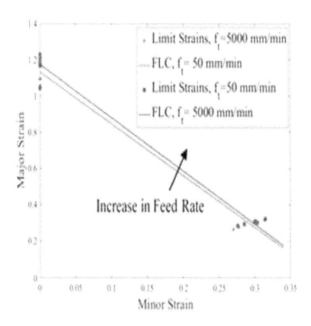

FIGURE 10.7   (Continued)

10.4   LUBRICANTS IN INCREMENTAL FORMING

Azevedo et al. [18] studied the influence of using lubricants on the surface finishes of AA1050 and DP780 sheets in SPIF. With the Al alloy sheet, enhanced surface quality was achieved by employing SAE 30 oil with AL-M grease, and for the dual-phase steel sheet, finarol B5746 oil with AS-40 grease performed better. Moreover, the harder material required lubricants with lower viscosity. Jawale et al. [19] studied the effects of different lubricants on surface finish, maximum forming angle, and fracture strains of copper sheet in incremental forming and concluded that mineral oil-based lubricants were more economical, with positive effects on surface quality and without affecting formability of the copper sheet.

Diabb et al. [20] investigated the friction and surface roughness of AA6061 sheet in SPIF using vegetable oil nano-lubricants comprising sunflower and corn oils and SiO_2 nanoparticles. Those authors found that surface wear reduced significantly using 0.025 wt% of SiO_2 in vegetable oils. Li et al. [21] investigated surface quality in ISF of aluminium alloy sheet using a novel lubricant that consisted of graphite grease and oil. Surface quality of the ISF part was observed to be better with lower forming force compared with that obtained using only oil or graphite.

Researchers have also investigated and reported on the effects of coatings on both tools and sheets to increase sheet formability and surface quality of the deformed sheets. Silva et al. [22] investigated the tool wear prediction or tool life

TABLE 10.1

Effects of Lubricants in Incremental Sheet and Bulk Metal Forming Processes

Sr no.	lubricants	Workpiece material	Forming results	Reference
1	SAE 30 oil and AL-M grease, B5746 oil and AS-40 grease	AA1050 and DP780	harder material required lubricants with lower viscosity	Azevedo et al. [18]
2	mineral oil	copper sheet	positive effects on the surface finish and no effect on formability	Jawale et al. [19]
3	sunflower/corn oils and SiO_2 nanoparticles	AA6061	significant reduction in surface wear	Diabb et al. [20]
4	graphite grease/oil	aluminium alloy sheet	better surface finish and formability	Li et al. [21]
5	grease	galvanized steel sheet	tool wear affects tool life	Silva et al. [22]
6	mineral oil and no lubricant	AA6061-T6 and 08F steel	better formability and surface finish for lubricated conditions	Wang et al. [23]
7	molybdenum disulphide	AA1050 Al alloy sheet	better surface quality and formability	Singh et al. [24]

in incremental forming of galvanized steel sheet using a SAE 1045 steel tool considering the process parameters and forming conditions. The developed models could predict forming tool wear and life within 33.4% and 35.9%, respectively, of the experimental values.

Wang et al. [23] investigated incremental punching of AA6061-T and 08F steel sheets using mineral oil and without use of lubricants and found that formability and surface finish were both higher under lubrication. Singh et al. [24] investigated suitable tool materials with lubricant for hot ISF and found that combining EN-13 steel tool and molybdenum disulphide as lubricant was better for the hot ISF of AA1050 sheets. An overview of different types of lubricants used in ISF with their effects on forming results is shown in Table 10.1.

10.5   COATING IN INCREMENTAL FORMING

Researchers have widely used coatings on both the sheet metals and the forming tools to improve formability in both conventional forming and ISF to reduce forming load and improve surface finish and forming tool life. For instance, Sieczkarek et al. [25, 26] showed that a structured surface tool with coated surface could decrease the forming forces and enhance the tool life. The forming tool with bionic structure and physical vapor deposition (PVD)-coated surface had longer tool life with better results in forming sheets of AlMg3 alloy and steel alloy DC04.

Sgarabotto et al. [27] found that good wear resistance could be obtained through applying a diamond-like coating (Ti-DLC) using PVD in the sheet metal forming tools. Li et al. [28] investigated the impacts of the tool–sheet interface on the formability and surface integrity of parts in the ISF of AA7075-O sheets. They found that using a roller contact tool delivered higher formability with better surface integrity than the sliding contact tool. Biermann et al. [29] also achieved longer forming tool life and better performance with a structured surface coated with CrAlN using PVD; the structure-coated tool surface had high wear resistance properties that improved the tool life and performance.

Tillmann et al. [30] obtained longer life of a PVD-coated forming tool with suitable pretreatment, specifically plasma nitriding and setting up magnetron-sputtering process parameters. Sieczkarek et al. [31] showed improved tool life of forming tools employing CrAlN PVD-coatings with different types of surface profiles. The forming tools had been observed to fail through abrasive and adhesive wear and crack formation, and these failures could be reduced by applying a coating of CrAlN on forming tools using PVD; tool life increased by about 84%.

Xu et al. [32] achieved the enhanced formability of aluminium alloy sheet in SPIF using a Cr/C-coated forming tool and molybdenum disulphide lithium base grease through delaying the sheet failure. Zhang et al. [33] investigated

TABLE 10.2
Effects of Coating in Incremental Sheet and Bulk Metal Forming Processes

Sr no.	Coating material	Workpiece material	Forming results	References
1	PVD coated surface	aluminium alloy (AlMg3) and steel alloy (DC04)	reduce the forming forces and improve the tool life	Sieczkarek et al. [25, 26]
2	Ti-DLC	C67S steel	wear resistance	Sgarabotto et al. [27]
3	Roller contact interface	AA7075-O	higher formability with better surface integrity	Li et al. [28]
4	CrAlN	DC04	higher tool life and better performance of forming tool	Biermann et al. [29]
5	CrAlN PVD-coatings	aluminium alloy and steel	significant increase of forming tool life	Sieczkarek et al. [31]
6	Cr/C	aluminium alloy	improve formability	Xu et al. [32]
7	amorphous hydrogenated carbon (a-C:H)	aluminium sheet	Dry forming	Abraham et al. [34]

temperatures at the coated tool–sheet contact area in ISF by analyzing influencing factors, namely sheet thickness, coating thickness, conductivities of the coating and sheet materials, etc. The authors found that temperature was significantly affected by the coating thickness and that conductivity influenced the formability of the sheet metal. Abraham et al. [34] showed successful dry forming of aluminium sheet deep drawing employing a fine diamond-like carbon (DLC) coating, amorphous hydrogenated carbon (a-C:H), on the forming tools. Effects of different coating materials in ISF performance are presented in Table 10.2.

10.6   CONCLUSIONS

This chapter presented a review of formability and surface integrity in the incremental sheet forming of different materials considering the parametric effects, forming temperature, and lubrication and coating on both tool and sheet materials. Some difficult-to-form sheet metals could be formed into complex shapes by hot incremental forming and the low-productivity ISF could be overcome through high-speed incremental forming employing suitable lubricants and coating material. Therefore, selecting suitable ISF process parameters, forming temperature, lubrication and coating materials, and other features, depending on the tool–sheet material combinations, can lead to the economically feasible fabrication of successful sheet metal components applicable for industries.

10.7   ACKNOWLEDGEMENT

The authors gratefully acknowledge the support of the Production and Industrial Engineering Department, National Institute of Technology Jamshedpur, Jharkhand, India, in drafting this book chapter.

REFERENCES

[1] W.C. Emmens, G. Sebastiani, and A.H. van den Boogaard, "The technology of incremental sheet forming—A brief review of the history," *J. Matr. Proces. Technol.*, vol. 210, no. 8, pp. 981–997, 2010, doi.org/10.1016/j.jmatprotec.2010.02.014.

[2] G. Kumar, "Formability analysis of tailored blanks in single point incremental forming," *PhD Thesis, NIT Patna*, India, 2021, doi.org/10.5281/zenodo.5884656.

[3] A. Kumar, V. Gulati, P. Kumar, V. Singh, B. Kumar, and H. Singh, "Parametric effects on formability of AA2024-O sheets in single point incremental forming," *J. Mater. Resear. Technol.*, vol. 8, no. 1, pp. 1461–1469, 2019, doi.org/10.1016/j.jmrt.2018.11.001.

[4] P. Gupta, A. Szekeres, and J. Jeswiet, "Design and development of an aerospace component with single-point incremental forming," *Intern. J. Advan. Manufact. Technol.*, vol. 103, pp. 3683–3702, 2019, doi.org/10.1007/s00170-019-03622-4.

[5] A.A. Riaz, N. Ullah, G. Hussain, M. Alkahtani, M.N. Khan, and S. Khan, "Experimental investigations on the effects of rotational speed on temperature and microstructure variations in incremental forming of T6-tempered and annealed AA2219 aerospace alloy," *Metals*, vol. 10, no. 809, pp. 1–17, 2020, doi.org/10.3390/met10060809.

[6] H. Wang, T. Wu, J. Wang, J. Li, and K. Jin, "Experimental study on the incremental forming limit of the aluminium alloy AA2024 sheet," *Inter. J. Advan. Manufact. Technol.*, vol. 108, pp. 3507–3515, 2020, doi.org/10.1007/s00170-020-05613-2.

[7] G. Kumar, and K. Maji, "Investigations into enhanced formability of AA5083 sheet in single point incremental forming," *J. Mater. Engin. Perform.*, vol. 30, pp. 1289–1305, 2021, doi.org/10.1007/s11665-021-05455-3.

[8] G. Kumar, and K. Maji, "Investigations on formability of tailor laminated sheets in single point incremental forming," *Proc. IMeche. Part B: J. Engine. Manuact. Perform.*, vol. 236, no. 10, pp. 1393–1405, 2022, doi.org/10.1177/09544054221076244.

[9] G. Kumar, and K. Maji, "Forming limit analysis of friction stir tailor welded AA5083 and AA7075 sheets in single point incremental forming," *Inter. J. Mater. Form.*, vol. 15, no. 20, pp. 1–23, 2022, doi.org/10.1007/s12289-022-01675-7.

[10] Z. Liu, "Heat-assisted incremental sheet forming: A state-of-the-art review," *Int. J. Advan. Manufact. Technol.*, vol. 98, no. 9–12, pp. 2987–3003, 2018, doi.org/10.1007/s00170-018-2470-3.

[11] J.D. Bressan, S. Bruschi, and A. Ghiotti, "Prediction of limit strains in hot forming of aluminium alloy sheets," *Inter. J. Mechan. Scien.*, vol. 115–116, pp. 702–710, 2016, doi.org/10.1016/j.ijmecsci.2016.07.040.

[12] P. Shrivastava, P. Kumar, P. Tandon, and A. Pesin, "Improvement in formability and geometrical accuracy of incrementally formed AA1050 sheets by microstructure and texture reformation through preheating, and their FEA and experimental validation," *J. Brazil. Soci. Mechani. Scien. Engine.*, vol. 40, no. 7, pp. 1–15, 2018, doi.org/10.1007/s40430-018-1255-9.

[13] P. Shrivastava, and P. Tandon, "Effect of preheated microstructure Vis-A'-vis process parameters and characterization of orange peel in incremental forming of AA1050 sheets," *J. Mater. Engine. Perform.*, vol. 28, pp. 2530–2542, 2019, doi.org/10.1007/s11665-019-04032-z.

[14] P. Gupta, and J. Jeswiet, "Effect of temperatures during single point incremental forming," *Inter. J. Advan. Manufact. Technol.*, vol. 95, pp. 3693–3706, 2018, doi.org/10.1007/s00170-017-1400-0.

[15] H. Khalatbari, A. Iqbal, X. Shi, L. Gao, G. Hussain, and M. Hashemipour, "High-speed incremental forming process: A trade-off between formability and time efficiency," *Mater. Manufact. Proce.*, vol. 30, no. 11, pp. 1354–1363, 2015, doi.org/10.1080/10426914.2015.1037892.

[16] A. Mulay, S. Ben, S. Ismail, A. Kocanda, and C. Jasiński, "Performance evaluation of high-speed incremental sheet forming technology for AA5754 H22 aluminum and DC04 steel sheets," *Archi. Civi. Mechan. Engine.*, vol. 18, no. 4, pp. 1275–1287, 2018, doi.org/10.1016/j.acme.2018.03.004.

[17] P. Shrivastava, and P. Tandon, "Microstructure and texture based analysis of forming behavior and deformation mechanism of AA1050 sheet during single point incremental forming," *J. Mater. Proce. Technol.*, vol. 266, pp. 292–310, 2019, doi.org/10.1016/j.jmatprotec.2018.11.012.

[18] N.G. Azevedo, J.S. Farias, R.P. Bastos, P. Teixeira, J.P. Davim, and R.J.A.D. Sousa, "Lubrication aspects during single point incremental forming for steel and aluminum materials," *Intern. J. of Preci. Engine. Manufact.*, vol. 16, no. 3, pp. 589–595, 2015, doi.org/10.1007/s12541-015-0079-0.

[19] K. Jawale, J.F. Duarte, A. Reis, and M.B. Silva, "Lubrication study in single point incremental forming of copper," *J. Phys. Conf. Ser.*, vol. 734, no. 032038, pp. 1–4, 2016, doi.org/10.1088/1742-6596/734/3/032038.

[20] J. Diabb, C.A. Rodriguez, N. Mamidi, J.A. Sandoval, J.T. Tijerina, O.M. Romero, and A.E. Zuniga, "Study of lubrication and wear in single point incremental sheet forming (SPIF) process using vegetable oil nanolubricants," *Wear*, vol. 376–377, pp. 777–785, https://doi.org/10.1016/j.wear.2017.01.045.

[21] Z. Li, S. Lu, T. Zhang, Z. Mao, and C. Zhang, "A simple and low-cost lubrication method for improvement in the surface quality of incremental sheet metal forming," *Tran. Indi. Inst. Met.*, vol. 71, no. 7, pp. 1715–1719, 2018, doi.org/10.1007/s12666-018-1305-0.

[22] P.J.D. Silva, and A.J. Alvares, "Investigation of tool wear in single point incremental sheet forming," *Proc. IMeche. Part. B: J. Engin. Manufact.*, vol. 234, pp. 1–19, 2019, https://doi.org/10.1177/0954405419844653.

[23] J. Wang, X. Wang, L. Li, X. Zhang, and N. Gu, "Feasibility of non-lubricated incremental sheet punching with sinusoidal tool path," *J. Brazil. Soci. Mechan. Scien. Engine.*, vol. 42, no. 234, pp. 1–8, 2020, doi.org/10.1007/s40430-020-02334-1.

[24] S.A. Singh, S. Priyadarshi, and P. Tandon, "Exploration of appropriate tool material and lubricant for elevated temperature incremental forming of aluminium alloy," *Intern. J. Preci. Engine. Manufact.*, vol. 22, pp. 217–225, doi.org/10.1007/s12541-020-00447-0.

[25] P. Sieczkarek, L. Kwiatkowski, A.E. Tekkaya, E. Krebs, D. Biermann, W. Tillmann, and J. Herper, "Improved tool surfaces for incremental bulk forming processes of sheet metals," *Key Engine. Mater.*, vol. 554–557, pp. 1490–1497, 2012, doi.org/10.4028/www.scientific.net/KEM.504-506.975.

[26] P. Sieczkarek, L. Kwiatkowski, A.E. Tekkaya, E. Krebs, P. Kersting, W. Tillmann, and J. Herper, "Innovative tools to improve incremental bulk forming processes," *Key Engin. Mater.*, vol. 554–557, pp. 1490–1497, 2013, doi.org/10.4028/www.scientific.net/KEM.554-557.1490.

[27] F. Sgarabotto, A. Ghiotti, and A.S. Bruschi, "Novel experimental set-up to investigate the wear of coatings for sheet metal forming tools," *Key Engine. Mater.*, vol. 554–557, pp. 825–832, 2015, doi.org/10.1007/s12289-014-1181-z.

[28] Y. Li, Z. Liu, W.J.T.B. Daniel, and P.A. Meehan, "Simulation and experimental observations of effect of different contact interfaces on the incremental sheet forming process," *Mater. Manufact. Proces.*, vol. 29, pp. 121–128, 2014, doi.org/10.1080/10426914.2013.822977.

[29] D. Biermann, D. Freiburg, R. Hense, W. Tillmann, and D. Stangier, "Influence of surface modifications on friction, using high-feed milling and wear resistant PVD-coating for sheet-metal forming tools," *Key Engine. Mater.*, vol. 639, pp. 275–282, 2015, doi.org/10.4028/www.scientific.net/KEM.639.275.

[30] W. Tillmann, D. Stangier, B. Denkena, T. Grove, and H. Lucas, "Influence of PVD-coating technology and pretreatments on residual stresses for sheet-bulk metal forming tools," *Prod. Eng. Res. Devel.*, vol. 10, pp. 17–24, 2016, doi.org/10.1007/s11740-015-0653-4.

[31] P. Sieczkarek, S. Wernicke, S. Gies, A.E. Tekkaya, E. Krebs, P. Wiederkehr, D. Biermann, W. Tillmann, and D. Stangier, "Wear behavior of tribologically optimized tool surfaces for incremental forming processes," *Tribol. Intern.*, vol. 104, pp. 64–72, 2016, doi.org/10.1016/j.triboint.2016.08.028.

[32] Q. Xu, M. Yang, and Z. Yao, "Influence of tool-sheet contact conditions on fracture behavior in single point incremental forming," *IOP Conf. Series: Mater. Scien. Engine.*, vol. 394, no. 032014, pp. 1–7, 2018, doi.org/10.1088/1757-899X/394/3/032014.

[33] X. Zhang, T. He, H. Miwa, T. Nanbu, R. Murakami, S. Liu, J. Cao, and Q.J. Wang, "A new approach for analysing the temperature rise and heat partition at the interface of coated tool tip-sheet incremental forming systems," *Inter. J. Heat Mass Transf.*, vol. 129, pp. 1172–1183, 2019, doi.org/10.1016/j.ijheatmasstransfer.2018.10.056.

[34] T. Abraham, G. Brauer, F. Flegler, P. Groche, and M. Demmler, "Dry sheet metal forming of aluminium by smooth DLC coatings-a capable approach for an efficient production process with reduced environmental impact," *Proce. Manufact.*, vol. 43, pp. 642–649, 2020, doi.org/10.1016/j.promfg.2020.02.140.

11 Optimizing Process Parameters in Novel SPIF to Reduce Excessive Thinning and Geometrical Inaccuracies for Aluminum Alloys

Viren Mevada, Jigar Patel, and Harit K. Raval

11.1 INTRODUCTION

In metal forming processes, the material is deformed plastically to achieve desired shape and size. The plastic deformation takes place under forming force that exceeds the yield strength of the material. There are many different metal forming processes, each with its own advantages and disadvantages.

For instance, in forging, a workpiece is hammered or pressed into the desired shape; in rolling, a workpiece is passed between two rotating rolls to reduce thickness and increase length; in extrusion, a workpiece is forced through a die of the desired shape; and in drawing, a workpiece is pulled through a die of the desired shape. However, the disadvantages of these processes—inability to form new-age hard-to-form materials or complex shapes, higher fixed and initial costs—gave birth to advanced metal forming techniques. One such technique that is gaining momentum in research as well as in industry is incremental sheet forming (ISF).

In ISF, the material is deformed layer by layer through the movement of tools of relatively smaller diameter, as shown in Figure 11.1. The tool movement is governed by the instructions (Duflou et al., 2007), which are usually a CNC programming file and which derive from the tool path; the tool path is generated from computer-aided manufacturing (CAM) software for complex parts by modeling the desired shapes (Attanasio, Ceretti, and Giardini, 2006). The forming tool follows the tool path, thus forming the material locally along the tool path. As the diameter of the forming tool is much smaller, the area to be deformed is also reduced drastically. Deforming a small amount of material requires less

DOI: 10.1201/9781003441755-11   **187**

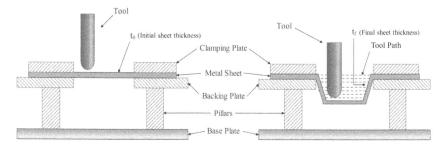

FIGURE 11.1   Working principle of ISF.

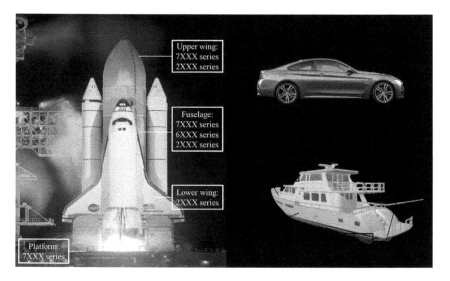

FIGURE 11.2   Applications of aluminum alloys in various industries.

force, which induces less stress in the material and increases the overall formability (Ham and Jeswiet, 2007). The operations are usually performed in the CNC machine; however, a robotic manipulator or a special-purpose ISF machine can also be used. ISF is derived from the spinning process and can form asymmetric shapes as well.

ISF was initially developed by Leszak, mainly to fulfill the need for automotive body parts in the early 1970s (Edward and Leszak, 1964; Matsubara Shigeo, 1994). However, following current advances in the process, it is being applied in vast areas like aerospace (Gupta and Jeswiet, 2019; Gupta, Szekeres, and Jeswiet, 2019), marine science (Mulay et al., 2017), medicine (in particular implants) (Ji and Park, 2008), complex-shaped shell elements, and rapid prototyping (Amino et al., 2015). Figure 11.2 highlights the uses of ISF in the aerospace, automotive, and marine industries for different grades of aluminum alloys. Unlike conventional forming, ISF can be a cold or hot forming process. However, there are a few limitations

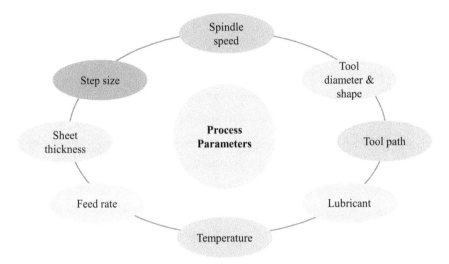

FIGURE 11.3 Classification of process parameters.

of ISF process such as longer forming time and geometrical accuracy that can be resolved through optimal tool paths or forming at an elevated temperature.

The ISF process can be classified according to geometrical parameters such as wall angle, major and minor dimensions, and forming depth and shape; further classification can be based on process parameters (step depth, feed rate, spindle speed, tool paths, etc.). The detailed classification is shown in Figure 11.3. The process parameters govern characteristics such as formability, surface quality, geometrical accuracy, and sheet thinning (McAnulty, Jeswiet, and Doolan, 2017). The objective of this book chapter is to present an idea of various process parameters and their influence on the ISF process, especially for aluminum alloys. Organizing the present literature will assist in developing the guidelines to identify the levels of these parameters to get the desired object through ISF. This will also help the researchers to identify potential research gaps and further enhance the ISF process.

11.2 INFLUENCE OF THE PROCESS PARAMETERS

Out of all the process parameters presented in Figure 11.3, the incremental step size, feed rate, spindle speed, and different tool profiles and diameters are reviewed in this section. The literature review is followed by a simulation case study of the effect of these parameters on the AA 6061 T6.

11.2.1 INFLUENCE OF STEP DEPTH

The range of step depth varies widely between 0.2 and 2 mm, as even the softer grades of aluminum can be deformed at higher step depth, although the

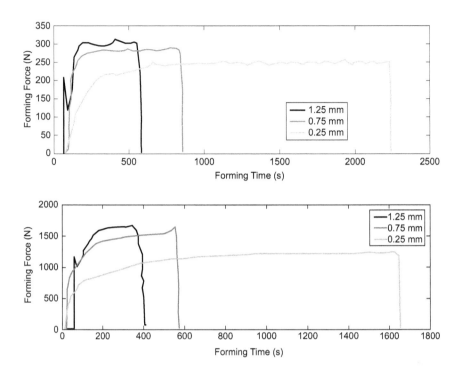

FIGURE 11.4   Effects of step depth on forming force for different sheet thicknesses.

hard-to-form grades are difficult to form at step depth more than 0.75 mm. With higher step depth, the force required to deform the material increases (Figure 11.4) as there is more material underneath the tool that can be deformed (Kumar and Suresh, 2019). The tool path generated with lower step depths reduces the staircase effect, thereby increasing the overall surface quality (Dodiya et al., 2021).

The effect of step size drastically increases with increased wall angle. At higher angles, it is difficult to form materials with higher step depths. This is because the higher forming force required to form the material generates even higher induced stresses (Kumar et al., 2019). The strain in all three direction peaks for higher step depths due to increased equivalent plastic strain is confirmed by finite element analysis (Li, Daniel, and Meehan, 2017). Higher step depths induce waviness in the formed area, increasing the surface roughness (Kumar, Gulati, and Kumar, 2018).

11.2.2   INFLUENCE OF FEED RATE

The feed rate in ISF defines the speed of forming and usually ranges from 50 mm/min to 4000 mm/min. The strength of the material formed increases with increasing feed rate, although after a certain limit (around 100 mm/min for harder grades of Al), the increase is nominal. This is because the material hardens during the

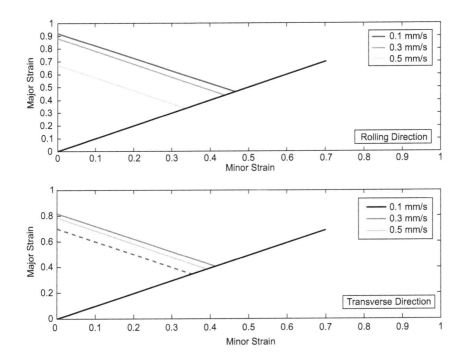

FIGURE 11.5   Variations in major & minor strain w.r.t. feed rate.

forming (Hussain et al., 2020). With increasing feed rate, the sum of the major and minor strain decreases, thus indicating that the lower feed rate helps increase the formability, albeit at the cost of higher forming force (Kim and Park, 2002), as shown in Figure 11.5.

11.2.3   INFLUENCE OF SPINDLE SPEED

Spindle speed in ISF denotes the rotational speed of forming tool. The tool can rotate in three different ways, i.e., free, locked, or at defined rotations per minute (RPM). The range of spindle speed in ISF varies between 0 RPM and 2000 RPM for aluminum alloys. The effect of spindle speed on the material properties is nominal, and the same is the case for the microstructure (Hussain *et al.*, 2020). However, the forming characteristics, in particular depth, increase at higher spindle speeds. This is because higher spindle speed increases the local friction at the tool–sheet interface and thereby generates a local heat flux. This heat flux helps material to flow easily, improving the formability through grain refinement as shown in Figure 11.6 (Kumar et al., 2019). In case of locked spindle, the heat generation is due to the sliding friction, which is much lesser compared to the rolling friction in case of higher spindle speed. Thus, at higher spindle speed excessive heating can be achieved to enhance the formability (Jeswiet et al., 2005).

11.2.4 Influence of Tool Diameter and Tool Profile

Researchers have explored various tool profiles such as spherical, flat with corner radii, and roller ball with wide-ranging diameters of 5–20 mm. As the process is still in the developing stage, the tools are not yet standardized. It is observed that with the increase in tool diameter, the force required to deform the material increases gradually because a larger area is in contact with the tool to be formed (Kumar and Suresh, 2019) as shown in Figure 11.7.

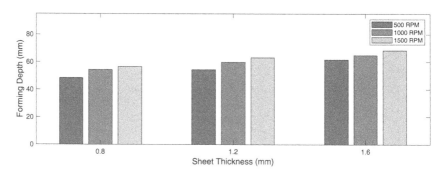

FIGURE 11.6   Effect of spindle speed and sheet thickness on the forming force.

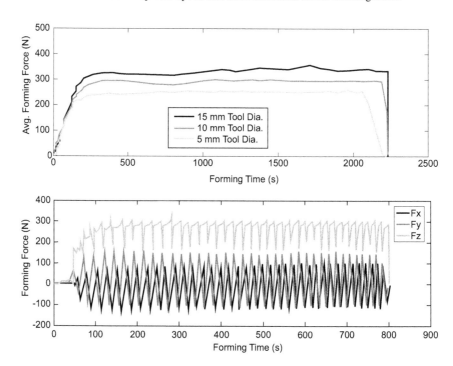

FIGURE 11.7   Effects of tool diameter on forming force: F_z = axial force; F_x and F_y = lateral forces.

Usually, the magnitude of axial force is drastically greater than the lateral forces. With a smaller tool diameter, the tool's penetration into the material results in chipping of the material and in turn reduces formability (Kumar et al., 2019). When the tool diameter increases, the higher contact area decreases the strain, increasing the formability (Kim and Park, 2002). Higher tool diameter also reduces the waviness effect in the formed area, resulting in decreasing average roughness (Kumar, Gulati, and Kumar, 2018). Larger tool diameter induces geometrical inaccuracies due to bending of the sheet at the starting contour.

11.3 INVESTIGATING THE EFFECTS OF ISF PROCESS PARAMETERS ON SHEET THICKNESS REDUCTION AND GEOMETRIC ACCURACY—A CASE STUDY

To investigate the influence of the process parameters on the section thickness and geometric accuracy while forming, we conducted a simulation analysis in Abaqus 2017 using AA6061 T6 with sheet dimensions of $177 \times 177 \times 1$ mm³. For the forming process, we employed a hemispherical tool and a spiral tool path generated in MATLAB® to shape the sheet into a truncated cone. To evaluate each process parameter's significance and contribution to the response variable, we analyzed the gathered data using statistical techniques. The accuracy and robustness of the simulation findings were improved by using the best factor settings from the DOE results, which took into account both major effects and interactions.

11.3.1 Design of Experiments

We optimized the ISF procedure using the design of experiments (DOE) created using the Taguchi technique and three levels for each of the critical process parameters that we identified. The generated DOE is shown in Table 11.1.

We used an L9 (3^4) orthogonal array in this work to effectively cover all combinations of factor levels with a minimum number of runs. By effectively covering all potential combinations, orthogonal arrays enable the examination of principal effects and interactions between components with the least testing. Following the L9 orthogonal array, we produced the DOE using MINITAB® software (Gundarneeya et al., 2022).

TABLE 11.1
Study Experimental Parameters

Tool diameter (mm)	Feed rate (mm/min)	Spindle speed (RPM)	Step depth (mm)
6	250	100	0.3
10	500	250	0.6
14	750	400	0.9

11.3.2 SIMULATION PROCESS

Based on the array results, the simulation research concentrated on using AA6061 T6, a well-known aluminum alloy with exceptional mechanical characteristics, formability, and corrosion resistance. With possible trace quantities of additional elements including chromium, copper, and zinc, the alloy is mostly composed of aluminum (about 97.9%), magnesium (about 1%), and silicon (about 0.6%) (Mirzakhani and Mansourinejad, 2011). The goal of the study was to evaluate the behavior and performance of AA6061 T6 throughout the forming process in the simulation, helping us to better understand its appropriateness for various manufacturing applications. The simulation required certain mechanical properties (Gómez-López et al., 2013) that we derived through tensile testing following ASTM E8 (Mirzakhani and Mansourinejad, 2011).

For the simulation, we used Abaqus/Explicit to generate a 2D deformable model of the AA6061 T6 sheet that was $177 \times 177 \times 1 \text{ mm}^3$ in size. The shaping tool had diameters of 6, 10, and 14 mm and was classified as an analytical rigid body. The sheet was supported by a fixture and secured at each of its four sides. The simulation model included AA6061's mechanical characteristics. Additionally, the friction coefficient between the sheet blank and the forming tool was set at 0.1 using Coulomb's friction law (Kumar and Reddy, 2016), and we used a mesh with linear quadrilateral components of the quad dominant type (Sajjad, Joy, and Jung, 2018) and a mesh size of 1 mm (Tayebi et al., 2019).

11.3.3 RESULTS AND DISCUSSION

The simulation study's findings will offer guidance on how to adjust the process variables to reach the specified section thickness and guarantee the accuracy of the generated truncated cone shape. We examined the sheet thickness reductions and geometric accuracy considering the process parameters, tool diameter, spindle speed, feed rate, and step depth. We ran 18 Abaqus simulations to calculate the parameters for the minimal thickness reduction of the sheet and the geometric correctness of the generated conical frustum.

To account for the variability in the simulations, we repeated the experimental runs at least twice. The created wall's thickness varied from 0.3 to 0.7 mm, while the deviation between the produced section's ideal and real geometry ranged from 17 to 29 mm (refer Table 11.2). The higher deviation is because the diametric offset of 30 mm was kept between the major diameter of the conical frustum and diameter of the backing plate.

11.3.4 INFLUENCES ON SHEET THICKNESS REDUCTION

Table 11.3 presents the influences of four factors (tool diameter, feed rate, spindle speed, and step depth) with corresponding levels represented by 1, 2, and 3 on the sheet thickness reduction. The rankings indicate that tool diameter had the highest impact, followed by spindle speed, feed rate, and step depth with the least.

TABLE 11.2

Simulation Results (rounded to three decimal places)

Run	Tool Diameter (mm)	Feed Rate (mm/min)	Spindle Speed (RPM)	Step-depth (mm)	Sheet thickness reduction (mm) Trial 1	Trail 2	Geometrical Deviation (mm) Trail 1	Trail 2
1	6	250	100	0.3	0.454	0.520	17.889	19.138
2	6	500	250	0.6	0.535	0.530	20.783	21.168
3	6	750	400	0.9	0.508	0.513	17.930	20.111
4	10	250	250	0.9	0.544	0.500	20.810	18.249
5	10	500	400	0.3	0.648	0.459	28.275	21.776
6	10	750	100	0.6	0.445	0.491	17.503	27.786
7	14	250	400	0.6	0.442	0.440	24.647	20.525
8	14	500	100	0.9	0.325	0.383	22.682	25.698
9	14	750	250	0.3	0.362	0.286	21.910	24.647

TABLE 11.3

SN Ratios for Sheet Thickness Reduction

Levels	Tool diameter (mm)	Feed rate (mm/min)	Spindle speed (RPM)	Step depth (mm)
1	−5.854	−6.336	−7.288	−7.059
2	−5.793	−6.544	−6.97	−6.393
3	−8.64	−7.408	−6.03	−6.836
Delta	2.847	1.072	1.258	0.665
Rank	1	3	2	4

The highest influence of tool diameter can be explained through the variations in contact area caused by tool diameter differences. Higher tool diameter increases the contact area and thus the local heat generation, which increases the formability of the material because thickness decreases more along the wall of the conical frustum. The same is confirmed by the main effect plot presented in Figure 11.8.

The step depth varied the least, with the best results obtained at moderate depths. These findings contradict those found by Kumar and Suresh (2019). This could be explained by the chipping of the material: At lower step depths, the tool's penetration is relatively limited, and due to insufficient heat and the interface, it is easier for the material to shear than flow.

The results for feed rate and spindle speed match well with the literature findings discussed in Sections 11.2.2 and 11.2.3. The spiral tool path is divided into several small straight segments, and the specified feed rate varies; higher feed rates result in increased thinning of the walls of the cone because the individual segments are much smaller. Similarly, the rolling friction is greater at higher spindle speeds than the sliding friction, which localizes the heating of the material, which leads material to flow until strain is higher.

11.3.5 INFLUENCES ON GEOMETRIC ACCURACY

Table 11.4 presents the influences of tool diameter, feed rate, spindle speed, and step depth on the geometric accuracy with corresponding levels represented by 1, 2, and 3. We measured the variations between the targeted geometry and the experimental results through the nodal displacement of individual elements in the simulation. The Taguchi analysis clearly shows that tool diameter had the greatest impact on geometric accuracy. The larger the tool diameter, the greater

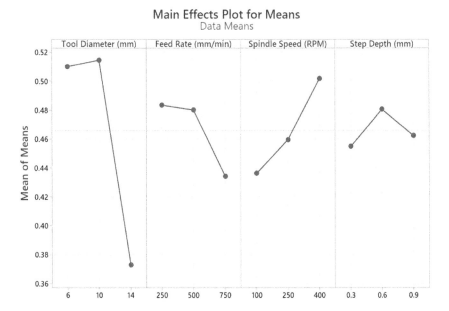

FIGURE 11.8   Main effects plot of sheet thickness reductions.

TABLE 11.4
SN Ratios for Geometric Accuracy

Level	Tool diameter (mm)	Feed rate (mm/min)	Spindle speed (RPM)	Step depth (mm)
1	−25.790	−26.080	−26.710	−26.89
2	−26.960	−27.360	−26.530	−26.87
3	−27.360	−26.670	−26.880	−26.36
Delta	1.570	1.280	0.350	0.53
Rank	1	2	4	3

the deviation from the targeted geometry. The effect of spindle speed on the geometric accuracy is lowest, following the step depth and feed rate. The feed rate achieved the highest geometric accuracy as denoted by the delta of 1.28. The optimal result can be achieved at the lowest levels of all the parameters.

11.4 CONCLUSION

We conducted a detailed literature review to understand the applications of aluminum alloys in different industries and their requirements, including better understanding the influences of the different ISF process parameters. To validate the literature, we conducted a study to form a conical cup simulating the material properties of AA6061 T6. We carried out finite element modeling simulations in Abaqus/Explicit to assess the decrease of sheet thickness and geometrical deviation.

To evaluate the impact of process factors, we used the Taguchi DOE approach. The results show that the sheet thickness reduction is greatly influenced by the tool diameter, while spindle speed and feed rate had the next-largest effects. In the case of geometric accuracy, the tool dimeter plays the significant role. These results provide valuable insights for optimizing the SPIF process and achieving desired part characteristics using different aluminum alloys, in particular AA6061 T6.

REFERENCES

Amino, M. *et al.* (2015) 'Single point "dieless" incremental forming', in *60 Excellent Inventions in Metal Forming*. Berlin, Heidelberg: Springer, pp. 155–159. doi: 10.1007/978-3-662-46312-3_24.

Attanasio, A., Ceretti, E. and Giardini, C. (2006) 'Optimization of tool path in two points incremental forming', *Journal of Materials Processing Technology*, 177(1–3), pp. 409–412. doi: 10.1016/j.jmatprotec.2006.04.047.

Dodiya, H. R. *et al.* (2021) 'Experimental investigation of surface roughness for AA 3003-0 aluminium alloy using single point incremental forming', *Materials Today: Proceedings*, 46, pp. 8655–8662. doi: 10.1016/j.matpr.2021.03.659.

Duflou, J. *et al.* (2007) 'Experimental study on force measurements for single point incremental forming', *Journal of Materials Processing Technology*, 189(1–3), pp. 65–72. doi: 10.1016/j.jmatprotec.2007.01.005.

Edward and Leszak (1964) 'Apparatus and process for incremental dieless forming', Patent No.- US3342051A United States.

Gómez-López, L. M. *et al.* (2013) 'Simulation and modeling of single point incremental forming processes within a solidworks environment', *Procedia Engineering*, 63, pp. 632–641. doi: 10.1016/j.proeng.2013.08.253.

Gundarneeya, T. P. *et al.* (2022) 'Optimization of single point incremental forming process through experimental investigation on SS 304 DDQ steel', *Materials Today: Proceedings*, 57, pp. 753–760. doi: 10.1016/j.matpr.2022.02.283.

Gupta, P. and Jeswiet, J. (2019) 'Manufacture of an aerospace component by single point incremental forming', *Procedia Manufacturing*, 29, pp. 112–119. doi: 10.1016/j.promfg.2019.02.113.

Gupta, P., Szekeres, A. and Jeswiet, J. (2019) 'Design and development of an aerospace component with single-point incremental forming', *The International Journal of Advanced Manufacturing Technology*, 103(9–12), pp. 3683–3702. doi: 10.1007/s00170-019-03622-4.

Ham, M. and Jeswiet, J. (2007) 'Forming limit curves in single point incremental forming', *CIRP Annals*, 56(1), pp. 277–280. doi: 10.1016/j.cirp.2007.05.064.

Hussain, G. *et al.* (2020) 'Mechanical properties and microstructure evolution in incremental forming of AA5754 and AA6061 aluminum alloys', *Transactions of Nonferrous Metals Society of China*, 30(1), pp. 51–64. doi: 10.1016/S1003-6326(19)65179-4.

Jeswiet, J. *et al.* (2005) 'Asymmetric single point incremental forming of sheet metal', *CIRP Annals*, 54(2), pp. 88–114. doi: 10.1016/S0007-8506(07)60021-3.

Ji, Y. H. and Park, J. J. (2008) 'Formability of magnesium AZ31 sheet in the incremental forming at warm temperature', *Journal of Materials Processing Technology*, 201(1–3), pp. 354–358. doi: 10.1016/j.jmatprotec.2007.11.206.

Kim, Y. H. and Park, J. J. (2002) 'Effect of process parameters on formability in incremental forming of sheet metal', *Journal of Materials Processing Technology*, 130–131, pp. 42–46. doi: 10.1016/S0924-0136(02)00788-4.

Kumar, A. *et al.* (2019) 'Parametric effects on formability of AA2024-O aluminum alloy sheets in single point incremental forming', *Journal of Materials Research and Technology*, 8(1), pp. 1461–1469. doi: 10.1016/j.jmrt.2018.11.001.

Kumar, A., Gulati, V. and Kumar, P. (2018) 'Investigation of surface roughness in incremental sheet forming', *Procedia Computer Science*, 133, pp. 1014–1020. doi: 10.1016/j.procs.2018.07.074.

Kumar, G. P. and Suresh, K. (2019) 'Experimental study on force measurement for AA 1100 sheets formed by incremental forming', *Materials Today: Proceedings*, 18, pp. 2738–2744. doi: 10.1016/j.matpr.2019.07.137.

Kumar, T. S. and Chennakesava Reddy, A. (2016) 'Single point incremental forming and significance of its process parameters on formability of conical cups fabricated from Aa1100-H18 alloy', *International Journal of Engineering Inventions*, 5(6), pp. 10–18.

Li, Y., Daniel, W. J. T. and Meehan, P. A. (2017) 'Deformation analysis in single-point incremental forming through finite element simulation', *The International Journal of Advanced Manufacturing Technology*, 88(1–4), pp. 255–267. doi: 10.1007/s00170-016-8727-9.

Matsubara Shigeo (1994) 'Incremental backward bulge forming of a sheet metal with a hemispherical head tool', *Journal of the Japan Society for Technology of Plasticity*, 35(406), pp. 1311–1316.

McAnulty, T., Jeswiet, J. and Doolan, M. (2017) 'Formability in single point incremental forming: A comparative analysis of the state of the art', *CIRP Journal of Manufacturing Science and Technology*, 16, pp. 43–54. doi: 10.1016/j.cirpj.2016.07.003.

Mirzakhani, B. and Mansourinejad, M. (2011) 'Tensile properties of AA6061 in different designated precipitation hardening and cold working', *Procedia Engineering*, 10, pp. 136–140. doi: 10.1016/j.proeng.2011.04.025.

Mulay, A. *et al.* (2017) 'Experimental investigation and modeling of single point incremental forming for AA5052-H32 aluminum alloy', *Arabian Journal for Science and Engineering*, 42(11), pp. 4929–4940. doi: 10.1007/s13369-017-2746-1.

Sajjad, M., Joy, J. A. and Jung, D. W. (2018) 'Finite element analysis of incremental sheet forming for metal sheet', *Key Engineering Materials*, 783, pp. 148–153. doi: 10.4028/www.scientific.net/KEM.783.148.

Tayebi, P. *et al.* (2019) 'Formability analysis of dissimilar friction stir welded AA 6061 and AA 5083 blanks by SPIF process', *CIRP Journal of Manufacturing Science and Technology*, 25, pp. 50–68. doi: 10.1016/j.cirpj.2019.02.002.

12 The Impacts of Generatrix Radius Variation and Temperature on Spifability in Warm Incremental Forming
A Comprehensive Study

*Sudarshan Kumar Choudhary,
Rahul Ramlal Gurpude, and Amrut Mulay*

12.1 INTRODUCTION

Incremental forming, also known as incremental sheet forming (ISF) or single-point incremental forming (SPIF), is a manufacturing method used to shape metal sheets into complex three-dimensional (3D) geometries. Comprising distinct traditional forming methods that use dies and moulds, incremental forming achieves the localized plastic deformation of different materials. In incremental forming, a computer-controlled tool, typically a CNC machine, moves a forming device (usually a small-diameter ball or roller) over the surface of a metal piece, applying localized force. The tool moves in a programmed path, incrementally deforming the sheet till the preferred shape is achieved.

Incremental forming can be performed on a variety of sheet metals, including aluminium, steel, copper, and titanium. The forming tool is usually a small-diameter ball or roller and can be attached to a robot support, a CNC machine, or a dedicated incremental forming machine. The tool moves in multiple directions to shape the sheet.

Incremental forming relies on the localized plastic deformation of sheet material. As the forming tool applies force, the material undergoes stretching, bending, and compressive deformation, resulting in shape changes. The process can be controlled to vary the amount of deformation and achieve the desired shape. Incremental forming offers great flexibility in shaping complex geometries. It allows the fabrication of prototypes, customized parts, and small lots without the

DOI: 10.1201/9781003441755-12

need for expensive and time-consuming tooling. An innovative method for testing thinning parameters of sheet metals in negative incremental forming is discussed as revealed in Figure 12.1.

SPIF involves forming an axisymmetric portion using a curve of a circle as the generatrix and using sine law to predict thickness and sheet metal thinning limit. It ensures accuracy and reduces processing time and cost. The process characteristics, cosine law of thickness scattering, and construction of formability measure thinning limit were previously discussed in detail [1].

The effect of the radius of curvature of a test specimen on the formability of aluminium sheet in negative incremental forming is debated, but researchers have shown that formability rises as the radius of curvature declines [2]. An empirical model was established and validated that can be used to forecast the formability of investigational material without conducting tests. Authors of the relevant study also highlighted the use of SPIF for customized small-batch production [2].

Other researchers have identified the outcomes of process factors on the extreme forming force, formability, and surface roughness in the ISF of thermoplastic materials [3]. The research highlights the significant influence of spindle speed on these factors, including the maximum depth reached. The results also demonstrate that temperature variation, caused by friction among tool and the sheet blank, plays a crucial part in altering the mechanical properties of thermoplastic polymers. Polyvinylchloride (PVC) was used as the material in this study due to its viscoelastic behaviour and rubber-like characteristics with increased temperature [3].

The force valuation in SPIF for flexible wall angle geometry was investigated in [4]. Various bending conditions were examined to recognize the impact of process parameters. The findings revealed that supreme forming strength rises with tool diameter and step depth, even though upper spindle speeds led to a decrease

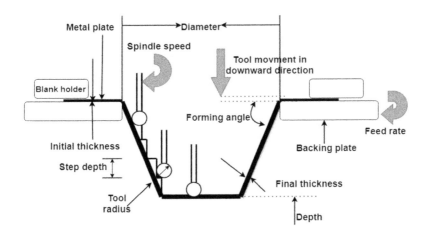

FIGURE 12.1   Single-point incremental forming setup with forming parameters.

in forming force due to increased friction and temperature. The development of forming force based on bending circumstances serves as a clue to preventing sheet failure [4].

Suresh and Regalla [5] investigated the forming ability of added deep drawing steel by numeric simulations using varying wall angle conical and pyramidal frustums as assessment geometries. The geometries had four changed generatrixes, circular, elliptical, parabolic, and exponential. The formability of the four geometries was verified using an analytical forming limit diagram curve in LS-DYNA. The determined wall angles found from the numeric simulations remained linked with investigational standards, showing decent associations among experimental and simulation outcomes.

Jin et al. [6] explored the use of SPIF on TRIP590 steel sheets, forming two parts with varying theoretical increases and reductions in wall thickness. Experimental results indicated that SPIF is a viable method for processing TRIP steel sheets, as evidenced by the division of wall thicknesses into three segments along the depth in both formed parts. Furthermore, the investigators determined the incremental forming limit angle for TRIP590 steel at a thickness of 0.67 mm to be approximately 66.5 degrees.

Erisov et al. [7] introduced an enhanced approach to determining limiting strains in incremental forming experiments. They suggested a method using a conical part with a radial generatrix to determine the limit angle of forming. However, this technique tends to overestimate limiting strains compared to measures with a rectilinear generatrix. The authors also offered a mathematical data processing technique that effectively eliminated overestimation.

Liu et al. [8] presented a new technique for evaluating the cylindricity fault of form parts built on the increment-simplex algorithm. Their method aimed to improve the productivity of analysis while maintaining accuracy. They analysed the existing cylindricity evaluation strategies and constructed mathematical models for fitting operations. They then compared their proposed method, based on incremental optimization, with traditional algorithms through experiments, demonstrating its effectiveness and feasibility in terms of accuracy and evaluation efficiency.

Hassan et al. studied the outcome of prerolling temperature on interfacial goods and formability of steel–steel bilayer sheets [9]. They conducted the roll bonding at three temperatures: 700 °C, 800 °C, and 950 °C, with 58% thickness decrease. The results displayed that increasing the prerolling temperature improved bond strength, critical strain energy release rate, and formability. The authors also discovered that heating the bilayer sheet beyond the critical temperature throughout manufacture facilitated decent attachment among the sheets.

Centeno et al. [10] provided an experimental examination of the impacts of process factors on the formability of AISI 304 metal pieces in SPIF that included determining conventional formability limits through extending and stretch-bending testing. The authors compared failure strains in SPIF and stretch-bending tests and discussed the respective failure modes. They also recorded the development of axial forming power to confirm the safe application of non-dedicated process equipment.

Suresh et al. [11] investigated the formability, thickness distribution, strain distribution, and microstructural variations of extra deep drawing steel in the SPIF process. The use of varying wall angle pyramidal frustum reduces the quantity of trials needed to determine the maximum wall angle associated with constant wall angle frustums, and the authors found the maximum formable wall angle to be 73 ± 2 degrees. Experimental and numeric results exhibited good correlation in terms of thickness, and numeric simulations revealed plane strain on the faces of the pyramid and biaxial elongating at the curves.

Mariem et al. [12] examined the result of generatrix profile on SPIF parameters in the manufacturing process of parts for minor series and samples. Specifically, they aimed to compare the effects of a straight generatrix and a circular generatrix on forming forces, thickness distribution, shape accuracy, and surface roughness of the formed profile. Honarpisheh et al. [13] studied the effects of different generatrix profiles on forming tool behaviour as the tool followed a predetermined spiral path defined by a CAD model. Specifically, the two profiles of interest were a truncated cone with a straight generatrix and another with a circular generatrix. The main parameters investigated included forming forces, thickness, shape correctness, and surface roughness of the formed profile. In order to estimate the properties of these generatrix profiles, the researchers introduced the factor of shape based on both experimental and analytical vertical forces. The investigational outcomes demonstrated that forming a straight generatrix leads to better surface roughness. On the other hand, shaping a circular generatrix resulted in extra-uniform thickness scattering of the blank beyond thinning and forming processes.

12.2 ROLE OF DIFFERENT FORMING PARAMETERS IN ISF

ISF involves shaping a sheet metal workpiece into final form through a series of small incremental deformations. Several forming parameters play a crucial role in the ISF process, and adjusting these parameters can affect the quality and efficiency of the forming operation. Following are some of the key forming parameters in ISF and their roles in forming operations.

12.2.1 THE ROLE OF STEP DEPTH IN INCREMENTAL FORMING

The step depth refers to the vertical space the forming tool travels during each incremental step or pass over the sheet material. It plays an important role in the process and has implications for the shape, accuracy, and efficiency of the formed part. The step depth directly influences the amount of material deformation that occurs during each pass and affects the time required to complete the forming process.

Larger step depths cover more surface area of the sheet per pass, potentially reducing the total number of required passes; this can lead to faster forming times. Choudhary and Mulay [14] described impacts of tool size and step depth on maximum wall angle achieved with SPIF along with presenting statistical analysis results and response surface methodology's D-optimal design to classify important parameters and relations. The findings indicate that tool radius is crucial for formability in AA1050 and AA7075-T6, while step depth is vital for AA6061-T6. A larger step depth will result in more significant deformation, while a smaller step depth will produce more gradual changes. By adjusting the step depth, the operator can control the amount of material displacement and achieve the desired shape.

12.2.2 THE ROLE OF FEED RATE IN INCREMENTAL FORMING

The feed rate denotes the rate at which the forming tool travels along the surface of the metal sheet during an incremental forming operation. It directly influences the deformation, forming time, accuracy, and surface quality of the desired portion. The feed rate determines the quantity of material that is incrementally shaped at each step of the forming process. A higher feed rate results in larger deformation, while a lower feed rate leads to smaller deformation.

Controlling the feed rate helps in achieving the desired form without causing excessive strain or failure of the material, and the feed rate affects the overall forming time. A higher feed rate allows for faster movement of the forming tool, resulting in shorter forming times. A lower feed rate allows for more precise control over the deformation, resulting in higher accuracy. The feed rate plays a role in ensuring process stability during incremental forming.

On-Uma Lasunon [15] investigated the outcomes of forming factors on arithmetic mean surface roughness (Ra) of AA5052 manufactured by a SPIF process. The three studied factors were feed rate (12.5, 25, and 50 in/min), depth increment (0.015 and 0.030 in), and wall angle (45° and 60°). The outcomes show that wall angle, depth, and their interaction play vital roles in the surface roughness, while feed rate has a minimal outcome. The ideal forming conditions for lowest surface roughness are feed rate of 25 in/min, depth of 0.015 in, and wall angle of 45°.

12.2.3 THE ROLE OF SPINDLE SPEED IN INCREMENTAL FORMING

The spindle speed affects the amount of material deformation during the forming process. Higher spindle speeds tend to induce more severe plastic deformation, resulting in greater material thinning and stretching. Controlling the spindle speed can help manage these forces to prevent excessive deformation or tool failure. Higher speeds can result in smoother surface finishes due to reduced tool marks and improved chip evacuation, and they generally lead to faster material removal and shorter forming times.

However, higher speeds may also increase the risk of vibration and instability, which can reduce process efficiency and accuracy. Spindle speed influences the tool life and wear during incremental forming. High spindle speeds can accelerate tool wear due to increased heat generation and mechanical stress. Bagudanch et al. [16] studied the influences of key method parameters (step down, spindle speed, feed rate, tool diameter, and sheet thickness) on the determined forming force in the ISF of polymeric constituents. They analysed maximum depth and surface roughness as indicators and identified significant impacts of spindle speed on forming force and formability; they also observed temperature variations triggered by friction between the tool and the sheet blank. The researchers used PVC, which exhibits rubber-like behaviour, due to its viscoelastic properties at higher temperatures.

12.2.4 THE ROLE OF TOOL SHAPE AND SIZE IN INCREMENTAL FORMING

The shape of the forming tool determines the deformation path of sheet metal. By controlling tool shape and movement, specific part geometries can be achieved. Proper tool shape design can help minimize wrinkling, tearing, and other undesired defects in the formed part, and different tool contours can result in variations in surface roughness and texture.

The tool shape also plays a role in process stability and feasibility. For instance, a larger tool diameter will result in more material being deformed during each incremental step, while a smaller tool diameter will produce less deformation. The tool diameter also affects the extent of forming forces and stresses experienced by material throughout the incremental forming. Bigger tool spans allocate the force over a larger area, reducing the localized stresses on the material, and this can help minimize the risk of material failure, such as tearing or wrinkling. Smaller tool diameters concentrate the force in a smaller area, resulting in localized stresses.

Hussain et al. [17] clarified that formability depends on tool size relative to sheet thickness using the response surface method. Although sheet thickness (t_0) and tool radius (r) individually lacked significance according to analysis of variance, they showed a strong interaction effect. Maximum spifability occurs at a specific r/t_0 for each sheet thickness under given conditions. Additionally, a tool radius smaller than the critical ratio ($r/t_0 < 2$) leads to material squeezing at the tool/sheet edge, adversely affecting spifability. The tool used in Hussain et al.'s experiment was made of high-speed steel (HSS) with a hemispherical tip. The diameter of the tool is 10 mm as shown in Figure 12.2.

FIGURE 12.2 Tool used in the experiments in [17].

12.2.5 The Role of Sheet Thickness in Incremental Forming

Incremental forming is a manufacturing method in which a sheet of material, typically metal, is gradually formed into a desired shape by a sequence of limited deformations using a forming tool, and the formability of sheet metal is influenced by its thickness. Thinner sheets generally exhibit higher formability, allowing for more complex shapes; thicker sheets, in contrast, may have limitations in terms of achievable forming depth and the complexity of the part due to increased material resistance. The forces required to deform a sheet metal also increase with increasing thickness. Heavier sheets need higher forming forces, which can impact the selection and design of forming tool and machine.

Springback refers to a material's elastic recovery after its deformation. Thinner sheets offer increased formability, require less force, and exhibit reduced springback. However, thicker sheets may be necessary for certain applications that require structural integrity or specific mechanical properties, even though they can present challenges related to forming forces, springback, and heat transfer. Sbayti et al. [18] performed numeric simulations using Box–Behnken design and response surface methodology to optimize SPIF parameters for titanium denture plates. Specifically, the authors used two algorithms, multi-objective genetic and global optimum determination, to seek optimal solutions to minimizing sheet thickness, final depth, and extreme forming force.

12.2.6 The Role of Lubricant in Incremental Forming

Lubricants play a crucial role in incremental forming processes in decreasing friction between the tool and the sheet metal, allowing for easier distortion and better surface finish of the shaped part. Incremental forming involves deforming the sheet metal with a tool, creating significant friction that generates heat in the process. Lubricants are applied to reduce this friction, making it easier for the tool to slide over the metal surface. This reduces the force required for deformation and prevents excessive heat generation, which can lead to tool wear and material damage.

Lubricants contribute to improving the material flow during the deformation and the final surface finish of the formed part, and they also help to maintain the uniformity of material thickness during the forming process. Different lubricants, such as oils, greases, and solids, may be used depending on the specific requirements of the forming operation. Azevedo et al. [19] proposed estimating the effect of kind of lubricant used, mainly on aluminium 1050 and DP780 steel sheets, on the surface quality of formed SPIF-formed pieces. The authors executed trials using a collection of different lubricants and conducted roughness tests to estimate surface quality. The results showed that the lubricants that improved aluminium qualities had poorer outcomes in steel and vice versa.

12.2.7 The Effects of Generatrix in Incremental Forming

The generatrix refers to the path a SPIF tool travels as it rotates around the workpiece. The generatrix in incremental forming can be significant and can influence several aspects of the forming process as it promotes controlled and continuous material flow during deformation. As the tool rotates around the workpiece, it applies pressure and radial forces, instigating the material to drift and gradually take the shape of the tool, as shown in Figure 12.3.

This results in a smooth and even distribution of material, reducing the risk of defects such as tearing or wrinkling. It helps to distribute the strain more evenly across the workpiece. The generatrix can impact forming limits; certain shapes or geometries are more suitable for incremental forming with a generatrix, while others may require additional tooling or different forming techniques as shown in Figure 12.4. Hussain et al. [20] conducted a study comparing the forming restrictions of an aluminium sheet metal using two types of parts with sideways depth,

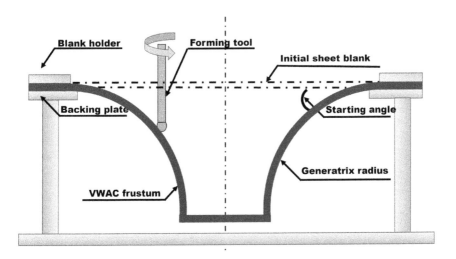

FIGURE 12.3   Representation sketch of an incremental forming setup.

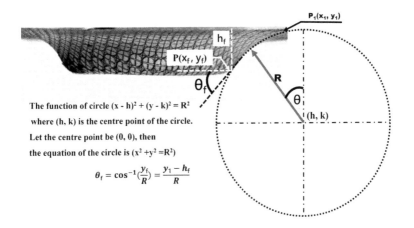

FIGURE 12.4   Schematic and description of generatrix.

with varying slopes and static slopes. The formability parameters were obtained by forming these parts until they fractured. The outcomes revealed that the formability of the sheet metal is influenced by slope scattering along the axisymmetric part to be formed. The parts with varying slope demonstrated higher forming limits compared to the parts with fixed slope.

12.3 MATERIAL AND METHODOLOGY

Warm incremental forming was applied to a 1 mm thick titanium grade 2 (Ti-Gr2) sheet that measured 105 mm × 105 mm in area. The selection of Ti-Gr2 was based on its superior strength-to-weight ratio. The chemical composition of the commercially pure Ti-Gr2 (wt %) is tabulated in Table 12.1. During the process, various generatrix radii of curvature and temperature ranges were employed. Subsequent runs were then conducted focusing on the radius of curvature that produced the highest forming depth.

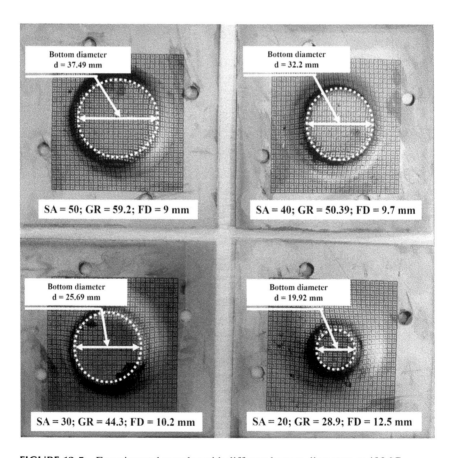

FIGURE 12.5 Experimental samples with different bottom diameters at 400 °C.

TABLE 12.1

Chemical Compositions of Commercially Pure Titanium Grade 2 (wt %)

Element	Ti	Fe	C	Cr	Ni	Cu	Si	Mn	Nb	V	Mo	Zr	Al
Wt. %	99.9	0.055	0.034	0.017	0.011	0.007	0.003	0.002	0.002	0.003	0.003	0.001	0.002

To obtain the highest forming depth, adjustments were made to the temperature, and a suitable mixture of MoS_2 and lithium-based grease in the ratio of 1:3 was applied as a lubricant; lubricant in warm ISF is crucial for achieving a smooth surface and minimizing tool wear. The lubricant was applied in a thin film by continually dipping the sheet being formed in a lubricant pool while the forming operation took place. The experimental generatrix was designed using the commercial CAD/CAM package NX12. The forming process was carried out on a CNC vertical milling machine with three axes (Batliboi Dart, Batliboi Ltd., Mumbai, India). Throughout the experiments, specific parameters—a constant feed rate of 275 mm/min, step depth of 0.2 mm, and spindle speed set at 90 rpm—remained unchanged. Experimental samples with different bottom diameters formed at 400 °C are shown in Figure 12.5.

12.4   RESULTS AND DISCUSSIONS

This section presents experimental outcomes for the impacts of generatrix radius and temperature on the formability of Ti-Gr2 sheets. The researchers aimed to analyse the influence of different factors on the formability and deformation behaviour of the material, providing valuable information for optimizing the SPIF process parameters for forming Ti-Gr2.

12.4.1   TEMPERATURE AND SPIFABILITY

Next, to examine the effect of temperature on the forming depth of Ti-Gr2 products, experiments were run at four different temperatures ranging from room temperature to 600 °C by keeping the radius of curvature and lubricant unchanged (Table 12.2), and the forming depth of the sheet metal varied according to the temperature. At room temperature, a generatrix radius of 28.9 mm resulted in a forming depth of 12.104 mm. When the temperature increased to 200 °C, the forming depth increased by 1.63% over room temperature. Similarly, at 400 °C, the forming depth reached 12.5 mm, a 3.27% increase in formability. The highest formability was achieved at 600 °C, with a forming depth of 13.096 mm at a generatrix radius of 28.99 mm. This represents an 8.19% increase in formability over room temperature.

The heat input was global heating, which targets achieving uniform heating across the entire sheet metal blank, while localized heating specifically focuses on the forming zone or its nearby region. Titanium and its alloys are known for their high flow stress at room temperature, which makes them tough to deform at room temperature. Heating Ti-Gr2 to 600 °C allowed the material to undergo plastic flow more easily. This increase in temperature encourages the flexibility of

TABLE 12.2

Experimental Conditions at Different Temperatures

S. no.	Starting angle (degrees)	Radius of curvature (mm)	Lubricant -	Temperature (°C)	Forming depth (mm)
01	20	28.9	Mixed	Room Temp	12.10
02	20	28.9	Mixed	200	12.30
03	20	28.9	Mixed	400	12.50
04	20	28.9	Mixed	600	13.09

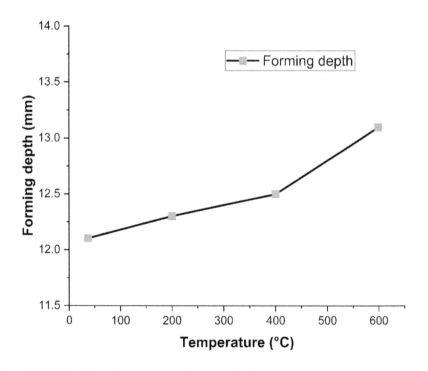

FIGURE 12.6 Plot of temperature vs forming depth.

dislocations and grain limits, leading to a decrease in flow stress and an improvement in formability [21]. Figure 12.6 plots the temperatures vs forming depths for the four experimental runs.

12.4.2 GENERATRIX RADIUS OF CURVATURE AND SPIFABILITY

To assess the impacts of generatrix radius on Ti-Gr2 sheet forming depth, we conducted multiple experiments at a constant temperature of 400 °C, and the lubricant we used was a mixture of pure MoS_2 and lithium-based grease in a ratio of 1:3. We measured the variations in the forming depth of the sheet metal corresponding to the different radii of curvature values tested. At the highest generatrix radius of 59.20 mm, the forming depth was 9 mm, while at the radius of 50.39 mm, the forming depth increased by 7.77%. Similarly, at radii of curvature of 44.30 mm and 28.90 mm, the forming depths increased by 13.33% and 38.88%, respectively. At 400 °C, Ti-Gr2 sheet metal material exhibits enhanced formability due to its plastic flow as it approaches its recrystallization point.

Due to this plastic flow and improved homogeneity, the material exhibits reduced resistance to forming forces. This means that it requires less force to deform material into the anticipated shape. Decreasing the radius of curvature further improves the spifability of the material [20] as depicted in Figure 12.7(a). Similarly, lowering the starting angle to 20° from 50° also enhances spifability by minimizing initial sheet bending and reducing localized strain, as depicted in Figure 12.7(b). Based on the aforementioned approach, the maximum forming depth was 12.10 mm at a radius of curvature of 28.90 mm, reflecting the highest formability. Further elaboration on the specific details of each run is tabulated in Table 12.3, and a pictorial representation is provided in Figure 12.8.

TABLE 12.3
Experiments Conducted with Varying Generatrix Radii of Curvature

S. no.	Starting angle (°)	Radius of curvature (mm)	Temperature (°C)	Lubricant -	Forming depth (mm)
01	50	59.20	400	Mixed	9.00
02	40	50.39	400	Mixed	9.70
03	30	44.30	400	Mixed	10.20
04	20	28.90	400	Mixed	12.10

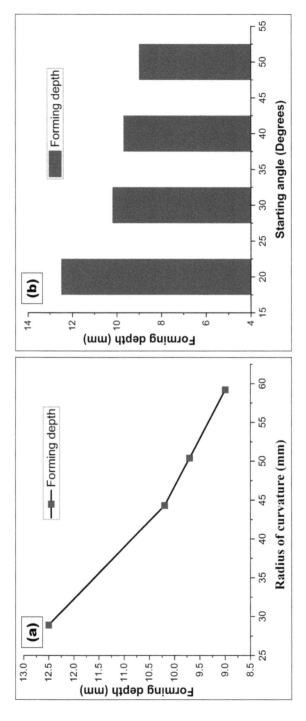

FIGURE 12.7   Plot of (a) radius of generatrix vs forming depth; (b) starting angle vs forming depth.

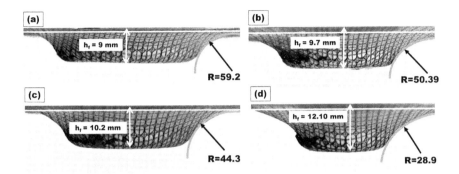

FIGURE 12.8   Experimental sample depicting generatrix radii and forming depths at (a) SA=50, GR = 59.2, FD = 9 mm; (b) SA = 40, GR = 50.39, FD = 9.7 mm; (c) SA = 30, GR = 44.3, FD = 10.2 mm; (d) SA = 20, GR = 28.9, FD = 12.5 mm.

12.5   CONCLUSIONS

This chapter concludes an experimental investigation on the effect of temperature and generatrix radius in the SPIF process which provides significant outcomes to the study.

- At a temperature of 600 °C, titanium grade 2 materials exhibit their highest formability, allowing for a forming depth of 13.096 mm with a generatrix radius of 28.9 mm. This temperature results in an 8.19% rise in formability over room temperature formability. The improved formability can be attributed to the reduced flow stress, which enhances the spifability of the titanium grade 2 materials.
- Formability of material can be further enhanced by decreasing the radius of curvature and the starting angle. Lowering the starting angle from 50° to 20° enhances spifability by minimizing the initial sheet bending and reducing localized strain compared with using higher starting angles.
- There is a correlation between the radius of curvature and forming depth: As the radius of curvature declines, the spifability increases due to the plastic flow and improved homogeneity.

12.6   ACKNOWLEDGEMENT

Authors would like to express thanks to SERB, Government of India, for the financial support under project SERB/2020/001355.

REFERENCES

[1] G. Hussain, and L. Gao, "A novel method to test the thinning limits of sheet metals in negative incremental forming," Int. J. Mach. Tools Manuf., vol. 47, no. 3–4, pp. 419–435, 2007, doi: 10.1016/j.ijmachtools.2006.06.015.

[2] G. Hussain, L. Gao, N. Hayat, and L. Qijian, "The effect of variation in the curvature of part on the formability in incremental forming: An experimental investigation," Int. J. Mach. Tools Manuf., vol. 47, no. 14, pp. 2177–2181, 2007, doi: 10.1016/j.ijmachtools.2007.05.001.

[3] I. Bagudanch, M. L. Garcia-Romeu, G. Centeno, A. Elías-Zúñiga, and J. Ciurana, "Forming force and temperature effects on single point incremental forming of polyvinylchloride," J. Mater. Process. Technol., vol. 219, pp. 221–229, 2015, doi: 10.1016/j.jmatprotec.2014.12.004.

[4] I. Bagudanch, G. Centeno, C. Vallellano, and M. L. Garcia-Romeu, "Forming force in single point incremental forming under different bending conditions," Procedia Eng., vol. 63, pp. 354–360, 2013, doi: 10.1016/j.proeng.2013.08.207.

[5] K. Suresh, and S. P. Regalla, "Analysis of formability in single point incremental forming using finite element simulations," Procedia Mater. Sci., vol. 6, no. Icmpc, pp. 430–435, 2014, doi: 10.1016/j.mspro.2014.07.055.

[6] W. Jin, W. Bao-ping, T. Long, and J. Hu-seng, "Study on possibility of processing TRIP steel sheet by single point incremental forming," Mater. Sci. Eng., no. 51205217, pp. 460–462, 2012, doi: 10.2991/mems.2012.76.

[7] Y. Erisov, S. Surudin, I. Petrov, and A. Kuzin, "An advanced method to test the formability in single point incremental forming," AIP Conf. Proc., vol. 2700, no. July 2005, 2023, doi: 10.1063/5.0124949.

[8] D. Liu, P. Zheng, J. Wu, H. Yin, and L. Zhang, "A new method for cylindricity error evaluation based on increment-simplex algorithm," Sci. Prog., vol. 103, no. 4, pp. 1–25, 2020, doi: 10.1177/0036850420959878.

[9] M. Hassan et al., "Effect of pre-rolling temperature on the interfacial properties and formability of steel-steel bilayer sheet in single point incremental forming," Proc. Inst. Mech. Eng. Part B J. Eng. Manuf., vol. 235, no. 3, pp. 406–416, 2021, doi: 10.1177/0954405420963004.

[10] G. Centeno, I. Bagudanch, A. J. Martínez-Donaire, M. L. García-Romeu, and C. Vallellano, "Critical analysis of necking and fracture limit strains and forming forces in single-point incremental forming," Mater. Des., vol. 63, pp. 20–29, 2014, doi: 10.1016/j.matdes.2014.05.066.

[11] K. Suresh, S. D. Bagade, and S. P. Regalla, "Deformation behavior of extra deep drawing steel in single-point incremental forming," Mater. Manuf. Process., vol. 30, no. 10, pp. 1202–1209, 2015, doi: 10.1080/10426914.2014.994755.

[12] M. Dakhli, A. Boulila, and Z. Tourki, "Effect of generatrix profile on single-point incremental forming parameters," Int. J. Adv. Manuf. Technol., vol. 93, no. 5–8, pp. 2505–2516, 2017, doi: 10.1007/s00170-017-0598-1.

[13] M. Honarpisheh, M. Mohammadi Jobedar, and I. Alinaghian, "Multi-response optimization on single-point incremental forming of hyperbolic shape Al-1050/Cu bimetal using response surface methodology," Int. J. Adv. Manuf. Technol., vol. 96, no. 9–12, pp. 3069–3080, 2018, doi: 10.1007/s00170-018-1812-5.

[14] S. Choudhary, and A. Mulay, "Influence of tool size and step depth on the formability behavior of AA1050, AA6061-T6, and AA7075-T6 by single-point incremental forming process," J. of Mater. Eng. Perform., pp. 1–16, 2023, doi: 10.1007/s11665-023-08231-7.

[15] O. U. Lasunon, "Surface roughness in incremental sheet metal forming of AA5052." Adv. Mat. Res., vol. 753, pp. 203–206, 2013, doi: 10.4028/www.scientific. net/amr.753-755.203.

[16] I. Bagudanch et al., "Forming force and temperature effects on single point incremental forming of polyvinylchloride", J. Mater. Process. Technol., vol. 219, pp. 221–229, 2015, ISSN 0924-0136, doi: 10.1016/j.jmatprotec.2014.12.004.

[17] G. Hussain et al., "Guidelines for tool-size selection for single-point incremental forming of an aerospace alloy," Mater. Manuf. Process., vol. 28, no. 3, pp. 324–329, 2013, doi: 10.1080/10426914.2012.700151.

[18] M. Sbayti et al., "Optimization techniques applied to single point incremental forming process for biomedical application," Int. J. Adv. Manuf. Technol., vol. 95, pp. 1789–1804, 2018, doi: 10.1007/s00170-017-1305-y.

[19] N. G. Azevedo et al., "Lubrication aspects during single point incremental forming for steel and aluminum materials," Int. J. Precis. Eng. Manuf., vol. 16, pp. 589–595, 2015, doi: 10.1007/s12541-015-0079-0A.

[20] G. Hussain et al., "A comparative study on the forming limits of an aluminum sheet-metal in negative incremental forming." J. Mater. Process. Technol., vol. 187, pp. 94–98, 2007, doi: 10.1016/j.jmatprotec.2006.11.112.

[21] G. Palumbo, and M. Brandizzi, "Experimental investigations on the single point incremental forming of a titanium alloy component combining static heating with high tool rotation speed." Mater. Des., vol. 40, pp. 43–51, 2012, doi: 10.1016/j.matdes. 2012.03.031.

13 Numeric Investigations to Improve the Final Sheet Thickness in the SPIF of DC04 Sheets

Viren Mevada and Amrut Mulay

13.1 INTRODUCTION

Incremental sheet forming (ISF) is a novel metal-forming process currently gaining momentum as a prototyping technique. In ISF, a sheet is deformed incrementally layer by layer, generally using a hemispherical tool. The tool's diameter is much smaller than a conventional punch, ranging from 6 to 20 mm. This hemispherical tool comes in contact with the sheet in a small region and deforms locally through bending, stretching, and shearing.

The tool follows a definite tool path generated from the CAD model of the desired geometry. The whole geometry is sliced into several layers, and the tool follows one layer at a time on a CNC machine, robotic manipulator, or particular ISF machine. The material deformation is highly localized, thus greatly reducing the deformation force. Due to fewer deformation forces, lower residual stresses are generated in the sheet, pushing the formability limits beyond those of traditional forming operations. Another significant advantage of the ISF process is its flexibility; it can even produce a lot size of one. All these advantages allow for applying ISF in biomedical implants and in the aerospace and automobile industries. However, limited geometric accuracy, surface quality, and excessive sheet thickness reduction limit its potential.

Researchers have attempted to reduce the excessive sheet thickness reduction, geometric inaccuracies, and surface roughness to scale the process for industry. Tera et al. (2019) proposed several strategies regarding formability along with implementing CAD/CAM for the SPIF process. The single-stage method using Archimedes' spiral strategy showed an acceptable reduction in sheet thickness and acceptable uniformity, but there were two-directional cracks in the finishing stage of the process.

Gipiela et al. (2017) studied the formability for concentric and spiral profiles in SPIF and compared the findings with conventional Nakajima test results for high-strength low-alloy 440. The concentric and spiral profiles could sustain much more strain than the Nakajima test. Khazaali and Fereshteh-Saniee (2019) presented a study on incremental forming at higher temperatures due to the low

DOI: 10.1201/9781003441755-13

formability of Mg alloy AZ31 at room temperature. The authors observed a significant increase in formability with the increasing ductility (faster slip plane movement) above the transition temperature.

Barnwal et al. (2018) presented a paper to understand the macro- and microstructural behavior of AA6061 Al-Mg-Si alloy. The direction of major true strain was observed to be always perpendicular to the tool path due to the flow of material in the normal course of tool movement. The zigzag nature of the major vs. minor strain graph showed higher formability with SPIF than with conventional forming. However, very few studies have been conducted to study the effect of tool path-related parameters on sheet thickness reduction, geometrical inaccuracies, and surface roughness.

Harish et al. (Nirala and Agrawal, 2022) improved the formability by developing an adaptive tool path that varied the increment concerning the layer's position, thereby improving the sheet thickness distribution. The value of increment decreases as the tool reaches the base of the geometry, making the process more localized and in the process increasing the formability. Fiorentino et al. (2015) developed an intuitive algorithm to modify the CAM-generated tool path using an artificial cognitive system that adjusted the tool path based on the error observed in the formed part. After five iterations, excellent agreement was found between the actual part and the CAM profile.

Dejardin et al. (2010) investigated the geometric inaccuracies in the formed part using finite element analysis (FEA). They concluded that these inaccuracies are very predictable and can be accommodated in the tool path before forming. Leem et al. (2022) developed a new strategy called regional plastic incremental bending (RPIB). In this method, a material is deformed in such a way that global spring-back can be avoided. The tool path made using RPIB predicts the springback and makes necessary changes in the tool path, which significantly improves the geometrical accuracies.

Ullah et al. (2023) determined the optimal position of the support tool for double-sided incremental forming (DSIF). The generated tool path manipulated the locations of tools in the circumferential direction in such a way that the two tools squeezed an equal amount of the sheet at the intersection. They also found that the support tool should be within $10°$ of the master tools, normally drawn perpendicular to the circumference. Controlling the position of the support tool within $10°$ coupled with DSIF drastically increased the precision of the formed part.

Durante, Formisano, and Lambiase (2018) studied unidirectional (UTP) and alternate directional (ATP) tool paths with fixed and rotating tools. They observed that the parts formed using UTP showed distortion at the wall for polycarbonated sheets. Alternatively, components formed using ATP do not display any distortions.

Ren et al. (2019) proposed an in situ springback compensation method to increase the geometric accuracy of the ISF process. The researchers identified a few control points from the formed geometries, compared those with the desired CAM profile, and calculated error; fewer control points reduces the computation time and improves the geometric accuracy. The feedback loop consisted of measurable force and the location of the tool to make the adaptive the tool path. The results showed significant improvement in geometric accuracy with compensated tool path without any iterations.

Carette, Vanhove, and Duflou (2019) created an automated feature-based tool path for double hemispherical geometry. However, the same algorithm can be used to improve the geometric accuracy and surface quality of previously formed parts. The results showed improved surface quality while maintaining the geometric accuracy. Lu et al. (2017) developed a SPIF algorithm to correct the vertical directional element of the generated tool path, which was based on model predictive control specifically designed for two-point incremental forming. However, in a later stage, they improved the algorithm, which can even modify the tool path in horizontal directions. This algorithm successfully increased the geometrical accuracy of non-axis symmetric shapes.

Fiorentino, Giardini, and Ceretti (2015) developed an intelligent artificial cognitive system for tool path corrections. The approach is iterative: A generated tool path is fed to the ISF machine to form the targeted geometry and compared with the CAD profile, and the tool path is corrected based upon calculated error. A non-axis symmetric object was formed using five iterations through this method with minimal deviation. Nirala and Agrawal (2020) developed a fractal geometry-based incremental tool path to improve formability while manipulating stresses. The tool path successfully formed a component with a better strength-to-weight ratio by applying excessive compressive stresses in the base region of the component.

Researchers have attempted to improve the geometric accuracy and surface quality of formed parts through tool path modifications but have given very little emphasis to the excessive thickness reduction. Our aim with this paper was to study nonconventional tool paths and their effects on sheet thickness reductions. We evaluated two nonconventional paths, zig with contour and trochoidal, and compared them with conventional constant depth tool paths. We used three tool profiles, flat with corner radii, hemispherical, and elliptical, in the experiments. We performed all the experiments using the FEA software Abaqus/Explicit.

13.2 METHODOLOGY

We performed 17 experimental structural simulations of the part to determine the sheet thicknesses for the material DC04. The material properties of the steel are shown in Table 13.1, obtained by the uniaxial tensile test. We calculated

TABLE 13.1
Material Properties of DC04 Steel

Density (kg/mm3)	Young's modulus (n/mm)	Anisotropy					
		R11	R22	R33	R12	R13	R23
7.87×10^{-6}	210×10^{3}	1.83	1.0243	1.231	1.0967	1	1

anisotropy using Abaqus formulas. The plastic properties of the material were given in the form of stress–strain curves generated from the tensile test.

The CAM program is shown in Figure 13.1(a). A sheet with geometrical dimensions 222 mm × 222 mm × 0.8 mm and a tool with a diameter of 10 mm with a shank length of 25 mm were modeled. The conical geometry was selected. The major diameter is 120 mm, minor diameter is 30 mm, total depth is 30 mm, and incremental step depth of 0.8 mm. Three types of tools were modeled for different kinds of experiments, i.e., flat end with a corner radius of 1 mm, ball end or spherical end, and elliptical tool with major diameter 10 mm and minor diameter of 5 mm. However, the tool was kept analytically rigid to save the processing power and simulation time. The interaction between tool and sheet is given surface-to-surface contact, while all the edges are fixed. The coefficient of friction was kept as 0.1 at the tool sheet interface. The design matrix for the proposed study is shown in Table 13.2.

The tool paths are generated with the help of Siemens NX® software, however, the software generates a G and M code files. A python program was written to get (x, y and z) coordinates of each steps out of the G and M code file. Several combinations of mass scaling factor and mesh size were run before selecting either of them. From the conclusion of this experimental run, mesh size was selected as 1 mm.

FIGURE 13.1   (a) Toolpath generation using Siemens UG NX®; (b) tools used.

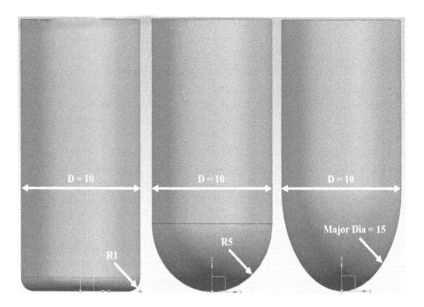

FIGURE 13.1   (Continued)

TABLE 13.2
List of Proposed Experiments

Simulation Expt. No.	Step over (%)	Tool shape	Tool path
1	40	Flat	Trochoidal
2	60	Flat	Trochoidal
3	40	Elliptical	Trochoidal
4	60	Elliptical	Trochoidal
5	40	Spherical	Constant Z level
6	60	Spherical	Constant Z level
7	40	Spherical	Zig with contour
8	60	Spherical	Zig with contour
9	50	Flat	Constant Z level
10	50	Elliptical	Constant Z level
11	50	Flat	Zig with contour
12	50	Elliptical	Zig with contour
13	50	Spherical	Trochoidal
14	50	Spherical	Trochoidal
15	50	Spherical	Trochoidal
16	50	Spherical	Trochoidal
17	50	Spherical	Trochoidal

13.3   RESULTS AND DISCUSSION

13.3.1   THE EFFECTS OF THE TOOL PATH ON THE SHEET THICKNESS

Our findings depict that stretching conditions prevail in SPIF, and significant sheet thinning occurs. The maximum allowable thinning of the sheet is an important response to the ISF process. The final sheet thickness of the component for a particular wall angle in the deformed area varies following the sine law. The final thickness of the sheet tends to decrease with the increase in wall angle and become 0 with a 90° wall angle, which results in sheet fracture.

If the material of the sheet can be considered incompressible, then the sheet is deformed at the expense of thickness. Hence, thickness reduction, thinning limit, and thickness distribution can be considered important criteria for predicting the feasibility of the ISF process. The sheet thickness at which failure occurs is known as the thinning limit of the sheet material. Typical sheet thicknesses are shown in Figure 13.2(a). The thicknesses show drastic reductions at the bending and initial stretching regions.

The sheet thickness is unaffected in the zone where the sheet is clamped. The thickness toward the center of the frustum starts reducing as the material flows from the edges to the center once the sheet starts deforming. Maximum thinning happens at the wall of the frustum due to the stretching, shearing, and squeezing of the sheet material. Similarly, the base of the frustum is unaffected as there is generally no contact between the tool and the sheet.

Figure 13.2(b) shows the comparison of minimum sheet thickness. A maximum sheet thickness of 0.6592 mm was observed in experiment 9, which had the parameters of 50% step-over, a flat tool with a corner radius of 1 mm, and a constant Z level tool path. In experiment 4, meanwhile, a minimum sheet thickness of 0.5858 mm was recorded with 60% step-over, an elliptical tool, and a trochoidal tool path.

We observed that experiments 1 and 2 followed close-to-typical distribution, and experiments 3, 4 and 13 showed completely different distribution patterns; in those three, the minimum thickness was observed at the base rather than at the walls where it is typically observed. This is because the trochoidal tool path starts contact at the center of the contour and expands to the walls; the direction of material flow is inward to outward, while in the conventional tool path, it is outward to inward.

Figure 13.3(a) depicts a simpler and more typical sheet thickness distribution. Zig with contour is a conventional tool path with a minor variation. These variations are mainly because of the other two variable parameters. Among these experiments, experiment 9 provides minimum sheet thickness reduction, where the step-over is 50%, and the tool is flat with a corner radius. Experiment 10 provides a similar sheet thickness with slightly more thinning, which is mainly because of the profile of the tool used, which was elliptical. While experiment 9 provides less sheet thinning, experiment 10 provides more uniform thickness. Experiments 5 and 6 gives almost the same thicknesses, which means that for spherical tools, percent of step-over is not the significant parameter. We also observed that the wall closer to the origin is thinner than the other half of the frustum. This is due to the location of the step depth for individual layers.

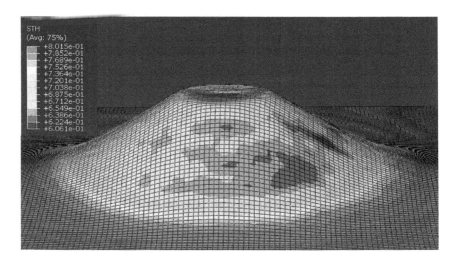

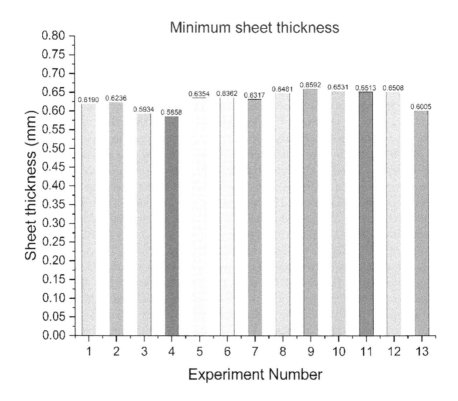

FIGURE 13.2   (a) Contour plot of a typical sheet thickness distribution (experiment 11); (b) minimum sheet thickness per the design matrix.

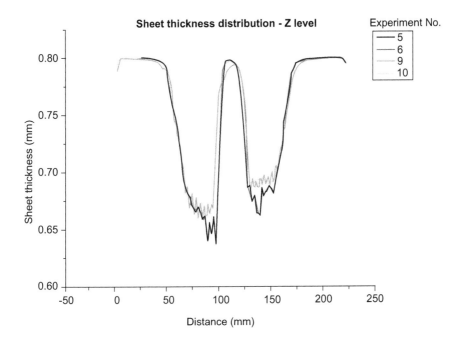

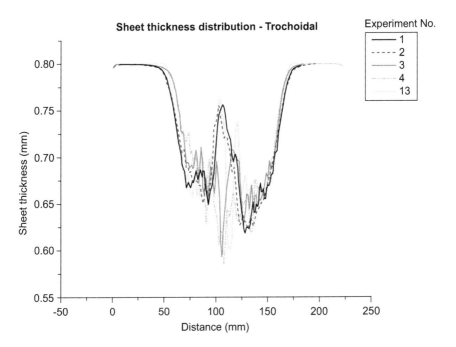

FIGURE 13.3   Sheet thickness distributions for (a) constant Z level tool path, (b) trochoidal tool path, and (b) zig with contour tool path.

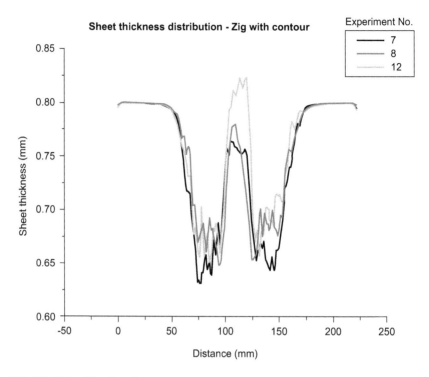

FIGURE 13.3   (Continued)

13.3.2   THE EFFECTS OF THE TOOL PROFILE ON THE SHEET THICKNESS

We used three tool profiles: flat with corner radius, elliptical, and spherical. For all the profiles, the tool diameter was constant (10 mm). The corner radius provided to the flat tool was 1 mm. Figure 13.4 shows the comparison of sheet thicknesses for all the tool profiles we used in the experiments. These graphs clearly show that minimum thickness depends more on the tool path than the profile of the tool, although tool profile is an important parameter for the uniformity of the sheet thickness distribution.

We also observed that the elliptical tool was responsible for material accumulation at the middle portion. The hemispherical and elliptical tools bent the sheet as it progressed toward the final geometry, but the flat tool stretched the sheet, which led to better thickness. The direction of the tool path is from the periphery to the center of the contour; hence, at the end of each contour the material is accumulated at the center of the frustum.

13.3.3   THE EFFECTS OF TOOL STEP-OVER (%) ON THE SHEET THICKNESS

Three types of percentage tool step over were used in this project i.e. 40%, 50% and 60% as shown in Figure 13.4. Percentage step over indicate that the overlap of the tool in the subsequent contour in the CAM programming. it is denoted by the percentage of the tool diameter. Figure 13.5 clearly shows that sheet thickness

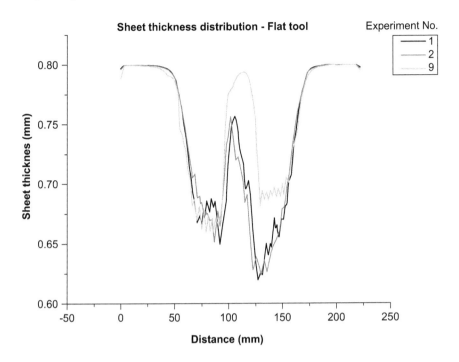

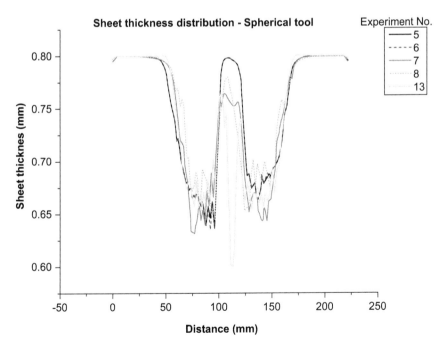

FIGURE 13.4   Sheet thickness comparisons for (a) flat, (b) spherical, and (3) elliptical profile tools.

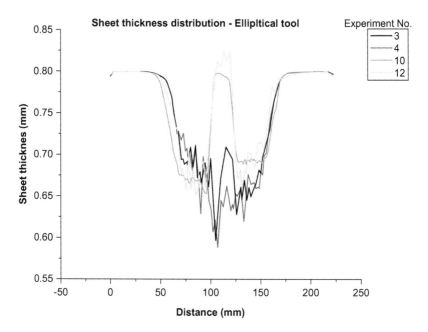

FIGURE 13.4   (Continued)

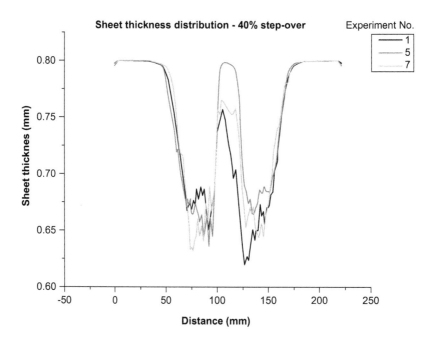

FIGURE 13.5   Sheet thickness comparison for (a) 40%, (b) 50%, and (c) 60% tool step-over.

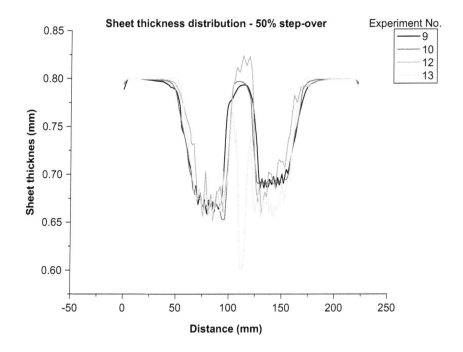

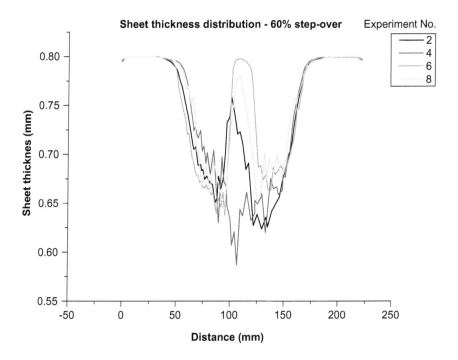

FIGURE 13.5 (Continued)

uniformity depends upon tool step-over, although it does not significantly affect the minimum sheet thickness.

We obtained the findings in Table 13.3 by response surface methodology, and the table shows the effects of individual variable parameters on minimum sheet thickness. Figure 13.6 clearly shows that the tool path is the dominant factor over tool profile and

TABLE 13.3

Effects of Individual Variable Parameters (Step-Over, Tool Profile, and Tool Path) on Minimum Sheet Thickness

Source	Sum of squares	df	Mean square	F	p	
Model	0.0091	9	0.0010	14.24	0.0010	Significant
A-Step over	0.0000	1	0.0000	0.3594	0.5677	
B-Tool profile	0.0006	1	0.0006	8.59	0.0220	
C-Tool path	5.513E−07	1	5.513E−07	0.0078	0.9323	
AB	0.0000	1	0.0000	0.5232	0.4929	
AC	0.0001	1	0.0001	0.8445	0.3887	
BC	7.728E−06	1	7.728E−06	0.1087	0.7513	
A^2	0.0001	1	0.0001	1.74	0.2284	
B^2	0.0005	1	0.0005	6.34	0.0400	
C^2	0.0077	1	0.0077	107.99	<0.0001	
Residual	0.0005	7	0.0001			
Lack of Fit	0.0005	3	0.0002			
Pure Error	0.0000	4	0.0000			
Cor Total	0.0096	16				

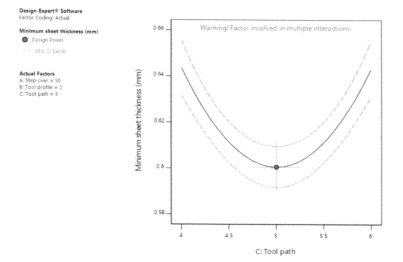

FIGURE 13.6   Effects of tool paths on minimum sheet thickness.

tool step-over; the trochoidal tool path performs more poorly than the other two paths. The trochoidal path initiates from the center of the geometry and pushes the material by elastoplastic deformation toward the wall of the cup. Although the thickness of the cup walls increases, thinning is found at the center of the cup.

The model F value of 14.24 reflects that the model is significant; there is only a 0.10% chance that an F value this large could occur due to noise. P values less than 0.05 indicate that model terms are significant. In this case, B (Tool profile), B^2, and C (Tool path)^2 are significant model terms. Values greater than 0.1 indicate that the model terms are not significant. The tool shape is the most influential parameter in the present investigation followed by step-over and tool path. The R^2 of 0.8817 explains the variance in the sheet thinning due to the parameters. Adeq. Precision of 10.59 measures the signal-to-noise ratio and indicates high precision. A ratio greater than 4 is desirable. This model can be used to navigate the design space.

13.4 RESPONSE SURFACE 3D GRAPHS FOR MINIMUM SHEET THICKNESS

Figure 13.7 shows the effects of combining the variable parameters on minimum sheet thickness. The differences between predicted and actual values are similar for variables tool profile and tool step-over. However, in the case of tool path, especially for the trochoidal tool path, the difference is more than 0.15 mm, which shows the complexity of the trochoidal tool path. The effect of the tool

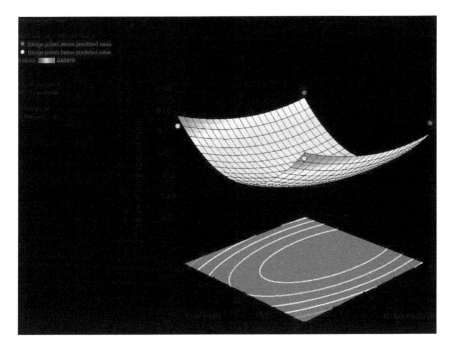

FIGURE 13.7   Response surface 3D graphs for tool paths and tool profile.

path is much more dominant than the effects of tool profile and tool step-over. For instance, the final thickness of the formed part is more sensitive to tool path than tool profile. The combination of Z-level tool path, a flat tool with a 1 mm corner radius, and 50% of step-over provides a thicker part. This finding supports selecting a flat tool for more formability. Moreover, the results depend on the sheet thickness. For the given tool profile, the thickness varies with the tool path in a parabolic manner.

13.5 CONCLUSION

The effects of nonconventional parameters, i.e., percentage step-over, tool profile, and nonconventional tool paths, were analyzed for sheet thickness and thinning. We performed ANOVA of the sheet thicknesses, and among all the parameters we analyzed, i.e., percentage stepover, tool profile, and nonconventional tool paths, the effect of the tool path was dominant, followed by the tool profile and step-over. We observed the maximum sheet thickness of 0.6592 with the constant Z-level tool path in experiment number 9 (50% step-over and flat tool with corner radius of 1 mm).

We observed the minimum thickness in experiment 4. The trochoidal tool path makes the material flow from the center of the frustum to the walls. This elastoplastic sheet deformation occurs because of the nature of the tool movement. We observed that the effect of percentage step-over was not significant for the thickness reduction. However, 50% overlap offers better thickness of the resultant part.

13.6 ACKNOWLEDGEMENT

The authors are thankful to the Government of India Science and Engineering Research Board for its support with project SRG/2020/001355.

REFERENCES

Barnwal, V. K. *et al.* (2018) 'Forming behavior and microstructural evolution during single point incremental forming process of AA-6061 aluminum alloy sheet', *The International Journal of Advanced Manufacturing Technology*, 95(1–4), pp. 921–935. doi: 10.1007/s00170-017-1238-5.

Carette, Y., Vanhove, H. and Duflou, J. (2019) 'Multi-step incremental forming using local feature based toolpaths', *Procedia Manufacturing*, 29, pp. 28–35. doi: 10.1016/j.promfg.2019.02.102.

Dejardin, S. *et al.* (2010) 'Experimental investigations and numerical analysis for improving knowledge of incremental sheet forming process for sheet metal parts', *Journal of Materials Processing Technology*, 210(2), pp. 363–369. doi: 10.1016/j.jmatprotec.2009.09.025.

Durante, M., Formisano, A. and Lambiase, F. (2018) 'Incremental forming of polycarbonate sheets', *Journal of Materials Processing Technology*, 253, pp. 57–63. doi: 10.1016/j.jmatprotec.2017.11.005.

Fiorentino, A., Giardini, C. and Ceretti, E. (2015) 'Application of artificial cognitive system to incremental sheet forming machine tools for part precision improvement', *Precision Engineering*, 39, pp. 167–172. doi: 10.1016/j.precisioneng.2014.08.005.

Fiorentino, A. *et al.* (2015) 'Part precision improvement in incremental sheet forming of not axisymmetric parts using an artificial cognitive system', *Journal of Manufacturing Systems*, 35, pp. 215–222. doi: 10.1016/j.jmsy.2015.02.003.

Gipiela, M. L. *et al.* (2017) 'A numerical analysis on forming limits during spiral and concentric single point incremental forming', *IOP Conference Series: Materials Science and Engineering*, 164, p. 012009. doi: 10.1088/1757-899X/164/1/012009.

Khazaali, H. and Fereshteh-Saniee, F. (2019) 'An inclusive experimental investigation on influences of different process parameters in warm incremental forming of AZ31 magnesium sheets', *Iranian Journal of Science and Technology, Transactions of Mechanical Engineering*, 43(2), pp. 347–358. doi: 10.1007/s40997-017-0122-0.

Leem, D. *et al.* (2022) 'A toolpath strategy for double-sided incremental forming of corrugated structures', *Journal of Materials Processing Technology*, 308, p. 117727. doi: 10.1016/j.jmatprotec.2022.117727.

Lu, H. *et al.* (2017) 'Part accuracy improvement in two point incremental forming with a partial die using a model predictive control algorithm', *Precision Engineering*, 49, pp. 179–188. doi: 10.1016/j.precisioneng.2017.02.006.

Nirala, H. K. and Agrawal, A. (2020) 'Residual stress inclusion in the incrementally formed geometry using Fractal Geometry Based Incremental Toolpath (FGBIT)', *Journal of Materials Processing Technology*, 279, p. 116575. doi: 10.1016/j.jmatprotec.2019.116575.

Nirala, H. K. and Agrawal, A. (2022) 'Adaptive increment based uniform sheet stretching in Incremental Sheet Forming (ISF) for curvilinear profiles', *Journal of Materials Processing Technology*, 306, p. 117610. doi: 10.1016/j.jmatprotec.2022.117610.

Ren, H. *et al.* (2019) 'In-situ springback compensation in incremental sheet forming', *CIRP Annals*, 68(1), pp. 317–320. doi: 10.1016/j.cirp.2019.04.042.

Tera, M. *et al.* (2019) 'Processing strategies for single point incremental forming—A CAM approach', *The International Journal of Advanced Manufacturing Technology*, 102(5–8), pp. 1761–1777. doi: 10.1007/s00170-018-03275-9.

Ullah, S. *et al.* (2023) 'A toolpath strategy for improving geometric accuracy in double-sided incremental sheet forming', *Chinese Journal of Aeronautics*, 36(1), pp. 468–479. doi: 10.1016/j.cja.2021.12.002.

14 Hydroforming
State-of-the-Art Developments and Future Trends

Sonanki Keshri and Sudha S.

14.1 INTRODUCTION TO HYDROFORMING

In hydroforming, fluid pressure is used in place of (or in addition to) conventional mechanical forces to form metal into complex shapes. Since the hydroforming method offers various benefits over conventional forming processes, it has become widely used for a variety of niche purposes. Several advantages can be observed, such as reduced thinning, enhanced mechanical properties, a more refined surface, fewer assembly components, and reduced rework due to the development of geometries that closely resemble the final shape. These benefits result from the working fluid capacity to provide uniform pressure throughout the workpiece whole surface and from the equipment capacity to adjust the working fluid pressure as needed during the forming cycle to achieve the best possible load path optimization.

Compared with stamping, forging, or casting, this method allows for producing more complicated structures that are both stronger and cheaper to produce. Because there is typically just one fabrication step involved in hydroforming, cost is reduced significantly: A sheet blank can be shaped into the final complex shape in a single step, whereas in stamping, multiple steps like blanking, drawing, restriking, trimming, and welding are required to complete a part. Hole piercing and trimming are two common stamping postprocessing operations.

Hydroforming has found extensive application across several industrial sectors, particularly in the automotive and aerospace domains, for the fabrication of components that would provide significant challenges, if not insurmountable obstacles, when employing alternative forming techniques [1–6].

The presence of fluid pressure facilitates a more uniform expansion of materials and enables the generation of pressure in orientations that differ from the direction of formation. This capacity allows for constructing supplementary or more pronounced features, consequently enhancing the overall quality and adaptability of the final product. By bridging the gap between standard cold

 DOI: 10.1201/9781003441755-14

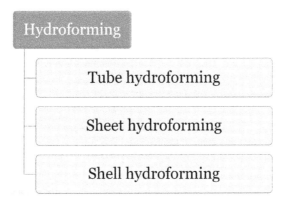

FIGURE 14.1 Categories of hydroforming.

forming and more advanced, unconventional sheet metal forming techniques like superplastic forming and creep forming, hydroforming helps manufacturers meet demand [6, 7]. Limit draw ratios for hydroformed materials can reach as high as 3.2, up from 2 for deep drawing [8]. Hydroforming can be broken down into three separate types, tube, sheet, and shell, as shown in Figure 14.1.

The geometry of the blank being worked on determines which type of hydroforming is applied. Different equipment and unique process variables characterize each of these three types of operations. Weight and cost savings can be achieved with hydroforming technology by consolidating parts, decreasing the need for post-forming activities like welding and piercing, although hydroforming process cycles are long. The cycle time, however, has been brought down to competitive levels with ongoing improvements to the hydraulic system and press designs. An improved metal forming technique known as Hydrogate™ was developed by Borit NV, a business headquartered in Belgium. The procedure incorporates into hydroforming an automated material feeding system from rolled sheets. This implementation significantly diminishes the cycle time, leading to a notable enhancement in output [9].

The ability to hydroform lightweight structures and parts utilizing lightweight materials is another fundamental premise of the hydroforming process. To fully achieve sustainable mobility and reduce energy consumption and negative impacts associated with transportation, the usage of lightweight structures is widely recognized as a prominent and long-term option [10]. In a car's whole life cycle, 80% of the energy it uses comes from when it's being driven. Lightweight construction improves automobiles' fuel efficiency and decreases pollutants, along with using alternative fuels and efficient power generation systems.

Lightweight structures are possible through the use of lightweight materials like aluminium, magnesium, high-strength steel, titanium, metal-matrix composites, and polymer composites and the development of low-cost, robust conversion processes that enable the effective use of these materials (i.e., novel manufacturing processes). According to existing literature, a decrease in vehicle weight by 10% improves fuel efficiency by 6% to 8% [11]. Hydroforming is favoured over stamping as a method of forming lightweight materials into complicated geometries because of its lower failure rate and greater potential for aesthetic improvement.

The most effective materials in lightweight structures are aluminium, magnesium, and high-strength steel, but these materials have very limited formability and are quite sensitive to production speeds. To increase the formability, producers can now use hydroforming to shape these materials by accurately controlling the fluid pressure inside the die. Adding selective heating to hydroforming (warm) can increase forming limits by an additional 100 to 300% [12]. Three different types of hydroforming are shown in Figure 14.2(a), 14.2(b), and 14.2(c) along with their way of processing.

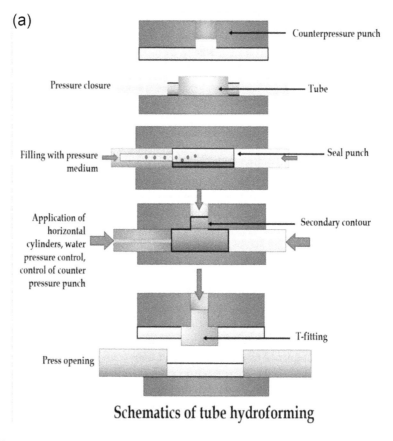

Schematics of tube hydroforming

FIGURE 14.2   Schematics of (a) tube, (b) sheet, and (c) shell hydroforming.

(b)

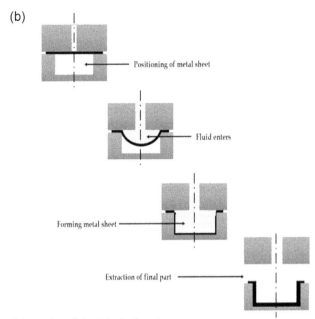

Positioning of metal sheet

Fluid enters

Forming metal sheet

Extraction of final part

Schematics of sheet hydroforming

(c)

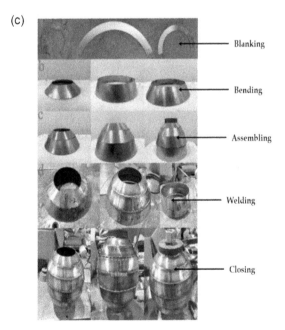

Blanking

Bending

Assembling

Welding

Closing

Schematics of shell hydroforming

FIGURE 14.2   (Continued)

In tube hydroforming (THF), to generate a shape of greater intricacy, a fluid substance is introduced into a metallic tube with a hollow interior and the tube is inflated. Rams are frequently employed in THF to facilitate the axial introduction of fresh material by exerting pressure on the tube edges. Axial feeding raises the fluid pressure in the tube and feeds new material to prevent the existing material from becoming too thin.

Sheet hydroforming (SHF) begins with a blank metal sheet to which fluid pressure is applied, often in conjunction with mechanical pressure, and the pressure moulds the sheet into the desired shape. SHF can use either a male or a female die, more frequently referred to as a punch or a cavity die, respectively. Both of these dies are commonly called punches. The procedure exhibits similarities to deep drawing, stamping (pressing), and rubber pad manufacturing.

Shell hydroforming or integrated hydro-bulge forming starts with flat pieces of sheet metal that are bent into place. The sheets are subsequently welded together to make an empty shape. The outside has a tip that is welded on. The inside is filled with fluid, which causes the form to bulge outward. Shell hydroforming doesn't use a die, so the shell grows in free space instead of against a die.

Koç presented a comprehensive survey on tube hydroforming, encompassing materials and applications, among other facets. The author also developed process design principles for THF [14]. The historical background and fundamental principles of sheet and tube hydroforming, dating back to the nineteenth century, were succinctly documented in a publication authored by Harjinder Singh [15] in 2003. Hydroforming of sheets, tubes, and shells were the subject of a 2004 review article by Lang et al. [16].

In 2005, Wang et al. [17] presented a review of the state of the art in shell hydroforming, discussing its history, current uses, and details on finite element simulations. The hydroforming process for tubes and sheets, together with the equations that control this process, was extensively examined in a 2008 edited volume authored by Koç et al. [18]. This publication offered a comprehensive analysis of the method, emphasized its present uses, and identified several study areas, such as heated hydroforming. Lee et al. [19] researched and reviewed the literature on tube and sheet hydroforming, including the most up-to-date technical studies and simulations. Similarly, Alaswad et al. [20] summed together the most recent experiments and literature on tube hydroforming in a review article published in 2012. Many authors have reviewed the advantages and disadvantages associated with hydroforming [21–28].

The goals of this chapter are to survey the current state of hydroforming and report new methods. Emphasis is laid on organizing the reported systems into a single taxonomy that makes the generic parallels and distinctions between the procedures. Development in the hydroforming process based on technologies, current status in the manufacturing sector, and the potential future applications are also discussed.

14.2   THE FUTURE OF HYDROFORMING

To predict where hydroforming will be used in the future, we must examine its past and present applications, as well as the market niches it now serves, active research areas, and the development of competing manufacturing techniques. This section will provide a high-level overview of the current market niches, their predicted evolution, and potential future applications of hydroforming techniques for tubes, sheets, and shells.

Hydroforming can replace superplastic and creep forming and allows for fabricating intricate geometries with deeper drawing capabilities. SHF has shown notable benefits compared with alternative manufacturing techniques, as evidenced by the growing attention to hot gas hydroforming in the production of automobile components [29] and of high-value components in general; the process is expensive and time-consuming, so it must be economically justified before being used to produce lower-value components. One competitor to THF was reported to be body in white, a protype of A-pillar reinforced by bending and hydroforming a conical tube, which can replace 10 sheet metal and bent tubes required for the actual product formation. This multifaceted process uses austenitic steel and was reported by EU venture. The benefits of employing higher component quality, tighter tolerances, increased part consolidation, decreased weight through more effective section design, enhanced structural rigidity and strength, and improved surface quality are evident. The advantages of this method include reduced secondary operations, such as the absence of welding or hole-making, as well as lower tooling costs due to fewer components and less waste [29].

From a hydroforming standpoint, the good news is that there are probably more opportunities for optimization due to the fact that hydroforming equipment has not been as heavily automated as other types of equipment. Hydroforming cycle times were cut in half for one firm thanks to the incorporation of automation techniques, including the use of sheet metal feeding from coiled sheets. In addition, Trzepieciński et al. created two parts in a single forming process, reducing the effective cycle time to just 4 seconds [30, 31]. Press makers like AIDA [32] and Schuler [33] are among those looking into more efficient ways to automate the process of pressing sheets of metal using transfer presses in the same physical footprint.

Automation improves productivity in both conventional and hydroforming processes, but due to longer and more prohibitive cycle times in hydroforming, the benefit of automation is more pronounced. Sales of tube hydroforming presses have levelled off during the past decade, and the predicted growth in demand for hydroforming machinery has not materialized [34]. However, THF shows promise in a number of contexts, most notably the production of high-value and lightweight components.

To save weight, eccentric components could be utilized more frequently in crankshafts, camshafts, and other rotating parts, as shown in a simulation by Zhang et al. [35], who used moveable-die tooling, a process that was suggested long ago as a possible area of research [13] and patented in 1959 [36]. Hydroforming has the potential to replace incremental manufacturing processes in the production of bleed valve ducts, fairings, and complex exhaust systems for gas-turbine engines. It is currently being investigated for an expanding variety of components (as stated in Table 14.1).

TABLE 14.1
Research Areas in the Field of Hydroforming

Technology	Description & technological maturity	References
Hybrid injection moulding & hydroforming	Allows for manufacturing complicated shapes using THF followed by injection moulding plastic around the resulting part while it is still in the die.	[37, 38]
Counterpunch	Having a counterpunch support the material while it is moving increases friction, preventing premature fracture and enhancing formability. This technology is already in place in THF applications.	[39]
Warm hydroforming	Hot gas, warmed fluid, steam, or conductive heat increases material formability. Many researchers have studied and published papers on warm hydroforming.	[10, 40–45]
Impulsive tube hydroforming	It has many benefits. Impulsive tube hydroforming is typically selected for applications in automotive suspension cradles, A-pillars, body structure components, exhaust components, rotating engine components, and bumpers. Components with varied cross sections can be produced using impulsive tube hydroforming without the requirement for additional procedures like welding. As a result, fewer parts are needed to make the finished product.	[46, 47]
Viscous pressure hydroforming	Using a high-viscosity fluid in hydroforming significantly enhances formability. Many authors have studied potential applications of the process.	[48–51]
Electro-hydraulic forming	Using an electrical impulse in the fluid, a large pressure spike is generated as fluid is vaporized, hydroforming metal with no moving parts and therefore without requiring a press.	[52, 53]
Impulsive sheet hydroforming	A pressure wave in the working media is commonly initiated by a quick discharge through an underwater spark gap, with or without an ignition wire.	[54, 55]
Multistage hydroforming	Work is ongoing on this well-known phenomenon in sheet and tube hydroforming.	[52, 56–58]

Technology	Description & technological maturity	References
Micro hydroforming	Research is ongoing in this area, and although there are multiple challenges, demand is increasing in the telecom, electronics, and medical device sectors.	[59]
Smart fluid hydroforming	Examines the use of smart hydroforming fluids instead of water and oil. Viscous fluids better support materials, and the local viscosity of a magnetorheological fluid can be controlled using a magnetic field.	[60–64]
Direct current on micro hydroformed tubes	Using electrical current during THF, pressures and axial forces required to form a component can be reduced as localized heat is applied to heat the work piece to below recrystallization.	[65]
Laser induced micro hydroforming	A pulsed laser is used to deform micro materials by inducing pressure changes in the adjacent fluid in both experimental and simulated settings, which show the process to be feasible.	[66]
Ultrasonically assisted hydroforming	By applying ultrasonic waves to the die during a THF simulation in MATLAB®, more corner filling is possible due to momentary surface conditions that occur during the process.	[67]

14.3   TUBE HYDROFORMING

THF dates to 1900, with a 1939 report by Gray [68] presenting its use for copper T fittings with internal pressure and axial feed; there were later publications on processing plumbing fittings, simple airplane blades, bicycle frame joints, and various other fittings. The commercial use of the technique began after 1986 with the making of large frame members, exhaust parts, instrument panels, support parts for automobiles like radiators, etc. During this time period, both theoretical and experimental investigations were performed to study stability, shapes, materials, performance, and shortcomings related to hydroforming in general and specifically about THF. Some automobile applications of tube hydroforming are displayed in Figure 14.3.

14.3.1   PROCESS DESCRIPTION

THF is a single-step solution for making product shapes which are otherwise formed by complex processes; the need for fewer steps reduces the cost, time, and tolerance control. THF consists of presses, tooling, a pressure system, and sealing methods. The process uses the presses, which are also referred to as clamping mechanisms, to open and close the die as well as to exert sufficient force during the forming stage to prevent elastic deflections and separation of the die. THF also uses tooling die holders, dies, and inserts [69]. Nevertheless, it is imperative

FIGURE 14.3   Tube hydroforming in automobile applications.

for THF tooling to adhere to a set of specified criteria [70], such as the ability to withstand the strains induced by elevated internal pressure and axial loading. THF systems must also incorporate interchangeable inserts to facilitate adaptability, efficient guiding systems, and a well-balanced design with minimal friction to increase formability and achieve superior surface finishes.

The failure modes that can happen during THF are buckling, bursting, wrinkling, and folding back. Buckling happens because of the use of longer/excessive tubes and shorter guidance in the die, and folding back commonly occurs because the die and tube parameters were not conditioned. Bursting is triggered because of constriction on the tube wall, which is usually controlled by the intermediate process of bulge, and wrinkling can be addressed by canalizing the internal pressure during the final phase of the process [71–73].

14.3.2 Materials Used in THF

Material selection in THF has a pivotal role; the factors that determine the material selection are the final product material, the production process, and the cost.

The quality of the material used for THF is determined by the material type/composition, tensile strength, flow characteristics, elongation percent, dimensions, and weld type. Selecting appropriate material is important for optimizing the process, and materials are tested with a biaxial test, which gives information about stress–strain relationships and material parameters. For instance, the authors of [74] report that slight increases in the tensile strength of the material increase formability, strain, and tube wall thickness and reduces corner thinning when the tube is converted to a square die.

Another parameter that affects thinning and expansion is longitudinal anisotropy, a function of internal pressure. Anisotropy (r) and the strain-hardening coefficient (n) have a huge bearing on the shape of the expanded tube. Researchers found that products with smaller die corners [75], efficient reproducibility requires low r and low axial displacement. Other properties to be evaluated before the process are elongation, with the cone test or expansion test, and hydraulic bulge. In the hydraulic bulge test, the tube is locked and stretched using hydraulic internal pressure which is close to the actual THF process and is a biaxial condition [19].

14.3.3 APPLICATIONS

THF is widely applied in the automobile and aircraft industries as well as in health care equipment and household appliances. In the automotive sector, THF is implemented to make camshafts, radiator frames, exhaust parts, rear axles, engine cradles, seat frames, and so on. Some well-known applications are the dashboard cowl of the Porsche Panamera, the rear axle of the BMW 500 series, Mercedes-Benz exhaust manifolds, and roof rails on the Jaguar XJ and Audi R8. Other parts include pillars, frames, front and rear axles, undercarriages cradles, bathroom faucet spouts, steel panic bars, and instrument panel beam. THF is so common in these applications because of the low product weights and costs [27, 29].

14.4 SHEET METAL HYDROFORMING

SHF involves placing a sheet workpiece into a die cavity and clamping it there (Figure 14.4 [76]) and using pressurized fluid to deform the sheet into the necessary shape without the need of a punch, as shown in Figure 14.4(a). As indicated in Figure 14.4(b), the pressurized fluid can also be employed to push the workpiece onto the punch as the punch goes downward.

SHF outperforms more conventional stamping processes in a variety of aspects, including a higher drawing ratio, a larger surface polish, superior dimensional stability, decreased tooling cost, and less tool wear [77, 78]. The superior formability of SHF is illustrated with circular cups [79, 80]. SHF, unlike THF, is typically employed in productions of modest scale. Its widespread use in industry has been hampered by its increased clamping force and longer cycle times.

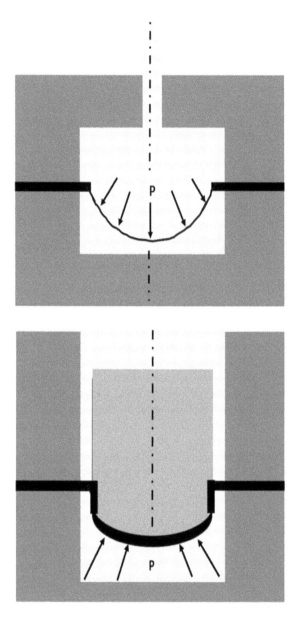

FIGURE 14.4.   (a) Water punch hydroforming; (b) water die hydroforming.

14.4.1   NEW APPROACHES TO SHF

By merging SHF with quick tooling, Palumbo [81] created a revolutionary method for producing an aluminium automotive component flexibly and affordably. The process consisted of assembling laser-cut layers in two dimensions for form a die with multiple layers. The proposed approach expedited the production of proto-types and streamlined the product design workflow.

To streamline SHF, Djavanroodi et al. [82] recommended employing a float-ing disk. With a floating disk, the normal blank holding force acts equally on all sides of the blank. This means that the punch force will be around half of what it would be if the regular blank holding force and chamber pressure were used. The proposed floating disk approach reduced the complexity of hydroforming equip-ment, which in turn can lower the price.

14.4.2 APPLICATIONS

Sheet hydroforming is used for a variety of vehicle components, such as door panels [77], roofs [87], and structural frames [88]. However, this technique is rarely used outside of niche markets due to its higher cost compared with con-ventional pressing methods. SHF is attractive because of its enhanced formability and superior surface smoothness; pieces made using sheet hydroforming are very similar in geometry to the final, prescribed geometry and therefore require very little finishing work. SHF can also be used to create semi-finished geometries, so hydroforming can be applied in a complex assembly while mitigating the higher costs associated with the longer cycle times.

14.4.2.1 Warm SHF

When it comes to high strain without fracture, warm SHF is a promising method for low-formability aluminium alloys. Koç et al. [83] investigated the optimal parameters, including temperature, pressure, and pressurization rate, for warm superplastic forming for the highest formability for AA5754-O. The optimal pro-cess conditions for simple stretch forming were found to be 268 °C, 25 MPa, and 0.22 MPa/s. These circumstances offered the best harmony between uniform thickness strain, cavity fill ratio, and radius sharpness.

14.4.2.2 SHF for Thin Sheets

The use of SHF in the production of thin sheets for fuel cell bipolar plates has received increased focus as of late. In fuel cells, the bipolar plates transport fuel and air while also conducting electricity and supporting the stack mechanically [84]. Because of their intricate designs, bipolar plates have traditionally been machined, but producing them using sheet forming can increase the productivity and efficiency. However, the typical stamping procedures' lack of malleability has prevented widespread use of sheet forming.

Mahabunphachai et al. [85] examined the formability and surface roughness of metallic bipolar plates produced using stamping and SHF (superplastic form-ing and hydroforming) and discovered that the dimensions of the bipolar plates made using hydroforming were more consistent than those made via stamping. In addition, the surface quality of hydroformed components was superior to that of stamped components. Hung and Lin [86] used hydroforming to create microflow channels in metallic bipolar plates. Using a custom-built HPH apparatus, they were able to increase the hydrostatic pressure to 0.468 at an aspect ratio of 0.468. Fuel cells' overall performance was enhanced by the channels' high aspect ratio.

14.5 COMPUTER SIMULATIONS IN HYDROFORMING

Manufacturing to numeric standards relies heavily on simulation to create new models. Simulation facilitates integrating production needs in the initial stages of production, and in conjunction with concurrent engineering has been empirically demonstrated to establish seamless connections between the design and manufacturing stages. Researchers such as Saran and Wagoner, Frey and Wenner, and M.P. Sklad created multiple finite element models for investigating sheet metal forming, specifically focusing on planar strain membrane conditions. The reason for this is that sheet metals have a higher propensity to fracture when subjected to plane strain [89–91]. These programs can accurately estimate deformation strains when shape tooling curvatures are absent but fail when tooling curvature is large enough to induce significant bending stresses [92].

Numeric simulation has quickly become one of the key techniques for solving engineering problems in research and industry applications, alongside theoretical and experimental studies. Several techniques commonly used in engineering and scientific disciplines include the finite difference method (FDM), the finite element method (FEM), the meshfree (or meshless) method, the slip line field method, and the upper bound method [93, 94]. Some, like finite element analysis (FEA), are much more widely employed. Table 14.2 lists some auto manufacturers that are currently implementing simulation in their production processes.

In this study, we first outline the anisotropy and yield criteria upon which numeric simulation approaches are based, and then we present recent advances in simulating sheet metal formation using several numeric methods. Because of its importance in FEA, sheet metal forming element selection is simulated. Researchers have reported on state-of-the-art sheet metal forming simulations in multiple phases. This chapter is meant to be an all-encompassing survey of the extant literature, and Table 14.3 presents a comprehensive overview of the literature, organized according to the specific issue of interest.

TABLE 14.2

Automakers Incorporating Simulation in Their Manufacturing Processes

European companies	Japanese companies	US companies
Daimler Benz AG	Mazda	Chrysler Corporation
Renault Automobile	Nissan Motor	Ford Motor Company
Volvo Car Corporation	Nippon Steel	National Steel
Sollac		US Steel

TABLE 14.3
The Extant Literature on Numeric Sheet Metal Forming

Area of interest	References
Review	Kaftanoglu et al. [95], Makinouchi [96], Makinouchi et al. [97], Tekkaya [98], Tisza [99], Wenner, [100] Ahmed et al. [101] Lee et al. [102], Banabic [103], Reddy et al. [104]
FMD	Woo [105], Doege et al. [106], TSENG [107]
FEM	Wifi [108], Gotoh et al. [109], N.-M. Wang et al. [110], Tang et al. [111], Toh et al. [112], Benson et al. [113], Belytschko et al. [114], Massoni et al. [115], N-M Wang et al. [116]
Formulation & Solution Type	Oniate et al. [117], Yang et al. [118], Nakamachi [119], Mamalis et al. [120], Jung [121], Carleer et al. [122], Finn et al. [123], Micari et al. [124], Tang et al. [125], Azizi [126]
Element Selection	Chung, et al. [127], Xu et al. [128], Papeleux et al. [129], Parente et al. [131], Alves de Sousa et al. [130], Lee et al. [132], Menezes et al. [133]
Meshfree method	Cueto et al. [134], Yoon et al. [135], Yoon et al. [136], Botkin et al. [137], Sidibe et al. [138], Liu et al. [139], Liu et al. [140]
Anisotropic yield criteria	Hill [141–143], Vial et al. [144], Barlat et al. [145], Barlat et al. [146], F. Barlat et al., [147, 148], Karafillis et al. [149], Gotoh [150], Budiansky [151], F. Barlat et al. [152], Banabic et al. [153], Cazacu et al. [154], Carleer et al. [155], Vegter et al. [156], Hu [157], Comsa et al. [158], Soare [159], Banabic et al. [160], Comsa et al. [161]
Formability	Emmens [162], Goodwin [163], Keeler [164], Havranek [165], Arrieux [166], Simha [167], Janssens et al. [168], Blek et al. [169], Hu et al. [170], Zimniak [171], Berstad et al. [172], Samuel [173], Duan et al. [174], Takuda et al. [175], Hajian et al. [176]
Springback	Baseri et al. [177], Tekiner [178], Moon et al. [179], Gomes et al. [180], Li et al. [181], Nakamachi et al. [182], Baseri et al. [183], Jamli et al. [184], Ank et al. [185]
Laser Forming	Geiger et al. [186], Shen et al. [187], Shichun et al. [188], Ji et al. [189], Shichun et al. [190], Hoseinpour et al. [191], Kheloufi et al. [192], Venkadeshwaran et al. [193], Shi et al. [194, 195], Yu et al. [196], Zohoor et al. [197], Gollo et al. [198], Pitz et al. [199], Che et al. [200], Lambiase [201], Paramasivan et al. [202], Guan et al. [203]
Incremental Sheet Forming	Leszak [204], Jeswiet et al. [205], Emmens et al. [206], Kumar et al. [207], Iseki [208], Shim et al. [209], Hirtl et al. [210], Yamashita et al. [211], Lequesne [212], Hadoush et al. [213], Ben et al. [214], Zhang et al. [215]

14.5.1 FINITE DIFFERENCE METHOD

The capacity to perform forming operations was initially limited to two-dimensional symmetric scenarios in early simulations of sheet metal. In the 1960s, the deep drawing of a 2D cylindrical cup was simulated using FDM

[97], marking the beginning of sheet metal forming simulation. After nonlinear FEM's correctness was established, it replaced FDM. In the 1990s, researchers attempted to apply the finite difference approach to 3D problems but failed due to the complexity of the problem's boundary conditions. In the case of sheet metal forming processes, however, FDM has been utilized to model the heat effect.

14.5.2 FINITE ELEMENT SIMULATION

The industrial application of numeric modelling in metal forming has significantly progressed, enabling the simulation of metal deformation and the determination of stress and strain states in complex scenarios. Simulation of the design using FEA allows for early error prediction and change prior to physical fabrication and testing. The time and money spent on labour can be cut down as a result. Therefore, advanced numeric simulations will gradually replace manual trial-and-error design iteration in favour of finite element methods. FEA (particularly in the pre-processing stage) is becoming increasingly popular in the manufacturing industry, and it's easy to see why: It dramatically improves productivity and reduces resource consumption.

Sheet metal forming simulation relies heavily on FEA. The finite element analysis of stretch forming and deep drawing process using elastoplastic, axisymmetric circular blank sheet was first presented by Wifi [108] in the late 1970s. FEA for a flange in deep drawing was reported by Gotoh and group [109], who based their work on the rigid-plastic material law and performed their analysis with the quadratic yield function and the fourth-degree yield function.

In their presentation of a comprehensive finite element approach for sheet metal stamping, Wang and group [110] considered a thin sheet material such that membrane theory could be used in the analysis. They assumed a rate-independent, J2-type flow rule and an elastic-plastic material law for the substance. The authors demonstrated that the strain distribution at the moment of material unloading was the same whether the material was elastic or rigid plastic.

By simulating a deformed auto body panel in the 1980s, Tang et al. [111] extended the use of finite element simulation from 2D to 3D. A 3D simulation of a flat metal sheet's shape was also provided by Toh and group [112]. Both the static implicit and static explicit methods were employed in these simulations, with elastoplasticity serving as the material model. Subsequently, Benson and his research group [113] implemented deformation mechanics into the DYNA3D simulation program. Belytschko et al. [114], implemented a dynamic explicit technique in the simulation. Massoni [115] first proposed the idea of utilizing an artificial force to replace draw bead, while Wang and group [116] accounted for viscosity issues. The design and development of robust, efficient, and accurate solution methods and algorithms for sheet metal simulation have been a primary emphasis of the field ever since the late 1990s, when researchers began focusing on precisely forecasting springback.

14.5.3 MESHFREE METHOD

Since the year 2000, the meshfree (meshless) method has been employed with FEM to simulate sheet metal forming. By proposing stabilized conforming nodal integration (SCNI), Chinesta et al. [134] offered a meshfree formulation for metal-forming simulation that avoided the high CPU cost of the Galerkin-based meshfree methods previously used. They demonstrated the operation's flanging and springback effects with examples.

Cylindrical punch and springback in tossing are examples of how Chen and Yoon's [135] generalization of SCNI can be applied to situations that are heavily reliant on the past. In order to keep the benefits of both FEM and the meshfree approach, Guo and his group suggested a coupled FEM/meshfree simulation [137]. Sidibe et al. [138] employed the replicating kernel particle method (RKPM) to illustrate the bending behaviour of sheet metal, and Liu [139] demonstrated the efficacy of the meshfree method by using the RKPM in the context of deep drawing and hemisphere drawing.

Liu et al. [140] employed adaptive multiple scale meshless simulation to examine the phenomenon of springback, using radial basis point interpolation (RKPM) at two distinct scales. The RKPM shape function's high and low scales were deconstructed and subsequently incorporated into a nonlinear elastoplastic formulation. This integration resulted in the determination of the high and low components of effective strain. After locating regions experiencing significant strain, the appropriate node refinement scheme was applied so that stresses could be calculated and springback could be predicted with great precision. Additionally, the authors compared results from an experiment, FEM (Abaqus), meshless methods, and adaptive meshless methods and demonstrated that the results from the adaptive meshless approach were most similar to those obtained from the experiments.

14.6 CONCLUSION

Organizations are exploring all possibilities for the less expensive production of lighter and tougher products. One technique for enhancing output is the integration of lubrication into the manufacture of tubes, for instance. In the quest for enhanced material qualities, researchers are looking at the feasibility of employing various welding techniques, including gas metal arc welding, laser welding, and electron beam welding. In some appropriate circumstances, rather than using traditional tube rolling mills, tube producing (forming) cells are being considered.

Despite the many benefits of hydroforming, it is still difficult to form lightweight materials into complex-shaped parts because there are still many unanswered questions regarding, for example, the impact of material behaviour under hydrostatic loading, the impact of surface topography on overall formability, the need for efficient lubricants, and the low formability of light alloys like aluminium and magnesium at room temperature. Hybrids of hydroforming with other techniques, including electromagnetic force and electrohydraulic forming, are also an area of development in this field. Overall, hydroforming has been shown to be a successful technology that can replace traditional stamping and forging processes due to its cost savings, tight dimensional tolerances, and formability of complex parts.

REFERENCES

[1] Zhang, S. H. and Danckert, J. (1998) "Development of hydro-mechanical deep drawing," *Journal of Materials Processing Technology*, 83(1–3), pp. 14–25. doi: 10.1016/s0924-0136(98)00039-9.

[2] We, L. W. and Yu, Y. (1996) Computer simulation of forming automotive structural parts by hydroforming process. In Proceedings of the Third International Conference Numisheet'96, Dearborn, Michigan, 29 September–3 October, pp. 324–329.

[3] Horton, F. (1997) Using forming simulation in development of complex hydroformed shapes. In TPA's Second Annual Automotive Conference, 13–14 May, Dearborn, MI, pp. 173–189.

[4] Neugebauer, R., Kunke, E. and Sterzing, A. (2003) Hydroforming of double blanks—A chance to increase the application area. In Proceedings of the SheMet 2003, Jordanstown, UK, pp. 107–114.

[5] Parsons W. J. (2007) The New Opel GT: Car body concept in development and manufacture. In The 25th European Car Body Conference, Bad Nauheim.

[6] Zhang, S. H. (1999) "Developments in hydroforming," *Journal of Materials Processing Technology*, 91(1–3), pp. 236–244. doi: 10.1016/s0924-0136(98)00423-3.

[7] Hirt, G. and Bambach, M. (2012) "Incremental sheet forming," *Sheet Metal Forming*, pp. 273–287. doi: 10.31399/asm.tb.smfpa.t53500273.

[8] Jeswiet, J. *et al.* (2008) "Metal forming progress since 2000," *CIRP Journal of Manufacturing Science and Technology*, 1(1), pp. 2–17. doi: 10.1016/j.cirpj.2008.06.005.

[9] *Hydrogate^{TM} Forming | Borit* (no date). Available at: www.borit.be/capabilities/hydrogate-forming.

[10] Koç, M. (ed.). (2008) Hydroforming for Advanced Manufacturing. Woodhead Publishing Limited.

[11] *Lightweight Materials for Cars and Trucks* (no date) *Energy.gov*. Available at: www.energy.gov/eere/vehicles/lightweight-materials-cars-and-trucks.

[12] Mildenberger, U. and Khare, A. (2000) "Planning for an environment-friendly car," *Technovation*, 20(4), pp. 205–214. doi: 10.1016/s0166-4972(99)00111-x.

[13] Koç, M. and Altan, T. (2001) "An overall review of the Tube Hydroforming (THF) technology," *Journal of Materials Processing Technology*, 108(3), pp. 384–393. doi: 10.1016/s0924-0136(00)00830-x.

[14] Koç, M. (1999) Development of design guidelines for part, tooling and process in the tube hydroforming technology (doctoral dissertation), Columbus Ohio: The Ohio State University.

[15] Singh, H. (2003) Fundamentals of Hydroforming. Society of Manufacturing Engineers.

[16] Lang, L. H. *et al.* (2004) "Hydroforming highlights: Sheet hydroforming and tube hydroforming," *Journal of Materials Processing Technology*, 151(1–3), pp. 165–177. doi: 10.1016/j.jmatprotec.2004.04.032.

[17] Wang, Z. R. *et al.* (2005) "Progress in shell hydroforming," *Journal of Materials Processing Technology*, 167(2–3), pp. 230–236. doi: 10.1016/j.jmatprotec.2005.05.045.

[18] Koç, M. (2008) Hydroforming for Advanced Manufacturing. Elsevier.

[19] Lee, M. G., Korkolis, Y. P. and Kim, J. H. (2014) "Recent developments in hydroforming technology," *Proceedings of the Institution of Mechanical Engineers, Part B: Journal of Engineering Manufacture*, 229(4), pp. 572–596. doi: 10.1177/0954405414548463.

[20] Alaswad, A., Benyounis, K. Y. and Olabi, A. G. (2012) "Tube hydroforming process: A reference guide," *Materials & Design*, 33, pp. 328–339. doi: 10.1016/j.matdes.2011.07.052.

[21] Zhang, S. H. and Danckert, J. (1998) "Development of hydro-mechanical deep drawing," *Journal of Materials Processing Technology*, 83(1–3), pp. 14–25. doi: 10.1016/s0924-0136(98)00039-9.

[22] Waters, T. F. (2017) Fundamentals of Manufacturing for Engineers. CRC Press.

[23] Olsson, S. (2016) Recent hydroform and flexform development. In New Developments in Sheet Metal Forming—Hydroforming of Sheets, Tubes and Profiles, Fellbach.

[24] Singh, S. K. and Ravi Kumar, D. (2008) "Effect of process parameters on product surface finish and thickness variation in hydro-mechanical deep drawing," *Journal of Materials Processing Technology*, 204(1–3), pp. 169–178. doi: 10.1016/j.jmatprotec.2007.11.060.

[25] Bell, C., Corney, J., Savings, D. and Storr, J. (2015) Assessing the potential benefits of manufacturing gas turbine components by utilizing hydroforming technology. In 13th International Cold Forming Congress, 2–4 September, Glasgow.

[26] Frith, R. and Stone, M. (2015) "Weld efficiency factors revisited," *Procedia Engineering*, 130, pp. 434–445. doi: 10.1016/j.proeng.2015.12.237.

[27] Ahmetoglu, M. and Altan, T. (2000) "Tube hydroforming: State-of-the-art and future trends," *Journal of Materials Processing Technology*, 98(1), pp. 25–33. doi: 10.1016/s0924-0136(99)00302-7.

[28] Hwang, Y. M., Lin, T. C. and Chang, W. C. (2007) "Experiments on T-shape hydroforming with counter punch," *Journal of Materials Processing Technology*, 192–193, pp. 243–248. doi: 10.1016/j.jmatprotec.2007.04.087.

[29] Tolazzi, M. (2010) "Hydroforming applications in automotive: A review," *International Journal of Material Forming*, 3(S1), pp. 307–310. doi: 10.1007/s12289-010-0768-2.

[30] Trzepieciński, T. (2020) "Recent developments and trends in sheet metal forming," *Metals*, 10(6), p. 779. doi: 10.3390/met10060779.

[31] Bell, C. et al. (2019) "A state of the art review of hydroforming technology," *International Journal of Material Forming*, 13(5), pp. 789–828. doi: 10.1007/s12289-019-01507-1.

[32] Aida, K., Rothenhagen, K. and Papaioanu, A. (2016) New developments in servo press technology. In New developments in Sheet Metal Forming—Hydroforming of Sheets, Tubes and Profiles, Fellbach.

[33] Früh, J. (2016) Efficient: Networked: Innovative. In New Developments in Sheet Metal Forming—Hydroforming of Sheets, Tubes and Profiles, Fellbach.

[34] Freytag, P. (2014) The renaissance of hydroforming—European market development, challenges and technical responses from the perspective of a serial manufacturer. In Hydroforming of Sheets Tubes and Profiles, Fellbach.

[35] Zhang, Q., Wu, C. and Zhao, S. (2012) "Less loading tube-hydroforming technology on eccentric shaft part by using movable die," *Materials Transactions*, 53(5), pp. 820–825. doi: 10.2320/matertrans.mf201121.

[36] Milton, O. M. (1959) Method of making cam shafts. United States of America patent US2892254, 30 June.

[37] Ruez, P. and Knoll, S. (2016) Hydroforming hybrid—Affordable hybrid light weight design. In New Developments in Sheet Metal Forming—Hydroforming of Sheets, Tubes and Profiles, Fellbach.

[38] Landgrebe, D., Kräusel, V., Rautenstrauch, A., Awiszus, B. and Boll, J. M. L. (2016) Energy-efficiency and robustness in a hybrid process of hydroforming and polymer injection molding. In Conference on Sustainable Manufacturing, Stellenbosch.

[39] Parsa, M. H. and Darbandi, P. (2008) "Experimental and numerical analyses of sheet hydroforming process for production of an automobile body part," *Journal of Materials Processing Technology*, 198(1–3), pp. 381–390. doi: 10.1016/j.jmatprotec.2007.07.023.

[40] Yuan, S., Tang, Z. and Liu, G. (2013) Simulation and experiment on warm hydroforming of AZ31 magnesium alloy tube. In International Manufacturing Science and Engineering Conference Collocated with the 41st North American Manufacturing Research Conference ASME, Volume 1: Processing. https://doi.org/10.1115/msec2013-1084.

[41] Landgrebe, D. and Schieck, F. (2015) Hot gas forming for advanced tubular automobile components: Opportunities and challenges. In ASME 2015 International Manufacturing Science and Engineering Conference, Charlotte.

[42] Aissa, S., Mohamed, S. and Tarek, L. (2017) "Experimental study of steam hydroforming of aluminum sheet metal," *Experimental Techniques*, 41(5), pp. 525–533. doi: 10.1007/s40799-017-0191-4.

[43] Takata, K. *et al.* (2000) "Formability of Al-Mg alloy at warm temperature," *Materials Science Forum*, 331–337, pp. 631–636. doi: 10.4028/www.scientific.net/msf.331-337.631.

[44] Maeno, T., Mori, K. I. and Unou, C. (2014) Improvement of die filling by prevention of temperature drop in gas forming of Aluminium alloy tube using air filled into sealed tube and resistance heating. In 11th International Conference on Technology of Plasticity, Nagoya.

[45] Türköz, M. *et al.* (2015) "Design, fabrication, and experimental validation of a warm hydroforming test system," *Journal of Manufacturing Science and Engineering*, 138(4). doi: 10.1115/1.4031498.

[46] Mori, K., Maeno, T. and Maki, S. (2007) "Mechanism of improvement of formability in pulsating hydroforming of tubes," *International Journal of Machine Tools and Manufacture*, 47(6), pp. 978–984. doi: 10.1016/j.ijmachtools.2006.07.006.

[47] Loh-Mousavi, M. *et al.* (2008) "Improvement of formability in T-shape hydroforming of tubes by pulsating pressure," *Proceedings of the Institution of Mechanical Engineers, Part B: Journal of Engineering Manufacture*, 222(9), pp. 1139–1146. doi: 10.1243/09544054jem1143.

[48] Wang, Z. J. *et al.* (2004) "Viscous Pressure Forming (VPF): State-of-the-art and future trends," *Journal of Materials Processing Technology*, 151(1–3), pp. 80–87. doi: 10.1016/j.jmatprotec.2004.04.010.

[49] Shulkin, L. B. *et al.* (2000) "Blank Holder Force (BHF) control in Viscous Pressure Forming (VPF) of sheet metal," *Journal of Materials Processing Technology*, 98(1), pp. 7–16. doi: 10.1016/s0924-0136(99)00300-3.

[50] Wang, Z. J. and Liu, Y. (2010) "Investigation on deformation behavior of sheet metals in viscous pressure bulging based on ESPI," *Journal of Materials Processing Technology*, 210(11), pp. 1536–1544. doi: 10.1016/j.jmatprotec.2010.04.014.

[51] Gao, T. *et al.* (2015) "The influence of length—Diameter ratio in forming area on viscous outer pressure forming and limit diameter reduction," *Journal of the Brazilian Society of Mechanical Sciences and Engineering*, 39(2), pp. 481–486. doi: 10.1007/s40430-015-0340-6.

[52] Golovashchenko, S. F., Bessonov, N. M. and Ilinich, A. M. (2011) "Two-step method of forming complex shapes from sheet metal," *Journal of Materials Processing Technology*, 211(5), pp. 875–885. doi: 10.1016/j.jmatprotec.2010.01.004.

[53] Soltanpour, M., Fazli, A. and Niaraki, R. J. (2017) "High speed hydroforming and direct quenching: An alternative method for production of hot stamped parts with high productivity," *Procedia Engineering*, 207, pp. 317–322. doi: 10.1016/j.proeng.2017.10.781.

[54] Maeno, T., Mori, K. and Hori, A. (2014) "Application of load pulsation using servo press to plate forging of stainless-steel parts," *Journal of Materials Processing Technology*, 214(7), pp. 1379–1387. doi: 10.1016/j.jmatprotec.2014.01.018.

[55] Xu, Y. *et al.* (2014) "Application of pulsating hydroforming in manufacture of engine cradle of austenitic stainless steel," *Procedia Engineering*, 81, pp. 2205–2210. doi: 10.1016/j.proeng.2014.10.309.

[56] Liu, W., Xu, Y. and Yuan, S. (2014) "Effect of pre-bulging on wrinkling of curved surface part by hydromechanical deep drawing," *Procedia Engineering*, 81, pp. 914–920. doi: 10.1016/j.proeng.2014.10.117.

[57] Chen, Y. Z. *et al.* (2017) "Analysis of wrinkling during sheet hydroforming of curved surface shell considering reverse bulging effect," *International Journal of Mechanical Sciences*, 120, pp. 70–80. doi: 10.1016/j.ijmecsci.2016.10.023.

[58] Hwang, Y. M. and Chen, Y. C. (2017) "Study of compound hydroforming of profiled tubes," *Procedia Engineering*, 207, pp. 2328–2333. doi: 10.1016/j.proeng.2017.10.1003.

[59] Hashemi, A., Hoseinpour Gollo, M. and Seyedkashi, S. M. H. (2015) "Process window diagram of conical cups in hydrodynamic deep drawing assisted by radial pressure," *Transactions of Nonferrous Metals Society of China*, 25(9), pp. 3064–3071. doi: 10.1016/s1003-6326(15)63934-6.

[60] Rösel, S. and Merklein, M. (2013) "Improving formability due to an enhancement of sealing limits caused by using a smart fluid as active fluid medium for hydroforming," *Production Engineering*, 8(1–2), pp. 7–15. doi: 10.1007/s11740-013-0496-9.

[61] Wang, Z., Wang, P. and Song, H. (2014) "Research on sheet-metal flexible-die forming using a magnetorheological fluid," *Journal of Materials Processing Technology*, 214(11), pp. 2200–2211. doi: 10.1016/j.jmatprotec.2014.04.016.

[62] Rösel, S. and Merklein, M. (2011) "Flow behaviour of magnetorheological fluids, considering the challenge of sealing in blank hydroforming in the flange area with rectangular leakage area cross-sections," *Key Engineering Materials*, 473, pp. 121–129. doi: 10.4028/www.scientific.net/kem.473.121.

[63] Merklein, M. and Rösel, S. (2010) "Characterization of a magnetorheological fluid with respect to its suitability for hydroforming," *International Journal of Material Forming*, 3(S1), pp. 283–286. doi: 10.1007/s12289-010-0762-8.

[64] Wang, Z. *et al.* (2017) "Property-adjustable forming medium induced extension of sheet metal formability under variable magnetic field," *Journal of Materials Processing Technology*, 243, pp. 420–432. doi: 10.1016/j.jmatprotec.2017.01.002.

[65] Wagner, S. W. *et al.* (2016) "Influence of continuous direct current on the microtube hydroforming process," *Journal of Manufacturing Science and Engineering*, 139(3). doi: 10.1115/1.4034790.

[66] Je, G. *et al.* (2017) "A study on micro hydroforming using shock wave of 355 nm UV-pulsed laser," *Applied Surface Science*, 417, pp. 244–249. doi: 10.1016/j. apsusc.2017.02.146.

[67] Eftekhari Shahri, S., Ahmadi Boroughani, S. Y., Khalili, K. and Kang, B. (2014) Ultrasonic tube hydroforming, a new method to improve formability. In 8th International Conference Inter disciplinarity in Engineering, Tirgu-Mures Romani.

[68] Gray, J. E. (1939) *US2203868A—Apparatus for Making Wrought Metal t's — Google Patents.* Available at: https://patents.google.com/patent/US2203868.

[69] Leitloff, F. (1997). Hydroforming—From feasibility analysis to series production. In Proceedings of 2nd International Conference on Innovations in Hydroforming Technology, Columbus, OH, September 17, 1997.

[70] Ahmed, M. and Hashmi, M. S. J. (1997) "Estimation of machine parameters for hydraulic bulge forming of tubular components," *Journal of Materials Processing Technology*, 64(1–3), pp. 9–23. doi: 10.1016/s0924-0136(96)02549-6.

[71] Fuchizawa, S. (1984) "Influence of strain hardening exponent on the deformation of thin-walled tube of finite length subjected to hydrostatic external pressure," *Advanced Technology of Plasticity*, 1, pp. 297–302.

[72] Dohmann, F. and Hartl, C. (1997) "Tube hydroforming—Research and practical application," *Journal of Materials Processing Technology*, 71(1), pp. 174–186. doi: 10.1016/s0924-0136(97)00166-0.

[73] Dohmann, F. and Hartl, C. (1994) "Liquid-bulge-forming as a flexible production method," *Journal of Materials Processing Technology*, 45(1–4), pp. 377–382. doi: 10.1016/0924-0136(94)90369-7.

[74] Lianfa, Y. and Cheng, G. (2008) "Determination of stress—Strain relationship of tubular material with hydraulic bulge test," *Thin-Walled Structures*, 46(2), pp. 147–154. doi: 10.1016/j.tws.2007.08.017.

[75] Koç, M., Billur, E. and Cora, Ö. N. (2011) "An experimental study on the comparative assessment of hydraulic bulge test analysis methods," *Materials & Design*, 32(1), pp. 272–281. doi: 10.1016/j.matdes.2010.05.057.

[76] Shahri, S. E. E. *et al.* (2015) "Ultrasonic tube hydroforming, a new method to improve formability," *Procedia Technology*, 19, pp. 90–97. doi: 10.1016/j.protcy.2015.02.014.

[77] Tolazzi, M. (2010) "Hydroforming applications in automotive: A review," *International Journal of Material Forming*, 3(S1), pp. 307–310. doi: 10.1007/s12289-010-0768-2.

[78] Zhang, S. H. *et al.* (2004) "Recent developments in sheet hydroforming technology," *Journal of Materials Processing Technology*, 151(1–3), pp. 237–241. doi: 10.1016/j.jmatprotec.2004.04.054.

[79] Bakhshi-Jooybari, M., Gorji, A. and Elyasi, M. (2012) "Developments in sheet hydroforming for complex industrial parts," *Metal Forming—Process, Tools, Design.* doi: 10.5772/48142.

[80] Pourboghrat, F., Venkatesan, S. and Carsley, J. E. (2013) "LDR and hydroforming limit for deep drawing of AA5754 Aluminum sheet," *Journal of Manufacturing Processes*, 15(4), pp. 600–615. doi: 10.1016/j.jmapro.2013.04.003.

[81] Palumbo, G. (2013) "Hydroforming a small scale Aluminum automotive component using a layered die," *Materials & Design*, 44, pp. 365–373. doi: 10.1016/j.matdes.2012.08.013.

[82] Djavanroodi, F. and Derogar, A. (2010) "Experimental and numerical evaluation of forming limit diagram for Ti6Al4V Titanium and Al6061-T6 Aluminum alloys sheets," *Materials & Design*, 31(10), pp. 4866–4875. doi: 10.1016/j.matdes.2010.05.030.

[83] Koç, M., Agcayazi, A. and Carsley, J. (2011) "An experimental study on robustness and process capability of the warm hydroforming process," *Journal of Manufacturing Science and Engineering*, 133(2). doi: 10.1115/1.4003619.

[84] Liu, Y. and Hua, L. (2010) "Fabrication of metallic bipolar plate for proton exchange membrane fuel cells by rubber pad forming," *Journal of Power Sources*, 195(11), pp. 3529–3535. doi: 10.1016/j.jpowsour.2009.12.046.

[85] Mahabunphachai, S., Cora, Ö. N. and Koç, M. (2010) "Effect of manufacturing processes on formability and surface topography of proton exchange membrane fuel cell metallic bipolar plates," *Journal of Power Sources*, 195(16), pp. 5269–5277. doi: 10.1016/j.jpowsour.2010.03.018.

[86] Hung, J.-C. and Lin, C.-C. (2012) "Fabrication of micro-flow channels for metallic bipolar plates by a high-pressure hydroforming apparatus," *Journal of Power Sources*, 206, pp. 179–184. doi: 10.1016/j.jpowsour.2012.01.112.

[87] Maki, T. and Walter, C. (2007) "Liquid curves—Sheet hydroforming helps the sporty solstice stand out," *Stamping Journal*, pp. 32–37, 8 May.

[88] Altan, T. (2002) "Sheet hydroforming in automotive applications," *Stamping Journal*.

[89] Saran, M. J. and Wagoner, R. (1991) A user's manual and verification examples, sheet stamping. Report no. ERC/NSM-S-91–17.

[90] Frey, W. H. (1987) "Development and application for a one dimensional finite element code for sheet metal forming," *Materials Processing*, 1, 307–320.

[91] Sklad, M. P. (1990), 16th IDDRG Congress, ASM International, Sweden. pp. 295.

[92] Pourboghrat, F. and Chandorkar, K. (1992) Springback calculation for plane strain sheet forming using finite element membrane solution, In Numerical Methods for Simulation of Industrial Metal Forming Processes, CED-Vol 5/AMD Vol 156, ASME, 1992, pp. 85–93.

[93] Assempour, A. and Emami, M. R. (2009) "Pressure estimation in the hydroforming process of sheet metal pairs with the method of upper bound analysis," *Journal of Materials Processing Technology*, 209(5), pp. 2270–2276. doi: 10.1016/j.jmatprotec.2008.05.020.

[94] Rubio, E. M. *et al.* (2009) "Analysis of plate drawing processes by the upper bound method using theoretical work-hardening materials," *The International Journal of Advanced Manufacturing Technology*, 40(3–4), pp. 261–269. doi: 10.1007/s00170-007-1347-7.

[95] Kaftanog̃lu, B. and Tekkaya, A. E. (1981) "Complete numerical solution of the axisymmetrical deep-drawing problem," *Journal of Engineering Materials and Technology*, 103(4), pp. 326–332. doi: 10.1115/1.3225023.

[96] Makinouchi, A. (1996) "Sheet metal forming simulation in industry," *Journal of Materials Processing Technology*, 60(1–4), pp. 19–26. doi: 10.1016/0924-0136(96)02303-5.

[97] Makinouchi, A., Teodosiu, C. and Nakagawa, T. (1998) "Advance in FEM simulation and its related technologies in sheet metal forming," *CIRP Annals*, 47(2), pp. 641–649. doi: 10.1016/s0007-8506(07)63246-6.

[98] Tekkaya, A. E. (2000) "State-of-the-art of simulation of sheet metal forming," *Journal of Materials Processing Technology*, 103(1), pp. 14–22. doi: 10.1016/s0924-0136(00)00413-1.

[99] Tisza, M. (2004) "Numerical modelling and simulation in sheet metal forming," *Journal of Materials Processing Technology*, 151(1–3), pp. 58–62. doi: 10.1016/j.jmatprotec.2004.04.009.

[100] Wenner, M. L. (2005) "Overview—Simulation of sheet metal forming," *AIP Conference Proceedings*. doi: 10.1063/1.2011187.

[101] Ahmed, M., Sekhon, G. S. and Singh, D. (2005) "Finite element simulation of sheet metal forming processes," *Defence Science Journal*, 55(4), pp. 389–401. doi: 10.14429/dsj.55.2002.

[102] Lee, M. G. *et al.* (2011) "Advances in sheet forming—Materials modeling, numerical simulation, and press technologies," *Journal of Manufacturing Science and Engineering*, 133(6). doi: 10.1115/1.4005117.

[103] Banabic, D. (2010) Sheet Metal Forming Processes. Springer Science & Business Media.

[104] Reddy, P., Reddy, G. and Prasad P. (2012) "A review on finite element simulations in metal forming," *International Journal of Modern Engineering Research*, 2(4), pp. 2326–2330.

[105] Woo, D. M. (1968) "On the complete solution of the deep-drawing problem," *International Journal of Mechanical Sciences*, 10(2), pp. 83–94. doi: 10.1016/0020-7403(68)90065-9.

[106] Doege, E. and Ropers, C. (1999) "Berechnung der Wärmeleitung in dreidimensional geformten Blechen mit der Finite-Differenzen-Methode während eines Umformprozesses," *Forschung im Ingenieurwesen*, 65(7), pp. 169–177. doi: 10.1007/pl00010874.

[107] Tseng, A. A. (1984) "A generalized finite difference scheme for convection-dominated metal-forming problems," *International Journal for Numerical Methods in Engineering*, 20(10), pp. 1885–1900. doi: 10.1002/nme.1620201009.

[108] Wifi, A. S. (1976) "An incremental complete solution of the stretch-forming and deep-drawing of a circular blank using a hemispherical punch," *International Journal of Mechanical Sciences*, 18(1), pp. 23–31. doi: 10.1016/0020-7403(76)90071-0.

[109] Gotoh, M. and Ishisé, F. (1978) "A finite element analysis of rigid-plastic deformation of the flange in a deep-drawing process based on a fourth-degree yield function," *International Journal of Mechanical Sciences*, 20(7), pp. 423–435. doi: 10.1016/0020-7403(78)90032-2.

[110] Wang, N. M. and Budiansky, B. (1978) "Analysis of sheet metal stamping by a finite-element method," *Journal of Applied Mechanics*, 45(1), pp. 73–82. doi: 10.1115/1.3424276.

[111] Tang, S. C., Chu, E. and Samanta, S. K. (1982) "Finite element prediction of the deformed shape of an automotive body panel during preformed stage," *Numer Methods Ind Form Process*, pp. 629–640.

[112] Toh, C. H. and Kobayashi, S. (1983) "Finite element process modeling of sheet metal forming of general shapes," *Berichte aus dem Institut für Umformtechnik der Universität Stuttgart*, pp. 39–56. doi: 10.1007/978-3-642-82186-8_2.

[113] Benson, D. J. and Hallquist, J. O. (1986) "A simple rigid body algorithm for structural dynamics programs," *International Journal for Numerical Methods in Engineering*, 22(3), pp. 723–749. doi: 10.1002/nme.1620220313.

[114] Belytschko, T. and Mullen, R. (1978) "Explicit integration of structural problems," *Finite Elements in Non-Linear Mechanics,* Edited by P. Bergan, et. al., Volume 2, (1977), pp. 697–720.

[115] Massoni, E., Bellet, M., Chenot, J. L., Detraux, J. M. and De Baynast, C. (1987) A Finite Element Modelling for Deep Drawing of Thin Sheet in Automotive Industry. Springer-Verlag, pp. 719–725.

[116] Wang, N. M. and Wenner, M. L. (1978) "Elastic-viscoplastic analyses of simple stretch forming problems," *Mechanics of Sheet Metal Forming*, pp. 367–402. doi: 10.1007/978-1-4613-2880-3_15.

[117] Oñate, E., Rojek, J. and García Garino, C. (1995) "Numistamp: A research project for assessment of finite-element models for stamping processes," *Journal of Materials Processing Technology*, 50(1–4), pp. 17–38. doi: 10.1016/0924-0136(94)01367-a.

[118] Yang, D. Y. *et al.* (1995) "Comparative investigation into implicit, explicit, and iterative implicit/explicit schemes for the simulation of sheet-metal forming processes," *Journal of Materials Processing Technology*, 50(1–4), pp. 39–53. doi: 10.1016/0924-0136(94)01368-b.

[119] Nakamachi, E. (1995) "Sheet-forming process characterization by static-explicit anisotropic elastic-plastic finite-element simulation," *Journal of Materials Processing Technology*, 50(1–4), pp. 116–132. doi: 10.1016/0924-0136(94)01374-a.

[120] Mamalis, A. G., Manolakos, D. E. and Baldoukas, A. K. (1997) "Simulation of sheet metal forming using explicit finite-element techniques: Effect of material and forming characteristics," *Journal of Materials Processing Technology*, 72(1), pp. 48–60. doi: 10.1016/s0924-0136(97)00128-3.

[121] Jung, D. W., Yoo, D. J. and Yang, D. Y. (1995) "A dynamic explicit/rigid-plastic finite element formulationand its application to sheet metal forming processes," *Engineering Computations*, 12(8), pp. 707–722. doi: 10.1108/02644409510104695.

[122] Carleer, B. D. and Hu, J. (1996) Closing the gap between the workshop and numerical simulations in sheet metal forming. In 2nd European Conference on Numerical Methods in Engineering, ECCOMAS 1996 – Paris, France, pp. 554–560.

[123] Finn, M. J. *et al.* (1995) "Use of a coupled explicit—Implicit solver for calculating spring-back in automotive body panels," *Journal of Materials Processing Technology*, 50(1–4), pp. 395–409. Doi: 10.1016/0924-0136(94)01401-l.

[124] Micari, F. *et al.* (1997) "Springback evaluation in fully 3-D sheet metal forming processes," *CIRP Annals*, 46(1), pp. 167–170. Doi: 10.1016/s0007-8506(07)60800-2.

[125] Tang, B., Li, Y. and Lu, X. (2010) "Developments of multistep inverse finite element method and its application in formability prediction of multistage sheet metal forming," *Journal of Manufacturing Science and Engineering*, 132(4). Doi: 10.1115/1.4001868.

[126] Azizi, R. (2009) "Different implementations of inverse finite element method in sheet metal forming," *Materials & Design*, 30(8), pp. 2975–2980. Doi: 10.1016/j.matdes.2008.12.022.

[127] Chung, W. *et al.* (2014) "Finite element simulation of plate or sheet metal forming processes using tetrahedral MINI-elements," *Journal of Mechanical Science and Technology*, 28(1), pp. 237–243. Doi: 10.1007/s12206-013-0959-0.

[128] "Proceedings of the 13th International Conference on Metal Forming" (2010) *Steel Research International*, 81(9), p. n/a-n/a. doi: 10.1002/srin.201190002.

[129] Papeleux, L. and Ponthot, J. P. (2002) "Finite element simulation of springback in sheet metal forming," *Journal of Materials Processing Technology*, 125–126, pp. 785–791. Doi: 10.1016/s0924-0136(02)00393-x.

[130] Alvesdesousa, R. *et al.* (2007) "On the use of a Reduced Enhanced Solid-Shell (RESS) element for sheet forming simulations," *International Journal of Plasticity*, 23(3), pp. 490–515. Doi: 10.1016/j.ijplas.2006.06.004.

[131] Parente, M. P. L. *et al.* (2006) "Sheet metal forming simulation using EAS solid-shell finite elements," *Finite Elements in Analysis and Design*, 42(13), pp. 1137–1149. Doi: 10.1016/j.finel.2006.04.005.

[132] Lee, M. C. *et al.* (2009) "Three-dimensional simulation of forging using tetrahedral and hexahedral elements," *Finite Elements in Analysis and Design*, 45(11), pp. 745–754. Doi: 10.1016/j.finel.2009.06.002.

[133] Menezes, L. F. and Teodosiu, C. (2000) "Three-dimensional numerical simulation of the deep-drawing process using solid finite elements," *Journal of Materials Processing Technology*, 97(1–3), pp. 100–106. Doi: 10.1016/s0924-0136(99)00345-3.

[134] Cueto, E. and Chinesta, F. (2013) "Meshless methods for the simulation of material forming," *International Journal of Material Forming*, 8(1), pp. 25–43. Doi: 10.1007/s12289-013-1142-y.

[135] Yoon, S. *et al.* (2000) "Efficient meshfree formulation for metal forming simulations," *Journal of Engineering Materials and Technology*, 123(4), pp. 462–467. Doi: 10.1115/1.1396349.

[136] Yoon, S. and Chen, J. S. (2002) "Accelerated meshfree method for metal forming simulation," *Finite Elements in Analysis and Design*, 38(10), pp. 937–948. Doi: 10.1016/s0168-874x(02)00086-0.

[137] Botkin, M. E., Guo, Y. and Wu, C. T. (2004) Coupled FEM/Mesh-Free shear-deformable shells for nonlinear analysis of shell structures. In Sixth World Congress on Computational Mechanics in Conjunction with the Second Asian-Pacific Congress, Proceedings of WCCM VI in Conjunction with APCOM, Volume 4, pp. 5–10.

[138] Sidibe, K. and Li, G. (2012) "A meshfree simulation of the draw bending of sheet metal," *International Journal of Scientific Engineering and Research*, 3(10), pp. 1–5.

[139] Liu, H. S., Xing, Z. W. and Yang, Y. Y. (2010) "Simulation of sheet metal forming process using reproducing kernel particle method," *International Journal for Numerical Methods in Biomedical Engineering*, 26(11), pp. 1462–1476. doi: 10.1002/cnm.1229.

[140] Liu, H. *et al.* (2011) "Adaptive multiple scale meshless simulation on springback analysis in sheet metal forming," *Engineering Analysis with Boundary Elements*, 35(3), pp. 436–451. doi: 10.1016/j.enganabound.2010.06.025.

[141] Hill, R. (1979) "Theoretical plasticity of textured aggregates," *Mathematical Proceedings of the Cambridge Philosophical Society*, 85(1), pp. 179–191. doi: 10.1017/s0305004100055596.

[142] Hill, R. (1990) "Constitutive modelling of orthotropic plasticity in sheet metals," *Journal of the Mechanics and Physics of Solids*, 38(3), pp. 405–417. doi: 10.1016/0022-5096(90)90006-p.

[143] Hill, R. (1993) "A user-friendly theory of orthotropic plasticity in sheet metals," *International Journal of Mechanical Sciences*, 35(1), pp. 19–25. doi: 10.1016/0020-7403(93)90061-x.

[144] Vial, C., Hosford, W. F. and Caddell, R. M. (1983) "Yield loci of anisotropic sheet metals," *International Journal of Mechanical Sciences*, 25(12), pp. 899–915. doi: 10.1016/0020-7403(83)90020-6.

[145] Barlat, F. and Richmond, O. (1987) "Prediction of tricomponent plane stress yield surfaces and associated flow and failure behavior of strongly textured f.c.c. polycrystalline sheets," *Materials Science and Engineering*, 95, pp. 15–29. doi: 10.1016/0025-5416(87)90494-0.

[146] Barlat, F., Lege, D. J. and Brem, J. C. (1991) "A six-component yield function for anisotropic materials," *International Journal of Plasticity*, 7(7), pp. 693–712. doi: 10.1016/0749-6419(91)90052-z.

[147] Barlat, F. *et al.* (1997) "Yielding description for solution strengthened Aluminum alloys," *International Journal of Plasticity*, 13(4), pp. 385–401. doi: 10.1016/s0749-6419(97)80005-8.

[148] Barlat, F. *et al.* (1997) "Yield function development for Aluminum alloy sheets," *Journal of the Mechanics and Physics of Solids*, 45(11–12), pp. 1727–1763. doi: 10.1016/s0022-5096(97)00034-3.

[149] Karafillis, A. P. and Boyce, M. C. (1993) "A general anisotropic yield criterion using bounds and a transformation weighting tensor," *Journal of the Mechanics and Physics of Solids*, 41(12), pp. 1859–1886. doi: 10.1016/0022-5096(93)90073-o.

[150] Gotoh, M. (1977) "A theory of plastic anisotropy based on yield function of fourth order (plane stress state)—II," *International Journal of Mechanical Sciences*, 19(9), pp. 513–520. doi: 10.1016/0020-7403(77)90044-3.

[151] Budiansky, B. (1984) "Anisotropic plasticity of plane-isotropic sheets," *Mechanics of Material Behavior—The Daniel C. Drucker Anniversary Volume*, pp. 15–29. doi: 10.1016/b978-0-444-42169-2.50008-5.

[152] Barlat, F. *et al.* (2003) "Plane stress yield function for Aluminum alloy sheets—Part 1: Theory," *International Journal of Plasticity*, 19(9), pp. 1297–1319. doi: 10.1016/s0749-6419(02)00019-0.

[153] Banabic, D., Balan, T. and Comsa, D. S. (2000) A new yield criterion for orthotropic sheet metals under plane-stress conditions. In Proceedings of 7th Cold Metal Forming Conference, D. Banabic (ed), May 11–12, 2000, Cluj Napoca, Romania, pp. 217–224. Available at: http://www.utcluj.ro/conf/tpr2000.

[154] Cazacu, O. and Barlat, F. (2001) "Generalization of drucker's yield criterion to orthotropy," *Mathematics and Mechanics of Solids*, 6(6), pp. 613–630. doi: 10.1177/108128650100600603.

[155] Carleer, B. D., Meinders. T. and Vegter, H. (1997) A planar anisotropic yield function based on multi axial stress states in finite elements. In COMPLAS V, 5th International Conference on Computational Plasticity, Barcelona, Spain.

[156] Vegter, H. and van den Boogaard, A. H. (2006) "A plane stress yield function for anisotropic sheet material by interpolation of biaxial stress states," *International Journal of Plasticity*, 22(3), pp. 557–580. doi: 10.1016/j.ijplas.2005.04.009.

[157] Hu, W. (2003) "Characterized behaviors and corresponding yield criterion of anisotropic sheet metals," *Materials Science and Engineering: A*, 345(1–2), pp. 139–144. doi: 10.1016/s0921-5093(02)00453-7.

[158] Comsa, D. S. and Banabic, D. (2007) "Numerical simulation of sheet metal forming processes using a new yield criterion," *Key Engineering Materials*, 344, pp. 833–840. doi: 10.4028/www.scientific.net/kem.344.833.

[159] Soare, S., Yoon, J. W. and Cazacu, O. (2008) "On the use of homogeneous polynomials to develop anisotropic yield functions with applications to sheet forming," *International Journal of Plasticity*, 24(6), pp. 915–944. doi: 10.1016/j.ijplas.2007.07.016.

[160] Banabic, D. (2005) "An improved analytical description of orthotropy in metallic sheets," *International Journal of Plasticity*, 21(3), pp. 493–512. doi: 10.1016/j.ijplas.2004.04.003.

[161] Comsa, D. and Banabic, D. (2008) "Plane-stress yield criterion for highly-anisotropic sheet metals," *Numisheet*, 0(1), pp. 43–48.

[162] Emmens, W. C. (2011) Formability. Springer Science & Business Media.

[163] Goodwin, G. M. (1968) "Application of strain analysis to sheet metal forming problems in the press shop," *SAE Technical Paper Series*. doi: 10.4271/680093.

[164] Keeler, S. P. (1968) "Circular grid system—A valuable aid for evaluating sheet metal formability," *SAE Technical Paper Series*. doi: 10.4271/680092.

[165] Havianek, J. (1977) "The effect of mechanical properties of sheet steels on the wrinkling behaviour during deep drawing of conical shells," *Journal of Mechanical Working Technology*, 1(2), pp. 115–129. doi: 10.1016/0378-3804(77)90001-8.

[166] Arrieux, R. (1981) Contribution to the determination of forming limit curves of titanium and aluminum. Proposal of an intrinsic criterion. PhD Thesis, INSA, Lyon (in French).

[167] Manoj Simha, C. H. *et al.* (2006) "Prediction of necking in tubular hydroforming using an extended stress-based forming limit curve," *Journal of Engineering Materials and Technology*, 129(1), pp. 36–47. doi: 10.1115/1.2400269.

[168] Janssens, K. *et al.* (2001) "Statistical evaluation of the uncertainty of experimentally characterised forming limits of sheet steel," *Journal of Materials Processing Technology*, 112(2–3), pp. 174–184. doi: 10.1016/s0924-0136(00)00890-6.

[169] Bleck, W. *et al.* (1998) "A comparative study of the forming-limit diagram models for sheet steels," *Journal of Materials Processing Technology*, 83(1–3), pp. 223–230. doi: 10.1016/s0924-0136(98)00066-1.

[170] Hu, P. *et al.* (2012) Theories, Methods and Numerical Technology of Sheet Metal Cold and Hot Forming. Springer Science & Business Media.

[171] Zimniak, Z. (2000) "Implementation of the forming limit stress diagram in FEM simulations," *Journal of Materials Processing Technology*, 106(1–3), pp. 261–266. doi: 10.1016/s0924-0136(00)00627-0.

[172] Berstad, T., Lademo, O. G., Pedersen, K. O. and Hopperstad, O. S. (2004) Formability modeling with LS-DYNA. In 8th International LS-DYNA Users Conference, 2(7465), pp. 6–53–64.

[173] Samuel, M. (2004) "Numerical and experimental investigations of forming limit diagrams in metal sheets," *Journal of Materials Processing Technology*, 153–154, pp. 424–431. doi: 10.1016/j.jmatprotec.2004.04.095.

[174] Duan, X., Jain, M. and Wilkinson, D. S. (2006) "Development of a heterogeneous microstructurally based finite element model for the prediction of forming limit diagram for sheet material," *Metallurgical and Materials Transactions A*, 37(12), pp. 3489–3501. doi: 10.1007/s11661-006-1044-4.

[175] Takuda, H. *et al.* (2009) "Forming limit prediction in bore expansion by combination of finite element simulation and ductile fracture criterion," *Materials Transactions*, 50(8), pp. 1930–1934. doi: 10.2320/matertrans.p-m2009817.

[176] Hajian, M. and Assempour, A. (2014) "Experimental and numerical determination of forming limit diagram for 1010 steel sheet: A crystal plasticity approach," *The International Journal of Advanced Manufacturing Technology*, 76(9–12), pp. 1757–1767. doi: 10.1007/s00170-014-6339-9.

[177] Baseri, H., Rahmani, B. and Bakhshi-Jooybari, M. (2012) "Predictive models of the spring-back in the bending process," *Applied Artificial Intelligence*, 26(9), pp. 862–877. doi: 10.1080/08839514.2012.726155.

[178] Tekıner, Z. (2004) "An experimental study on the examination of spring-back of sheet metals with several thicknesses and properties in bending dies," *Journal of Materials Processing Technology*, 145(1), pp. 109–117. doi: 10.1016/j.jmatprotec.2003.07.005.

[179] Moon, Y. H. *et al.* (2003) "Effect of tool temperature on the reduction of the spring-back of Aluminum sheets," *Journal of Materials Processing Technology*, 132(1–3), pp. 365–368. doi: 10.1016/s0924-0136(02)00925-1.

[180] Gomes, C., Onipede, O. and Lovell, M. (2005) "Investigation of springback in high strength anisotropic steels," *Journal of Materials Processing Technology*, 159(1), pp. 91–98. doi: 10.1016/j.jmatprotec.2004.04.423.

[181] Li, K. P., Carden, W. P. and Wagoner, R. H. (2002) "Simulation of springback," *International Journal of Mechanical Sciences*, 44(1), pp. 103–122. doi: 10.1016/s0020-7403(01)00083-2.

[182] Nakamachi, E. *et al.* (2014) "Two-scale finite element analyses for bendability and springback evaluation based on crystallographic homogenization method," *International Journal of Mechanical Sciences*, 80, pp. 109–121. doi: 10.1016/j.ijmecsci.2014.01.011.

[183] Baseri, H., Rahmani, B. and Bakhshi-Jooybari, M. (2011) "Selection of bending parameters for minimal spring-back using an ANFIS model and simulated annealing algorithm," *Journal of Manufacturing Science and Engineering*, 133(3). doi: 10.1115/1.4004139.

[184] Jamli, M. R., Ariffin, A. K. and Wahab, D. A. (2015) "Incorporating feedforward neural network within finite element analysis for L-bending springback prediction," *Expert Systems with Applications*, 42(5), pp. 2604–2614. doi: 10.1016/j.eswa.2014.11.005.

[185] Ank, R. and Barauskas, R. (2006) "Finite element investigation on parameters influencing the springback during sheet metal forming," 5(5), pp. 57–62.

[186] Geiger, M. and Vollertsen, F. (1993) "The mechanisms of laser forming," *CIRP Annals*, 42(1), pp. 301–304. doi: 10.1016/s0007-8506(07)62448-2.

[187] Shen, H. and Vollertsen, F. (2009) "Modelling of laser forming—An review," *Computational Materials Science*, 46(4), pp. 834–840. doi: 10.1016/j.commatsci.2009.04.022.

[188] Shichun, W. and Jinsong, Z. (2001) "An experimental study of laser bending for sheet metals," *Journal of Materials Processing Technology*, 110(2), pp. 160–163. doi: 10.1016/s0924-0136(00)00860-8.

[189] Ji, Z. and Wu, S. (1998) "FEM simulation of the temperature field during the laser forming of sheet metal," *Journal of Materials Processing Technology*, 74(1–3), pp. 89–95. doi: 10.1016/s0924-0136(97)00254-9.

[190] Shichun, W. and Zhong, J. (2002) "FEM simulation of the deformation field during the laser forming of sheet metal," *Journal of Materials Processing Technology*, 121(2–3), pp. 269–272. doi: 10.1016/s0924-0136(01)01241-9.

[191] Hoseinpour Gollo, M., Mahdavian, S. M. and Moslemi Naeini, H. (2011) "Statistical analysis of parameter effects on bending angle in laser forming process by pulsed Nd:YAG laser," *Optics & Laser Technology*, 43(3), pp. 475–482. doi: 10.1016/j.optlastec.2010.07.004.

[192] Kheloufi, K. and Amara, E. H. (2008) Numerical simulation of steel plate bending process using thermal mechanical analysis, In Laser and Plasma Applications in Materials Science: First International Conference on Laser Plasma Applications in Materials Science – LAPAMS-08, El-Hachemi Amara, Said Boudjemai, and Djamila Doumaz. *AIP Conference Proceedings* 1047. Conference Location and Date: Algiers, Algeria, 23-26 June 2008. Published November 2008., p. 176–179.

[193] Venkadeshwaran, K., Das, S. and Misra, D. (2010) "Finite element simulation of 3-D laser forming by discrete section circle line heating," *International Journal of Engineering, Science and Technology*, 2(4). doi: 10.4314/ijest.v2i4.59284.

[194] Shi, Y. *et al.* (2006) "Numerical investigation of straight-line laser forming under the temperature gradient mechanism," *Acta Metallurgica Sinica (English Letters)*, 19(2), pp. 144–150. doi: 10.1016/s1006-7191(06)60036-7.

[195] Shen, H. *et al.* (2006) "Fuzzy logic model for bending angle in laser forming," *Materials Science and Technology*, 22(8), pp. 981–986. doi: 10.1179/174328406x100725.

[196] Yu, G. *et al.* (2000) "FEM simulation of laser forming of metal plates*," *Journal of Manufacturing Science and Engineering*, 123(3), pp. 405–410. doi: 10.1115/1.1371930.

[197] Zohoor, M. and Zahrani, E. G. (2012) "Experimental and numerical analysis of bending angle variation and longitudinal distortion in laser forming process," *Scientia Iranica*, 19(4), pp. 1074–1080. doi: 10.1016/j.scient.2012.06.020.

[198] Hoseinpour, G. M., Moslemi, N. H. and Mostafa Arab, N. B. (2011) "Experimental and numerical investigation on laser bending process," *Journal of Computational and Applied Research in Mechanical Engineering*, 1(1), pp. 45–52.

[199] Pitz, I., Otto, A. and Schmidt, M. (2010) "Simulation of the laser beam forming process with moving meshes for large Aluminium plates," *Physics Procedia*, 5, pp. 363–369. doi: 10.1016/j.phpro.2010.08.063.

[200] Che Jamil, M. S., Sheikh, M. A. and Li, L. (2011) "A study of the effect of laser beam geometries on laser bending of sheet metal by buckling mechanism," *Optics & Laser Technology*, 43(1), pp. 183–193. doi: 10.1016/j.optlastec.2010.06.011.

[201] Lambiase, F. (2012) "An analytical model for evaluation of bending angle in laser forming of metal sheets," *Journal of Materials Engineering and Performance*, 21(10), pp. 2044–2052. doi: 10.1007/s11665-012-0163-x.

[202] Paramasivan, K., Das, S. and Misra, D. (2014) "A study on the effect of rectangular cut out on laser forming of AISI 304 plates," *The International Journal of Advanced Manufacturing Technology*, 72(9–12), pp. 1513–1525. doi: 10.1007/s00170-014-5695-9.

[203] Guan, Y. *et al.* (2005) "Influence of material properties on the laser-forming process of sheet metals," *Journal of Materials Processing Technology*, 167(1), pp. 124–131. doi: 10.1016/j.jmatprotec.2004.10.003.

[204] Leszak, E. (1967) Apparatus and process for incremental dieless forming. Patent no. US3342051A. https://patents.google.com/patent/US3342051A/en.

[205] Jeswiet, J. *et al.* (2005) "Asymmetric single point incremental forming of sheet metal," *CIRP Annals*, 54(2), pp. 88–114. doi: 10.1016/s0007-8506(07)60021-3.

[206] Emmens, W. C., Sebastiani, G. and van den Boogaard, A. H. (2010) "The technology of incremental sheet forming—A brief review of the history," *Journal of Materials Processing Technology*, 210(8), pp. 981–997. doi: 10.1016/j.jmatprotec.2010.02.014.

[207] Kumar, Y. and Kumar, S. (2015) "Incremental Sheet Forming (ISF)," *Advances in Material Forming and Joining*, pp. 29–46. doi: 10.1007/978-81-322-2355-9_2.

[208] Iseki, H. (2001) "An approximate deformation analysis and FEM analysis for the incremental bulging of sheet metal using a spherical roller," *Journal of Materials Processing Technology*, 111(1–3), pp. 150–154. doi: 10.1016/s0924-0136(01)00500-3.

[209] Shim, M. S. and Park, J. J. (2001) "The formability of Aluminum sheet in incremental forming," *Journal of Materials Processing Technology*, 113(1–3), pp. 654–658. doi: 10.1016/s0924-0136(01)00679-3.

[210] Hirt, G. *et al.* (2004) "Forming strategies and process modelling for CNC incremental sheet forming," *CIRP Annals*, 53(1), pp. 203–206. doi: 10.1016/s0007-8506(07)60679-9.

[211] Yamashita, M., Gotoh, M. and Atsumi, S. Y. (2008) "Numerical simulation of incremental forming of sheet metal," *Journal of Materials Processing Technology*, 199(1–3), pp. 163–172. doi: 10.1016/j.jmatprotec.2007.07.037.

[212] Lequesne, C., Henrard, C. and Bouffioux, C. (2008) "Adaptive remeshing for incremental forming simulation," *Numer Simul*, 32, pp. 4–8.

[213] Hadoush, A. and van den Boogaard, A. H. (2009) "Substructuring in the implicit simulation of single point incremental sheet forming," *International Journal of Material Forming*, 2(3), pp. 181–189. doi: 10.1007/s12289-009-0402-3.

[214] Ben Ayed, L. *et al.* (2014) "Simplified numerical approach for incremental sheet metal forming process," *Engineering Structures*, 62–63, pp. 75–86. doi: 10.1016/j. engstruct.2014.01.033.

[215] Zhang, M. H. *et al.* (2014) "Selective element fission approach for fast FEM simulation of incremental sheet forming based on dual-mesh system," *The International Journal of Advanced Manufacturing Technology*, 78(5–8), pp. 1147–1160. doi: 10.1007/s00170-014-6723-5.

15 Applying Adaptive and Arlequin Meshing to Tensile Testing and Metal Forming

C. Anand Badrish, Baloji Dharavath, and Swadesh Kumar Singh

15.1 INTRODUCTION

Finite element methods (FEM) have been widely employed effectively in the whole domain of engineering to replicate realistic work conditions using partial differential equations and preprocessing techniques. One of the major challenges in FE simulation is to predict accurate results with less computation time; various numeric methods, equations, and modeling strategies have been implemented grounded on error estimation. To achieve the prerequisite considerations, adaptive mesh refinement methodology is carried out [1]. Adaptivity is primarily employed to solve preprocessed models with equitable CPU time.

Adaptive mesh techniques are used to optimize element size and mesh flow along with element transition [2]. By increasing the order of approximations, adaptive meshing has better convergence than simple first order of elements. From a widespread literature survey, various adaptive mesh techniques have been recommended, including hierarchical (*h*), polynomial *(p)* and relocation *(r)* adaptivity. The latter refers to relocating nodes with the same number of degrees of freedom *(dof)* and element connectivity with low computation cost.

In *r*-adaptive mesh, elements can be adjusted by relocating nodes in a complete domain without adding a degree of freedom, and optimum mesh is attained by distributing error quantity equally [3]. In *h*-adaptive mesh, *dof* and element connectivity change while polynomial degree and shape function remain unchanged. Implementing *h*-adaptive mesh is simple, with less preprocessing time but longer computation time [4]. To obtain uniform mesh refinement in critical areas when the entire system features high nonlinearity, *h* adaptivity is taken up in preprocessing; this method helps in refining mesh in areas with many errors. Polynomial adaptivity involves varying the polynomial degree locally by aggregating the order of the trial function under identical mesh [5].

The Arlequin method was initially used to analyze engineering structures; it entails combining two dissimilar mechanical states to conduct multi

 DOI: 10.1201/9781003441755-15

scale or multi model simulations [6]. It boasts exceptional versatility in effectively integrating two distinct mechanical conditions. Applying the Arlequin method to change the overall structure facilitates attaining both flexibility and reduced computation time, particularly with intricate nonlinear structures. This approach is enhanced by incorporating FE tearing and interconnecting [7]. Incorporating the Arlequin method into existing commercial finite element software has become essential for the reliable, flexible, efficient multi scale analysis of large complex structures. However, the extensive literature survey revealed little scholarly effort at testing different meshing techniques and their effects on formability with highly complex materials and nonlinear metal forming.

Inconel 625 (IN625) is a combination of diverse alloys comprising mostly nickel followed by chromium and molybdenum. Based on its extraordinary mechanical properties, along with good weld ability and resistance to creep, thermal shock led to the usage of Inconel 625 in risky environmental conditions. It is widely used in sheet metal forming, primarily in aerospace but also in the automotive, marine and petrochemical industries. IN625 exhibits exceptional resistance to deformation, primarily due to its limited workability, intricate microstructure characterized by high strength and inherent resistance to deformation. However, machining this alloy poses significant challenges owing to the presence of hard abrasive carbides, work-hardening tendencies and remarkable toughness, even when subjected to elevated temperatures [8].

Thus, the focus of this study was to investigate and numerically demonstrate *h*-, *p*- and *r*-adaptive meshing and the Arlequin method to couple the shell element S4R with the solid element C3D8R in typical metal-forming processes in commercial Abaqus CAE software. We evaluated the meshing techniques and the Arlequin method (2D-3D) based on accuracy, forming behavior at both room and higher temperatures and solving or CPU time. We compared all iterations with normal automated mesh, which is generated only by considering geometrical dimensions of the components with default element size and element flow; we had observed that normal mesh showed lower accuracy and longer CPU time.

15.2 MATERIAL AND METHODS

In the present work, we conducted all the experiments on 1 mm thick sheets of IN625 mostly comprising nickel (61.51%), chromium (21.75%) and molybdenum (9.49%), which give it strength and ductility at higher temperatures. Niobium at a concentration of 3.29% and aluminum at 0.069% play a significant role in the age-hardening process of IN625. This is primarily attributed to the formation of $Ni3(Al, Nb)$ precipitates, which consistently take on an ellipsoidal γ' phase distribution throughout the entire metallic sheet. This phenomenon effectively restricts creep deformation at elevated temperatures during testing [9].

Table 15.1 provides the detailed chemical composition of the IN625. Figure 15.1(a) illustrates a computer-controlled Zwick/Roell universal testing machine equipped with a maximum loading capacity of 100 kN, a box furnace and a non-contact laser extensometer for conducting high-temperature tensile tests. For deep drawing and stretching operations, we employed a computer-controlled hydraulic press setup with a maximum capacity of 40 tons, depicted in Figure 15.1(b). To maintain the desired operational temperatures, a water-cooling tower was integrated, and temperature measurement was facilitated through the use of a K-type optical cable probe and a PID temperature controller [10]. Figure 15.1(c) displays a compression-testing machine with capacity of 100 kN to find spring back angle using V-bending process for IN625 [11]. The die was constrained at all *dof*, with nose angle of 60° and punch radius of 3 mm. Finally, we conducted V-bending analysis in isothermal test conditions with rectangular plates of 80 mm × 40 mm.

TABLE 15.1

Chemical Composition of IN625

Element	Ni	Cr	Nb	Mo	Ti	Al	Fe
wt.%	62.495	20.739	3.291	9.489	0.168	0.065	3.314
Element	C	Si	Mn	P	S	Co	Cu
wt.%	0.024	0.103	0.125	0.002	0.003	0.085	-

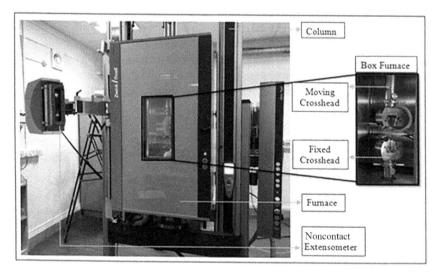

FIGURE 15.1(A)   100 KN capacity universal testing machine with noncontact laser extensometer.

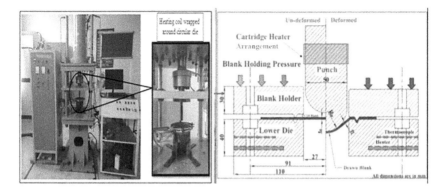

FIGURE 15.1(B) Hydraulic press setup with magnified view of stretch forming and deep drawing dies along with induction heating arrangement.

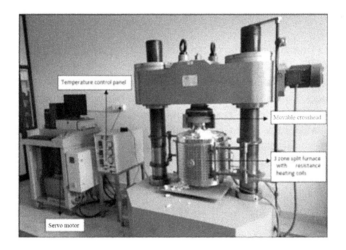

FIGURE 15.1(C) Compression machine used for V bending.

15.3 METHODOLOGIES AND PROCESS PLAN

Flow chart 15.1 shows the process flow of this study implementing adaptive meshing techniques and the Arlequin method in various forming processes.

15.3.1 STANDARD METHOD

Standard meshing is customarily used for symmetrical components because it maintains uniform element size and mesh flow [12]. However, capturing small or highly curved features can be challenging and leads to node averaging, which reduces accuracy in results by increasing computation time. This method is

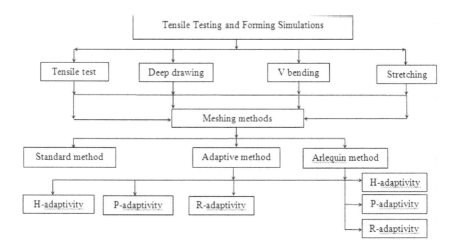

FLOWCHART 15.1   Study methodology adopted.

restricted to simple and trivial geometries and offers error-free CAD models with no projections, gaps or duplicate surfaces.

To build a geometric or CAD model, we require certain entities: points, lines, curves, areas, surfaces, volumes and solids. For standard meshing, we need elements and nodes to build FEA entities as shown in Figure 15.2 (a, b). Automatic or standard meshing is widely used because the models are simpler to build; here, we could create the mesh automatically by only defining the mesh density along the edges or surfaces of the CAD model [13]. However, this method has major limitations with respect to mesh quality and solution accuracy. Standard solid meshing is also limited in that it can only be used for some regular shapes like cubes. Figure 15.2 (c, d) shows the major difference between manual and standard or automated meshing methods.

15.3.2   Adaptive Meshing

Adaptive mesh refinement is a technique of familiarizing the accuracy of a solution within certain limits [14]. Numerically calculated solutions are often restricted to predetermined quantified grids in a cartesian plane, which determines the mesh, or grid. Adaptive meshing allows for exact numeric calculations by considering requirements in specific areas. This method helps in maintaining element connectivity to achieve high mesh quality. In the present work, we evaluated adaptive meshing by implementing h, p adaptivity and r adaptivity.

15.3.3   Hierarchical Adaptivity

As previous specified, h-adaptive meshing fluctuates the refinement level or density of the mesh. It allows for manually defining the number of elements necessary to obtain not worthy outcomes at each locality [15]. In this study, we selected fine

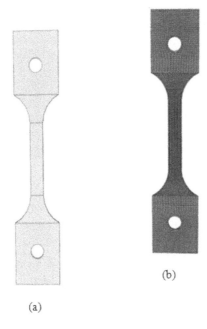

FIGURE 15.2   (a) CAD and (b) FEA models.

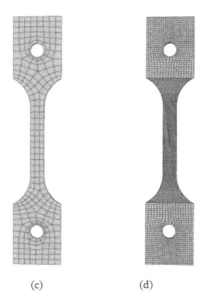

FIGURE 15.2   (c) Automated and (d) manual FEA models.

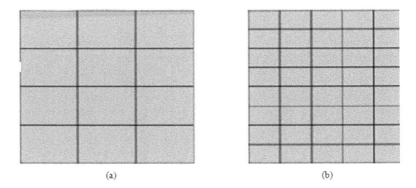

FIGURE 15.3   *H*-adaptive (a) coarse and (b) fine mesh.

or ultra-fine mesh for critical areas and coarse mesh for noncritical areas. Figure 15.3 (a, b) depicts the coarse and fine mesh used. Hierarchical adaptive meshing has the following advantages:

- It moderates mesh size in areas where calculated strain energy error is high.
- It can coarsen or increase mesh size where strain energy is too low.
- Mesh can be refined until either the target accuracy is achieved or the maximum specified loops are run.
- Uses local mesh control to achieve convergence where required.

15.3.4   POLYNOMIAL ADAPTIVITY

Polynomial adaptivity does not entail increasing the number of nodes or elements; rather, it upsurges the order of the polynomial approximation used within each element. It eliminates the need for re meshing, instead solving the problem during each iteration cycle (denoted as "p") until the desired level of accuracy is achieved. This technique involves adding mid-side nodes to element edges, thereby augmenting solution accuracy.

Similar to the *h*-adaptive method, error estimates are utilized at various points within the mesh. Polynomial order is increased in regions exhibiting high errors until the results meet the user-defined precision criteria. However, it's worth noting that this advantage diminishes in the presence of severe stress concentrations. Furthermore, significantly elevating the polynomial order (referred to as "p") can lead to increased computation costs. Figure 15.4 (a, b) shows *p*-adaptive fine and coarse mesh. One of the silent features of *p*-adaptive meshing is that it does not modify the mesh but changes the polynomial order to approximate the displacement field. Second-order polynomial adaptivity adapts to distortion with larger elements, which in turn results in more computation time than *h*-adaptive mesh. Complete mesh should maintain good quality and mesh flow. Because the displacement and von mises condition normally will not show any improvement in accuracy, it is imperative to use the total strain energy as the stopping criteria.

15.3.5 RELOCATION ADAPTIVITY

In the context of r adaptivity, nodes can be repositioned, enabling the adjustment of element sizes across the entire domain. However, it's important to note that degrees of freedom cannot be introduced. The ideal mesh configuration is achieved by evenly distributing the error quantity, meaning that the product of error and element size remain consistent for each element; in this approach, the user can generate the mesh that is best suited for requirements. Figure 15.5 (a, b) illustrates r-adaptive fine and coarse mesh.

To achieve realistic solutions at minimal computation costs, r-adaptive structures make use of optimal mesh topology and are cheap to optimize for given

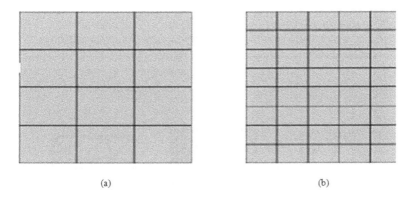

(a)   (b)

FIGURE 15.4   *P*-adaptive (a) coarse and (b) fine mesh.

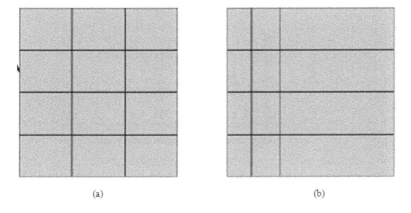

(a)   (b)

FIGURE 15.5   *R*-adaptive (a) coarse and (b) relocated mesh.

finite element configurations. As alluded to earlier, r refinement moves nodes either to deform a mesh boundary or to diminish concentrated error in the mesh. Among its advantages, r adaptivity allows for effectively improving Lagrangian calculations, which involves a restrained amount of deformation. Additionally, node movement can control element deformation and significantly increase the time required to achieve numeric stability [16].

The art of implementing FEM lies in selecting the accurate mesh density to solve the problem; striking a balance is crucial. If the mesh is excessively coarse, it may not yield an accurate solution, but overly fine mesh can significantly increase computation times. Establishing a suitable mesh requires considering variables such as stress, temperature and pressure within the component. In cases where high parameter gradients and strain variations exist, a finer mesh is necessary. Conversely, for less demanding scenarios, a coarser mesh suffices.

A well-suited mesh strikes the right balance by enabling precise resolution of primary physical phenomena while also facilitating a rapid solution. Linear elements necessitate a finer mesh than parabolic (quadratic) elements, which in turn require a finer mesh than cubic elements. Consequently, users of FE software must possess a fundamental understanding of the underlying physical phenomena and their behavior to make informed decisions about mesh refinement. Figure 15.6(a–c) shows pictorial representation of meshing techniques of course, fine mesh along with percentage increase in node count. Figure 15.6(d) shows mesh enhancement at precise locality.

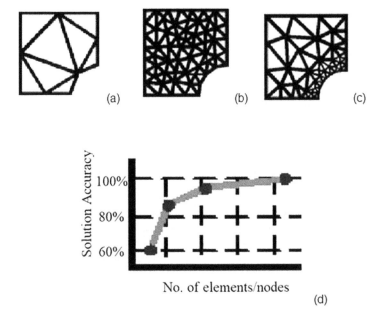

FIGURE 15.6 Representations of mesh: (a) coarse, (b) fine, (c) % increase in node count, (d) mesh at a particular location.

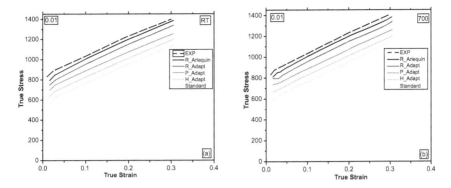

FIGURE 15.7   The (a) shell and (b) solid elements to be coupled under the Arlequin method.

(Source: Practical FEA and Abaqus documentation).

15.3.6   Arlequin Method

The Arlequin method demonstrates remarkable versatility and adaptability in connecting distinct mechanical states or elements. It has become increasingly essential to incorporate the method into established commercial FE software for the analysis of large-scale civil structures, intricate mechanical systems and scenarios requiring high reliability and efficiency. The transparency and simplicity of the coupling matrix presentation significantly enhance comprehension of the Arlequin framework.

Surprisingly, there is a dearth of literature on this subject. With this paper, we focus on integrating the Arlequin method into Abaqus and experimentally validating the procedure. Arlequin coupling pairs coarse and fine elements within the Arlequin framework using the Lagrange multiplier field. This field introduces a "virtual element" that inherently aligns with the coarse element in cases of coarse coupling. The virtual element also eases implementing the Arlequin framework within Abaqus. In the present work, we compare standard and adaptive meshing techniques for coupling a 2D shell to a 3D solid. Figure 15.7 (a, b) shows representations of the shell and solid elements we coupled for this study.

15.4   NONLINEAR FE ANALYSIS

In the present work, we conducted a comparative study of an experimental metal-forming process with FE analysis by incorporating *p, r* and *h* adaptivity along with the Arlequin method. We performed the FE analysis at room temperature and at 400 °C and 700 °C, and we developed all CAD models based on the dimensions of the experimental and forming specimens. We performed all simulations following the standard implicit method considering all nonlinearity effects, i.e., material, geometric and boundary.

TABLE 15.2
Material Properties of IN625 Applied in FE Analysis

S	Room temperature (RT)
Density (Kg/m³)	8440
Yield strength (Mpa)	812.07
Poisson's ratio	0.28

We applied gradual load to address the nonlinearities, wherein the total load is divided into multiple small load increments. Within each load increment, the model is treated as behaving linearly, and the corresponding changes in the model's shape are calculated. Stresses are continuously updated from one increment to the next until the complete applied load is reached. We coupled shell (S4R) and solid elements (C3D8R) in tensile and metal-forming processes.

We performed all the simulations considering relevant load and boundary conditions in both the experimental and forming strategies. We performed all tests in three iterations and took the averages as the final values. We analyzed and interpreted all data in Abaqus Viewer and origin software. The material properties used for numerical simulations are presented in Table 15.2.

15.4.1 Newton–Rapson Method

The Newton–Raphson method, often referred to as the Newton method, stands as a robust numerical approach for solving equations. In the context of nonlinear analyses, where loads are incrementally applied instead of being applied all at once, and where the stiffness matrix's elements are influenced by the displacement matrix, solving equilibrium equations directly becomes impractical. Consequently, iterative techniques are essential, and among these techniques, the Newton–Raphson method holds significant prominence and is widely employed for addressing nonlinear problems.

The Newton–Raphson method operates on the concept that when the root of the equation $f(x)=0$ is initially approximated at x_i, drawing a tangent to the curve at the point $f(x_i)$ will intersect the x-axis at a new point, x_{i+1}. This new point serves as an enhanced estimate for the root.

Utilize the definition of the slope of the function, at $x = x_i$:

$$f'(x_i) = \tan\theta \tag{15.1}$$

$$f'(x_i) = \frac{f(x_i) - 0}{x_i - x_{i+1}} \tag{15.2}$$

These give

$$x_{i+1} = x_i - \frac{f(x_i)}{f'(x_i)} \tag{15.3}$$

Equation (15.3) is commonly referred to as the Newton-Raphson formula, specifically designed for solving nonlinear equations of the given form $f(x) = 0$. To progress from an initial estimate x_i, the next approximation, x_{i+1}, can be determined by applying equation (15.3).

The Newton–Raphson method follows a sequence of steps for locating the root of $f(x) = 0$:

1. Calculate the derivative, $f'(x)$, of the function symbolically.
2. Begin with an initial guess for the root x_i and compute the new approximation for the root x_{i+1} as follows:

$$x_{i+1} = x_i - \frac{f(x_i)}{f'(x_i)} \tag{15.4}$$

3. Find the absolute error of approximation $|\epsilon_a|$ as

$$|\epsilon_a| = \left| \frac{x_{i+1} - x_i}{x_{i+1}} \right| \times 100 \tag{15.5}$$

4. Compare $|\epsilon_a|$ with the predefined relative error tolerance, ϵ_s. If $|\epsilon_a| > \epsilon_s$, then go to step 2 and terminate the algorithm.

15.5   RESULTS AND DISCUSSION

15.5.1   TENSILE TESTING

Figure 15.8 (a, b) illustrates flow stress behavior under different adaptive meshing strategies and the Arlequin method at room temperature (RT) and at 700 °C at the specific deformation rate of 0.01 s⁻¹. The figure shows that the Arlequin method combined with r-adaptive mesh delivers more accurate results than h- and p-adaptive mesh with increased flow stress and results close to experimental findings. The computation time also reduced with increased node count, which further strengthens the capability of combined Arlequin r-adaptive meshing.

In r adaptivity, the nodes are relocated to a minimal distance of 0.1–0.3 mm established from previous iterations. Figure 15.9 (a–e) shows the preprocessing models of all the adaptivity meshing techniques we investigated. Figure 15.9 (e) shows r-adaptive mesh adopted with the Arlequin method to couple the shell and the solid elements. Figure 15.9 (f–j) demonstrates variation in stress pattern of all adaptive meshing techniques carried out. Figure 15.9 (j) shows the corresponding post processing model of combined *(r)* adaptivity with the Arlequin method. The results in Table 15.3 demonstrate that combined Arlequin method and r-adaptive meshing provides more accurate results based on stress, CPU time, and numbers of nodes and elements. Figure 15.10 (a, b) shows our strategy for coupling shell S4R with solid C3D8R.

TABLE 15.3
Effects of Adaptive and Arlequin Methods on Tensile Testing

Method	No. of nodes	No. of elements	CPU Time (Sec)	Stress Values (Mpa)
Standard	440	365	39	47
R_Adaptivity	1171	1013	42	45
H_Adaptivity	1171	1013	41	42
P_Adaptivity	1971	1013	46	40
R_Alerquin	3061	2081 (S4R+C3D8R)	28	38

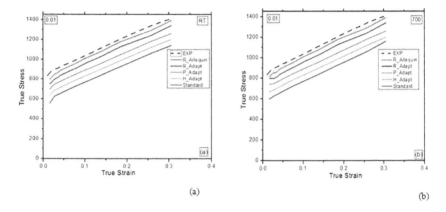

(a)

(b)

FIGURE 15.8 Flow stress behavior with various adaptive approaches: (a) RT and (b) 700 °C, both at strain rate 0.01 s⁻¹.

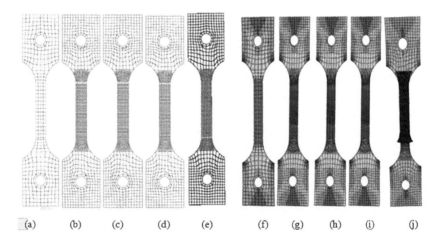

FIGURE 15.9 Outputs from preprocessing models: (a) standard, (b) h adaptivity, (c) p adaptivity, (d) r adaptivity. Outputs from (e) Arlequin method and (f–j) post processing models.

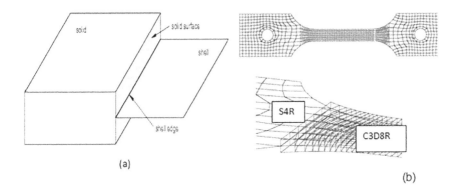

FIGURE 15.10   Arlequin coupling: (a) shell S4R to (b) solid C3D8R.

15.6   DEEP DRAWING

In deep drawing, certain parameters such as limiting draw ratio, thickness, maximum thinning rate and thickness deviation must exhibit consistency for a drawn cup to qualify as high quality. For this study, we ran numeric simulations of circular deep drawing processes at various temperatures while maintaining a constant punch speed of 5 mm/min and a consistent blank holding force of 15 bar. We developed FE simulations to authenticate the experimental results in terms of average thickness and drawn height using adaptive and Arlequin methods.

The sample we considered for the drawing cup was of 58 diameter, and we employed a discrete rigid model to represent the punch, die and holder in the simulation. The samples motion was controlled by a reference node assigned to each rigid body, and the blank was modeled with S4R. Then we assessed each meshing technique for its forming of deep-drawn cups.

In r adaptivity, the nodes were relocated to a minimum distance of 0.06–0.1 mm, which we established from previous iterations. In critical regions where maximum thinning is bound to happen, the nodes were relocated along walls where we observed maximum stress levels. Figure 15.11 shows the deep drawing assembly setup modeled with standard meshing methods.

Figure 15.12(a–e) shows the preprocessing models we ran for the simulations. The thickness and the deformation zone of the fully deep-drawn cup are shown in Figure 15.13. The drawn heights of the fully formed cup are presented in Figure 15.14(a–e). The average thickness and % increase in height at room temperature and 400 °C are shown in Figure 15.15(a–d).

The findings in Table 15.4 demonstrate that the Arlequin method combined with r adaptivity gives the best deep-drawn height and average thickness at both room temperature and 400 °C, which is because solid elements can capture more than shell elements. Furthermore, relocating nodes in failure regions strengthens the structure and increases the material's formability. Merging solid and shell elements in critical regions increases cup height and improves average thickness.

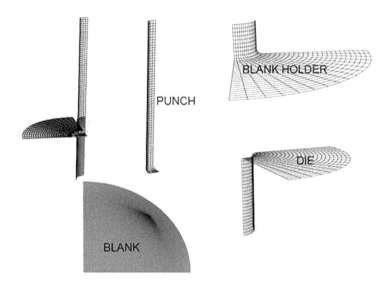

FIGURE 15.11   Deep drawing FE assembly.

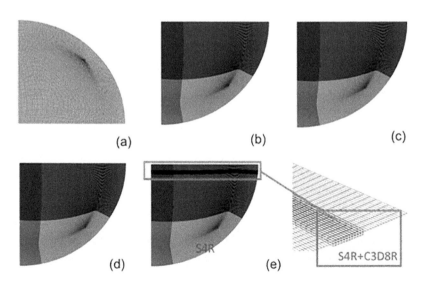

FIGURE 15.12   Preprocessing models: (a) standard method, (b) *h* adaptivity, (c) *p* adaptivity, (d) *r* adaptivity, (e) Arlequin method.

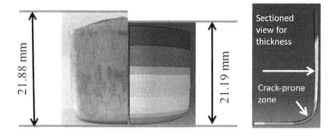

FIGURE 15.13 Thickness and crack zone of the deep-drawn cup.

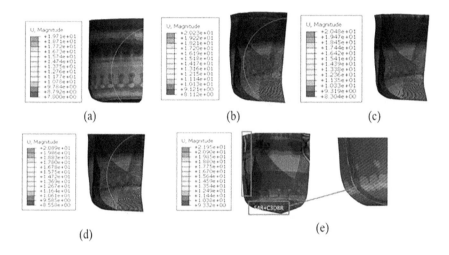

FIGURE 15.14 Drawn cup heights: (a) standard method, (b) *h* adaptivity, (c) *p* adaptivity, (d) *r* adaptivity, (e) Arlequin method.

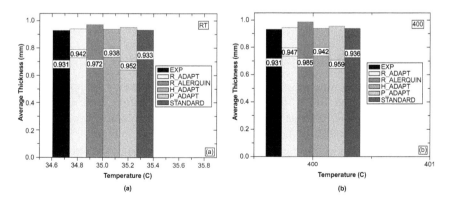

FIGURE 15.15A Average thickness with adaptive and Arlequin methods at (a) room temperature and (b) 400°C.

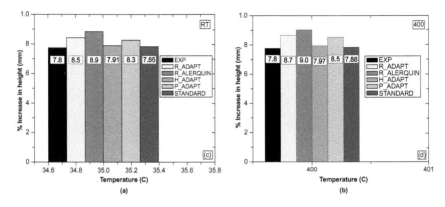

FIGURE 15.15B   Average % increase in height with adaptive and Arlequin methods at (a) room temperature and (b) 400 °C.

TABLE 15.4
Influence of Meshing Methods on Deep-Drawn Cup Parameters

Method	No. of nodes	No. of elements	CPU time (sec)	Drawn height (mm) (RT)	Average Thickness (mm) (RT)	Drawn height (400 °C)	Average thickness (400 °C)
Standard	9142	8968	4487	19.71	0.933	19.91	0.936
R_Adaptivity	34560	26022	4771	20.23	0.942	20.46	0.947
H_Adaptivity	34560	26022	4668	20.48	0.938	20.34	0.942
P_Adaptivity	44560	26022	5741	20.86	0.952	20.85	0.959
R_Alerquin	34560	26022 (S4R+C3D8R)	4132	21.19	0.972	22.15	0.985
Experiment				21.88	0.983	22.86	0.991

15.7   STRETCH FORMING

We ran stretch forming simulations with adaptive and Arlequin methods at room temperature and at 400 °C to learn the influence of meshing method on forming limit curve and limiting dome height. We used a circular specimen with a diameter of radius R75 for the numeric simulations shown in Figure 15.16(a). The stretch forming assembly we used for the numeric simulations is represented in Figure 15.16(b).

Figure 15.17(a–e) exhibits the stretch forming preprocessing adaptive and Arlequin method models, and the limiting dome height models are represented in Figure 15.18(a–e).

We plotted the true major and minor strain as represented in Figure 15.19a for stretch forming at (a) room temperature and (b) 400 °C. All meshing strategies showed similar trends to the experimental results, but r adaptivity combined with the Arlequin method showed the capturing capability that was closest to the experimental results.

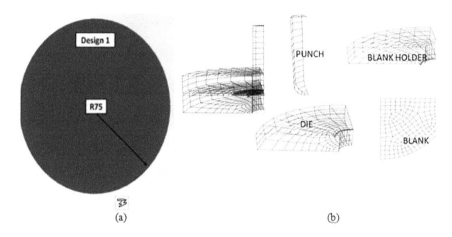

FIGURE 15.16   Stretch forming (a) circular specimen and (b) FE assembly.

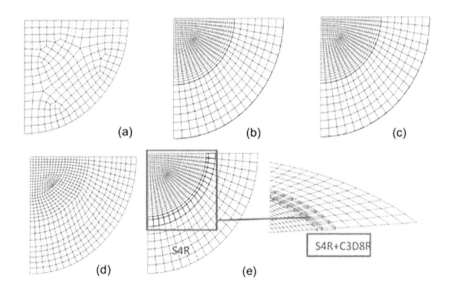

FIGURE 15.17   Stretch forming preprocessing models: (a) standard method, (b) h adaptivity, (c) p adaptivity, (d) r adaptivity, (e) Arlequin method.

Figure 15.19b shows the limiting dome heights at (a) room temperature and (b) 400 °C. The figure findings indicate that r adaptivity combined with the Arlequin method provides the best height. The better capturing capacity of solid C3D8R combined with the greater *dof* from the shell elements contributed to solving highly nonlinear forming problems with better convergence, time and accuracy. Table 15.5 also supports that limiting draw height increases with r-adaptive mesh and the Arlequin approach with shorter solving time than with other the methodologies.

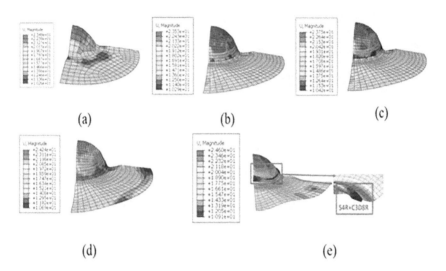

FIGURE 15.18 Stretch forming limiting dome height: (a) standard method, (b) h adaptivity, (c) p adaptivity, (d) r adaptivity, (e) Arlequin method.

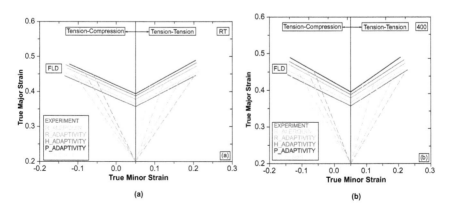

FIGURE 15.19A Influence of adaptive and Arlequin methods on major and minor strain at (a) room temperature and (b) 400 °C.

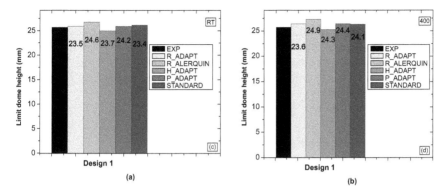

FIGURE 15.19B Influence of adaptive and Arlequin methods on limiting dome height at (a) room temperature and (b) 400 °C.

TABLE 15.5
Impacts of Adaptive and Arlequin Methods on Limiting Dome Height

Method	No. of nodes	No. of elements	CPU time (sec.)	Drawn height (mm)	Drawn height (°C)
Standard	210	183	30	23.40	24.10
R_Adaptivity	618	575	28	23.53	23.59
H_Adaptivity	618	575	31.5	23.75	24.34
P_Adaptivity	765	575	32	24.02	24.41
R_Alerquin	982	700 (S4R+C3D8R)	24	24.60	24.95
Experiment				25.11	26.32

15.8 V BENDING

For the V-bending analyses, we conducted the numeric simulations on a rectangular specimen of 80 × 40 mm with thickness of 1 mm. The multistep FEA involved loading, holding and unloading. We conducted the numeric simulation by incorporating adaptive and Arlequin method, specifically by coupling shell S4R and solid C3D8R at room temperature and at 400 °C at punch speed of 1 mm/min held for 90 sec to calculate the lowest spring back angle. Figure 15.20 (a–f) represents V-bending assembly along with preprocessing models with adaptive and Arlequin meshing techniques, and deformation behavior models are shown in Figure 15.21 (a–e). In the case of *(r)* adaptivity method nodes are relocated with a minimal distance of 0.01 to 0.003 based on earlier iterations. Table 15.6 shows spring back calculations at room temperature and 400 °C and specifically shows that the spring back angle was lower with *r*-adaptive mesh and the Arlequin approach.

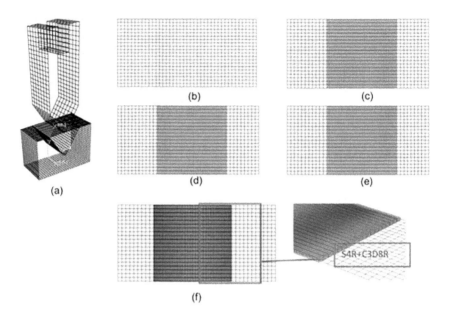

FIGURE 15.20 (a) V-bending assembly; preprocessing models with (b) standard method, (c) h adaptivity, (d) p adaptivity, (e) r adaptivity, (f) Arlequin method.

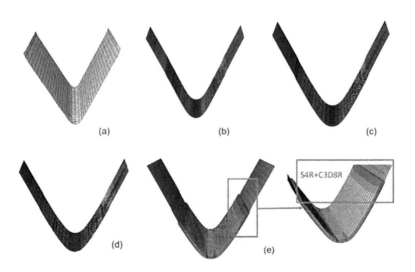

FIGURE 15.21 Deformation behavior: (a) standard method, (b) h adaptivity, (c) p adaptivity, (d) r adaptivity, (e) Arlequin method.

TABLE 15.6

Spring back Angle Calculations Based on Adaptive and Arlequin Methods

Method	No. of nodes	No. of elements	CPU time (sec)	Spring back angle(°C)	Spring back angle(°C)
Standard	861	800	285 (Loading+Unloading)	3.85	4.09
R_Adaptivity	2541	2400	450 (Loading+Unloading)	2.85	3.04
H_Adaptivity	2541	2400	489 (Loading+Unloading)	3.47	3.81
P_Adaptivity	2987	2567	579 (Loading+Unloading)	3.05	3.48
R_Alerquin	8841	6360	410 (Loading+Unloading)	2.47	2.89
		(S4R+C3D8R)			
Experiment				2.72	2.95

15.9   CONCLUSIONS

In the present work, we combined various adaptive meshing strategies like hierarchical, polynomial and relocation adaptivity with the Arlequin approach to couple shell and solid) and performed experimental (tensile) testing. We also applied forming methods such as deep drawing, stretch forming and V bending, and combined the findings to determine the optimal combination for the best formability with the shortest computation time. We arrived at the following findings.

Integrating shell (S4R) and solid (C3D8R) elements is effective because solid elements have greater capturing capacity in failure regions, and combining adaptive meshing with the Arlequin method provided noteworthy results with highly nonlinear materials, including improved convergence rate and time. From all implemented meshing techniques, combining relocation-adaptive meshing with the Arlequin approach provided more precise results with less CPU time. Relocating nodes in failure or critical regions affects both stress and material formability.

The main reason we implemented the current strategy is that whenever two geometric dimensions exceed a third one, it is compulsory to use shell elements, but this process has limited capture rates and long solving time, and incorporating a 3D (solid) element can improve accuracy. We highly recommend combining adaptive meshing with the Arlequin method to perform numeric simulations of the forming of sheet metal or plastic components, especially in fracture, thinning and deformation regions.

REFERENCES

[1] Zienkiewicz, O.C. and Zhu, J.Z., 1991. Adaptivity and mesh generation. *International Journal for Numerical Methods in Engineering*, *32*(4), pp. 783–810. https://doi.org/10.1002/nme.1620320409.

[2] Askes, H. and Sluys, L.J., 2000. Remeshing strategies for adaptive ALE analysis of strain localisation. *European Journal of Mechanics-A/Solids*, *19*(3), pp. 447–46. https://doi.org/10.1016/S0997-7538 (00)00176-5.

[3] Díez, P. and Huerta, A., 1999. A unified approach to remeshing strategies for finite element h-adaptivity. *Computer Methods in Applied Mechanics and Engineering*, *176*(1–4), pp. 215–229. https://doi.org/10.1016/S0045-7825(98)00338-7.

[4] Askes, H. and Rodríguez-Ferran, A., 2001. A combined rh-adaptive scheme based on domain subdivision: Formulation and linear examples. *International Journal for Numerical Methods in Engineering*, *51*(3), pp. 253–273. https://doi.org/10.1002/NME.142.

[5] Urekew, T.J. and Rencis, J.J., 1990. Absolute p-refinement of two-dimensional elasticity problems in the vicinity of boundary solution singularities. In *Boundary Element Methods in Engineering: Proceedings of the International Symposium on Boundary Element Methods: Advances in Solid and Fluid Mechanics East Hartford, Connecticut, USA, October 2–4, 1989* (pp. 193–199). Berlin, Heidelberg: Springer. https://doi.org/10.1007/978-3-642-84238-2_25.

[6] Qiao, H., Yang, Q.D., Chen, W.Q. and Zhang, C.Z., 2011. Implementation of the Arlequin method into ABAQUS: Basic formulations and applications. *Advances in Engineering Software*, *42*(4), pp. 197–207. https://doi.org/10.1016/j.advengsoft.2011.02.005.

[7] Dhia, H.B., 2006. Global-local approaches: The Arlequin framework. *European Journal of Computational Mechanics/Revue Européenne de Mécanique Numérique*, *15*(1–3), pp. 67–80. https://doi.org/10.3166/remn.15.67-80.

[8] Ikumapayi, O.M., Afolalu, S.A., Kayode, J.F., Kazeem, R.A. and Akande, S., 2022. A concise overview of deep drawing in the metal forming operation. *Materials Today: Proceedings*, *62*, pp. 3233–3238.

[9] Badrish, C.A., Kotkunde, N., Mahalle, G., Singh, S.K. and Mahesh, K., 2019. Analysis of hot anisotropic tensile flow stress and strain hardening behavior for Inconel 625 alloy. *Journal of Materials Engineering and Performance*, *28*, pp. 7537–7553. https://doi:10.1007/s11665-019-04475-4.

[10] Badrish, A., Morchhale, A., Kotkunde, N. and Singh, S.K., 2020. Influence of material modeling on warm forming behavior of nickel based super alloy. *International Journal of Material Forming*, *13*(3), pp. 445–465. https://doi.org/10.1007/s12289-020-01548-x.

[11] Badrish, A., Morchhale, A., Kotkunde, N. and Singh, S.K., 2020. Parameter optimization in the thermo-mechanical V-bending process to minimize springback of inconel 625 alloy. *Arabian Journal for Science and Engineering*, *45*, pp. 5295–5309. https://doi.org/10.1007/s13369-020-04395-9.

[12] Manual, A.S.U.S., 2012. Abaqus 6.11. *http://130.149*, *89*(2080), p. v6.

[13] Gokhale, N.S., 2008. *Practical Finite Element Analysis*. India: Finite to Infinite. ISBN:9788190619509, 8190619500 Page count:416.

[14] Dharavath, B., Morchhale, A., Singh, S.K., Kotkunde, N. and Naik, M.T., 2020. Experimental determination and theoretical prediction of limiting strains for ASS 316L at hot forming conditions. *Journal of Materials Engineering and Performance*, *29*, pp. 4766–4778.

[15] Raizer, A., Hoole, S.R.H., Meunier, G. and Coulomb, J.L., 1990. P-and H-type adaptive mesh generation. *Journal of Applied Physics*, *67*(9), pp. 5803–5805. https://doi.org/10.1063/1.345969.

[16] Badrish, C.A., Kotkunde, N., Salunke, O. and Singh, S.K., 2019. Study of anisotropic material behavior for inconel 625 alloy at elevated temperatures. *Materials Today: Proceedings*, *18*, pp. 2760–2766. https://doi.org/10.1016/j.matpr.2019.07.140.

16 Edge Formability in Sheet Metal Forming

*André Rosiak, Luana de Lucca de Costa,
and Lirio Schaeffer*

16.1 INTRODUCTION

In today's industry, marked by fierce competition, companies face a growing demand for high-quality products, driving them to minimize losses to the lowest possible level. In this context, the edge formability of metal sheets plays a crucial role. Understanding and controlling this property has become essential for ensuring efficient manufacturing processes, minimizing losses, and maximizing the quality of the final products.

The consequences of manufacturing losses are diverse and impactful. In addition to the material waste itself, waste consumes considerable energy and ultimately increases greenhouse gas emissions. Understanding and controlling the edge formability of metal sheets is essential for addressing these challenges and ensuring the efficiency and sustainability of industrial processes. Minimizing losses by reducing failures and waste not only enhances energy efficiency and preserves natural resources but also directly impacts the profitability of companies by reducing costs associated with rework and waste. This chapter aims to delve deeply into the edge formability of metal sheets, providing valuable insights on how to comprehend, evaluate, and control this fundamental property.

16.2 FLANGING

Modern sheet metal manufacturing methods primarily involve cutting, bending, and stretching. Among these processes is bending around holes. This operation, known as flanging, is incorporated into designs to reinforce edge regions. Flanged holes find application in bolted connections or serve as strengthening components within mechanical joints and bushings [1]. The conventional flanging process is schematically represented in Figure 16.1. This operation typically consists of a die, a punch, and a blank holder. The blank-holder pressure is fine-tuned to effectively grip the sheet, while the punch impels the hole's outer edge into the die.

Flanging can be executed with or without an ironing stage. In flanging without ironing (Figure 16.1(a)), the sheet is held in place between the die and the blank holder using suitable force, and the flange is shaped by stretching the edge. In contrast, flanging with ironing (Figure 16.1(b)) involves regulating a minor gap

DOI: 10.1201/9781003441755-16

between the punch and the die. Through this approach, metal is extruded between the punch and the die, yielding elongated flanges [2].

The edges of a part can exhibit different geometric arrangements as a result of flanging (Figure 16.2); during strain, the flange edge can experience tension or compression or remain undeformed. Concave bending is particularly prominent because it is commonly incorporated in stamping processes and because of the possibility of crack formation. Failure is usually linked to the material's elongating tangentially to the open edge, concomitant with contraction in the direction perpendicular to the open edge [3].

Forming a flange in conventional mild steels is relatively simple, but flanging high-strength steels is a more complex task that requires precautions; they have a high tendency to crack, which has restricted their use. Over the past few decades, these materials have gradually increased in use, leading to a heightened focus on fracture analysis within edge regions. Researchers have extensively explored the subject.

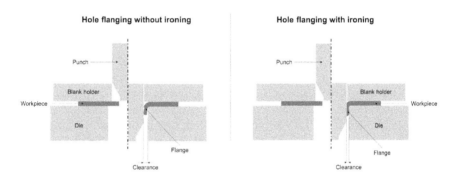

FIGURE 16.1   Conventional hole flanging: (a) without ironing; (b) with ironing.

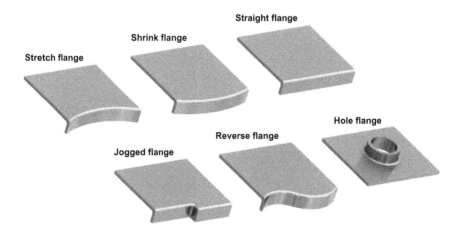

FIGURE 16.2   Different flanging geometries.

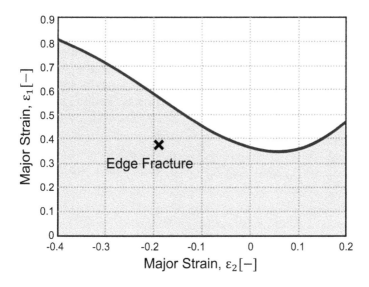

FIGURE 16.3 Comparison of edge fracture results with the FLC of AISI 1008 steel.

16.3 EDGE FORMABILITY

Sheet formability can be classified as bulk or edge formability. Bulk formability, also known as global formability, refers to the ability of a sheet to be deformed without failure in the body of the part. Edge formability, also referred to as local formability, encompasses the material's capacity to undergo deformation in edge regions or perforations without experiencing failure.

Global formability can be ascertained through techniques such as forming limit curves (FLC), deep-drawing experiments, stretching tests, and ductility assessments derived from standard tensile tests. However, these tools are not efficient for analyzing edge formability [4]. In Figure 16.3, the results of edge fracture are plotted on the forming limit diagram of AISI 1008 steel. The fracture of the punched-hole sheet occurs below the FLC, demonstrating that the FLC cannot successfully predict fracture in flanging.

16.4 EDGE FORMABILITY TEST

The hole expansion test is the prevalent approach for quantifying edge formability. The strain state imposed by the test is equivalent to that developed in concave flanging under production conditions. This makes the test ideal for predicting the behavior of a material subjected to edge stretching [5]. The test quantifies the edge formability of a material class with a specific edge condition. ISO 16630 is the commonly used standard for this test, but other test standards have also been developed such as JFS T 1001-96 and IISI–AHSS Guidelines Hole Expansion-2004.

The hole expansion test encompasses subjecting a flat sheet featuring a circular hole to the progressive penetration of a conical punch until cracks emerge at the edge (Figure 16.4). The test is halted at the initial indication of a crack extending through the sheet's entire thickness (Figure 16.5). The outcome of this test yields the hole expansion ratio (λ), a widely employed parameter for appraising the edge formability of sheet metal. This ratio is expressed by Equation 16.1:

$$\lambda = \frac{d_1 - d_0}{d_0} \cdot 100 \tag{16.1}$$

where d_0 and d_1 correspond, respectively, to the initial hole diameter and the diameter after test.

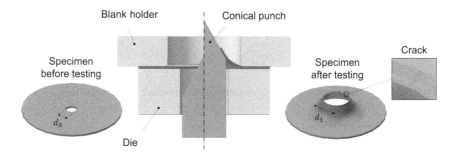

FIGURE 16.4 Schematic drawing of the hole expansion test.

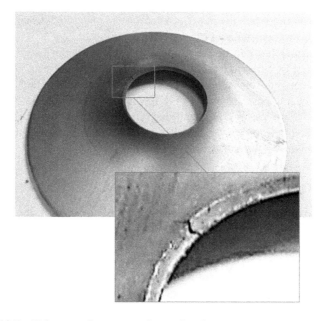

FIGURE 16.5 Hole expansion test specimen after the test.

Sample preparation involves shearing a hole through the sheet, and as previously mentioned, the cutting operation for hole fabrication influences edge formability. For this reason, the standard defines specific conditions for fabricating the hole in the samples. The standard also stipulates the clearance between the punch and the cutting die, along with the permissible tolerances for tool dimensions.

16.5 PARAMETERS THAT INFLUENCE EDGE FORMABILITY

Edge formability is a function of sheet properties and process design. Steel mills have made continuous efforts to enhance hole expansion capability by developing new high-performance products. At the same time, designers must evaluate whether a specific steel grade is compatible with the desired manufacturing route. If they are not compatible, the material may need to be replaced or the process modified, or possibly both. Edge formability is influenced by a number of factors: material microstructure; mechanical properties; hole preparation (shearing, laser cutting, water jet, etc.); hole diameter; burr position relative to the punch; and punch shape (conical, hemispherical, flat, etc.) [6].

16.5.1 MICROSTRUCTURE

From a microstructural perspective, the volumetric fraction and, more importantly, the type of microconstituents significantly affect the hole expansion capability. As a result, the same steel grade produced by different steel mills may exhibit significantly different edge formability. This aspect is critical in modern high strength steels. The complex microstructure of these steels generates an optimized balance of properties and influences edge formability [7].

Materials such as dual-phase (DP) steels, which differ greatly in hardness owing to their microconstituents (martensite and ferrite), tend to exhibit inferior performance. This is due to the concentration of strain in the softer and more ductile phase and the occurrence of microcracks at the ferrite/martensite interface (Figure 16.6). The edge formability of DP steels can be improved by achieving a fine and uniform distribution of martensite colonies [8].

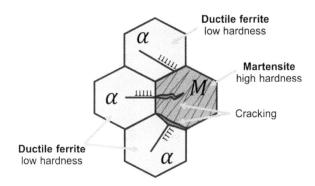

FIGURE 16.6 Failure mechanism of DP steels subjected to hole expansion.

Complex-phase (CP) steels exhibit better edge formability than DP steels. In CP steels, the stronger phase consists of a mixture of bainite and tempered martensite, and the relatively smaller difference in strength between the phases results in higher λ values. Due to bainite's higher ductility compared with martensite, in steels comprising this microconstituent, strain doesn't localize exclusively in the ferrite. Rather, the strain dispersion becomes more even, resulting in enhanced hole expansion potential.

In TRIP (transformation-induced plasticity) steels, the prematurely retained austenite that transforms into martensite during strain is detrimental to formability. The retained austenite tends to transform into martensite during the hole execution through the punching process. Consequently, the presence of low stability and a significant fraction of retained austenite tends to diminish the λ values.

16.5.2 Mechanical Properties

Experimental and numeric results from the hole expansion test of AISI 1008 carbon steel are used to enhance understanding of the stress and strain states developed during the forming process. The distribution of equivalent stress in the sample at the moment of failure is illustrated in Figure 16.7(a), while Figure 16.7(b) shows the distribution of the algebraically larger principal stress. Upon test completion, the stress condition at the edge of the central hole within the sample is primarily tensile in nature, analogous to the stress state generated during uniaxial tensile testing [9].

Figure 16.8(a) displays the pattern of equivalent strain dispersion within a hole expansion sample made from AISI 1008 steel posttest. Meanwhile, Figure 16.8(b) depicts a side-by-side representation of the experimental and numeric outcomes showcasing the fluctuations in principal strains, ε_1 and ε_2, across the sample.

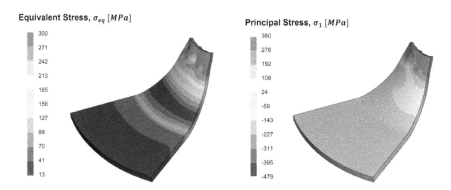

FIGURE 16.7 Distributions of (a) equivalent stress, σ_{eq}, and (b) algebraically larger principal stress, σ_1, in the sample at the moment of failure obtained through finite element analysis using FORGE software.

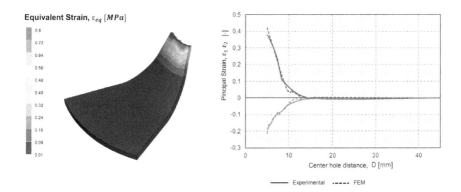

FIGURE 16.8 (A) Distribution of equivalent strain, ε_{eq}, in the AISI 1008 sample at the moment of failure obtained through finite element analysis using FORGE software. (b) Variation of principal strains, ε_1 and ε_2, along the sample.

It is evident that in the hole expansion process, strains are concentrated in the edge region, while the periphery of the sheet undergoes minimal plastic deformation. The maximum strain coincides with the location of material failure, namely at the edge of the sheet. The analysis of principal strains demonstrates that during hole expansion, the strain near the hole edge follows the path of uniaxial tension, where $\varepsilon_1 = -2\,\varepsilon_2$. As a result of the tensile stress and strain states in the edge region, material stretching occurs, leading to failure.

Figure 16.8(a) shows that expansion generates a strain gradient along the geometry. Therefore, despite the fundamentally tensile stress and strain states and the deformation path that coincides with uniaxial tension, the hole expansion test differs from the uniaxial tensile test due to the heterogeneous distribution of strain along the material. In a tensile test, at the macro scale, strain is homogeneous until the onset of necking. Furthermore, in a tensile test, fracture becomes noticeable solely within the necking zone, whereas within the hole expansion sample, multiple cracks can become apparent along the central edge [10]. In other words, even though the edge presents a deformed state analogous to uniaxial tension, tensile tests and hole expansion tests have a significant difference that prevent obtaining precise correlations between uniform elongation and hole expansion ratio. An example is TRIP steels, which, despite exceptional uniform elongation in tension, exhibit limited formability in terms of hole expansion [11].

Several researchers have sought to relate other properties to the hole expansion. Recent studies have shown that nonuniform elongation is a relevant parameter for edge formability. Materials showcasing heightened elongation post-necking typically yield enhanced outcomes in the hole expansion test. Moreover, increased hole expansion should correlate with a higher strain-hardening exponent and a higher coefficient of normal anisotropy. These properties restrict deformation in the thickness direction, enhancing resistance to necking and delaying premature failure in the thickness direction induced by flanging.

16.5.3 HOLE PREPARATION

The procedure for creating the central hole in the samples assumes a pivotal role in shaping the outcomes of the hole expansion test. Diverse hole preparation techniques are utilized in industrial applications, including punching, milling, wire-EDM cutting, laser cutting, and more. Each of these methods imparts varying degrees of damage to the hole edge, subsequently impacting the material's local formability [12–14].

Typically, for reasons of cost-efficiency, components are fabricated from steel sheets derived from coil cutting. Within these sheets, a single edge is shaped during the shearing stage, followed by subsequent stamping. In the course of punching, the material is essentially subjected to catastrophic shearing failure. However, the process doesn't perfectly achieve pure shear; instead, it engenders stress and strain gradients. The heterogeneous distribution of strain in the edge region results in strain hardening. Consequently, compared with other cutting processes, punching results in more damage to the material edge [15].

A punching operation yields a cutting edge characterized by three distinct zones and the presence of a burr (Figure 16.9), and the extension of each of these regions fluctuates based on punching conditions and material mechanical properties [16–18]. The slope of shear and the sheared surface are influenced by the cut's clearance. The extent of the fracture region and the height of the burr are regulated by the sheet's tensile strength.

In contemporary steels, an edge featuring a well-defined shear zone and a seamless transition to the fracture zone yields improved hole expansion outcomes. The fracture zone should be sleek and devoid of secondary shearing

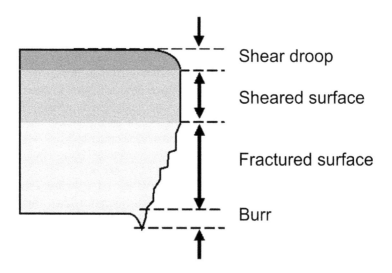

FIGURE 16.9 Schematic drawing showing the characteristic zones of a punched cutting edge; AHSS: advanced high-strength steel.

or impairment. If clearances are too low, secondary shearing might transpire, escalating fracture susceptibility due to the intricate microstructure. Likewise, a nonuniform shift from the shear zone to the fracture zone, as depicted in Figure 16.10, is undesirable.

To ensure the quality of the cutting surface and prevent degradation of the edge formability, control of tool wear is crucial. Over time, the quality of the punch edge decreases and the clearance changes due to wear mechanisms. The progression of tool wear damages the edge of the holes. The edges become unevenly work-hardened, and the flangeability is dramatically affected.

16.5.4 HOLE DIAMETER

While ISO 16630 designates a 10 mm diameter hole for hole expansion tests, it's important to note that the initial hole diameter directly impacts the achievable expansion extent. That is, the initial diameter should be considered when selecting the work material for a specific process. Tests should be conducted on sheets with hole diameters close to those used in the production process [19].

16.5.5 BURR POSITION

The positioning of the cutting burr in relation to the punch significantly influences the outcomes of a hole expansion test. The negative impacts are mitigated when the burr is situated on the punch side (as illustrated in Figure 16.11). This effect is primarily due to the reduced tension exerted on the surface near the punch during expansion. Correct burr positioning can increase edge expansion capability by up to 20%. This factor should be duly acknowledged in designs that are susceptible to edge expansion issues.

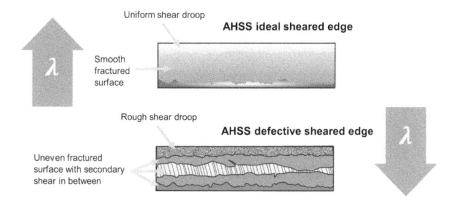

FIGURE 16.10 Schematic drawings of cut surface of advanced high-strength steel sheets.

In the standardized test according to ISO 16630, the test specimens are positioned with the burr on the opposite side of the punch. In this way, the material is evaluated under the most critical condition in terms of edge formability.

16.5.6 PUNCH GEOMETRY

The geometry of the flanging punch affects the metal flow during forming and, consequently, the extent to which the steel can be deformed before fracture [20]. Figure 16.12 shows the numeric results of Cockcroft–Latham damage obtained in QFORM software for different punch geometries with the same hole expansion.

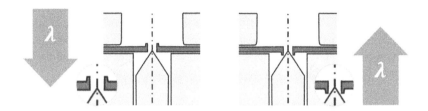

FIGURE 16.11   Effects of burr position on the hole expansion ratio.

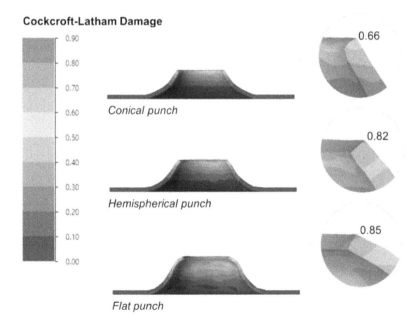

FIGURE 16.12   Numeric result of Cockcroft–Latham damage distribution for hole expansion with different punch geometries.

The use of a conical punch results in bending and stretching of the material edge. On the other hand, the use of a flat or hemispherical punch eliminates bending, and all deformation occurs only in the form of edge stretching. These differences in strain state lead to different levels of damage; sheets deformed with a conical punch show the least damage. This indicates that conical punches can expand holes farther than can other geometries.

While hole expansion tests utilizing hemispherical and flat punches are not standardized techniques, they have been explored in research. Investigations indicate that in these punch geometries, the point of failure occurs slightly removed from the edge. This observation suggests that the edge condition's impact is diminished when employing flat and hemispherical punches.

16.6 CONCLUSION

Understanding the edge formability of sheet metal is crucial for the design and manufacturing of stamped components. Accurate quantification of this property allows for proper material selection and optimization of forming processes, resulting in high-quality end products.

When analyzing edge formability, the cutting process plays a crucial role in maximizing local formability characteristics. The geometry and cutting conditions can influence the hole expansion capability of the material, directly impacting edge formability. Therefore, careful design of preflanging operations is necessary to maximize the deformation properties of sheet edges.

Another important factor to consider is the influence of the microstructural characteristics of each type of steel on hole expansion capability. Variations in chemical composition and heat treatments can significantly alter edge formability, making a detailed analysis of the microstructural properties of the materials used essential. The success of a flanging operation depends on the precise control of factors from the metallurgical properties of the materials to the adopted manufacturing route. In summary, in quantifying edge formability, processing conditions, proper design of the cutting process, analysis of microstructural characteristics, and careful control of involved factors are key to achieving excellence in flanging operations.

REFERENCES

[1] S.K. Paul, "A critical review on hole expansion ratio", *Materialia*, vol 9, 2020, https://doi.org/10.1016/j.mtla.2019.100566.

[2] Y. Dewang, R. Purohit and N. Tenguria, "A study on sheet metal hole-flanging process", *Materials Today: Proceedings*, vol 4, Issue 4, Part D, pp. 5421–5428, 2017, https://doi.org/10.1016/j.matpr.2017.05.053.

[3] C.T. Wang, "Mechanics of bending, flanging, and deep drawing and a computer-aided modeling system for predictions of strain, fracture, wrinkle, and springback in sheet metal forming", Ph.D. Dissertation, The Ohio State University, 1993.

[4] N. Pathak, C. Butcher and M. Worswick, "Assessment of the critical parameters influencing the edge stretchability of advanced high-strength steel sheet", *J. of Materi Eng and Perform*, vol 25, pp. 4919–4932, 2016, https://doi.org/10.1007/s11665-016-2316-9.

[5] S.K. Paul, "Non-linear correlation between uniaxial tensile properties and shear-edge hole expansion ratio", *J. of Materi Eng and Perform*, vol. 23, pp. 3610–3619, 2014, https://doi.org/10.1007/s11665-014-1161-y.

[6] A. Karelova, C. Krempaszky, E. Werner, P. Tsipouridis, T. Hebesberger and A. Pichler, "Hole expansion of dual-phase and complex-phase AHS steels—Effect of edge conditions", *Materials Technology*, 2008, https://doi.org/10.2374/SRI08SP110.

[7] M. Shi and X. Chen, "Prediction of stretch flangeability limits of advanced high strength steels using the hole expansion test", *SAE Technical Paper*, 2007-01-1693, 2007, https://doi.org/10.4271/2007-01-1693.

[8] O. Terrazas, K.O. Findley and C.J. Van Tyne, "Influence of martensite morphology on sheared-edge formability of dual-phase steels", vol 57, Issue 5, pp. 937–944, 2017, https://doi.org/10.2355/isijinternational.ISIJINT-2016-602.

[9] C. Butcher, D. Anderson and M. Worswick, "Predicting failure during sheared edge stretching using a damage-based model for the shear-affected zone", *SAE Int. J. Mater. Manf.*, vol 6, Issue 2, pp. 304–312, 2013, https://doi.org/10.4271/2013-01-1166.

[10] C. Chiriac and G. Chen, "Local formability characterization of AHSS-digital camera based hole expansion test development, best in class stamping", *International Deep Drawing Research Group IDDRG 2008 International Conference*, 16–18 June 2008, Olofström, Sweden, pp. 1–11.

[11] D. Hyun, S. Oak, S. Kang and Y. Moon, "Estimation of hole flangeability for high strength steel plates", *Journal of Materials Processing Technology*, vol 130–131, Issue20, pp. 9–13, 2002, https://doi.org/10.1016/S0924-0136(02)00793-8.

[12] A. Konieczny and T. Henderson, "On formability limitations in stamping involving sheared edge stretching", *SAE Technical Paper*, 2007-01-0340, 2007, https://doi.org/10.4271/2007-01-0340.

[13] N. Pathak, C. Butcher and M. Worswick, "Influence of the sheared edge condition on the hole expansion of dual phase steel", *IDDRG 2013 Conference*, Swedish Deep Drawing Research Group, 2013.

[14] H.-C. Shih, C. Chiriac and M.F. Shi, "The effects of AHSS shear edge conditions on edge fracture", *ASME 2010 International Manufacturing Science and Engineering Conference*, 2010, 10.1115/MSEC2010-34062.

[15] K.I. Mori, Y. Abe and Y. Suzui, "Improvement of stretch flangeability of ultra high strength steel sheet by smoothing of sheared edge", *J. Mater. Proc. Technol.*, vol 210, pp. 653–659, 2010, https://doi.org/10.1016/j.jmatprotec.2009.11.014.

[16] J.I. Yoon, J. Jung, J.G. Kim, S.S. Sohn, S. Lee and H.S. Kim, "Key factors of stretch-flangeability of sheet materials", *J Mater Sci*, vol 52, 7808–7823, 2017, https://doi.org/10.1007/s10853-017-1012-y.

[17] M. Li, C.J. Vantyne and Y.H. Moon, "The effect of mechanical properties on hole flangeability of stainless steel sheets", *J. Mech. Sci. Technol.*, vol 29, pp. 5233–5239, 2015, https://doi.org/10.1007/s12206-015-1123-9.

[18] X. Chen, H. Jiang, Z. Cui, C. Lian and C. Lu, "Hole expansion characteristics of ultra high strength steels", *Procedia Eng.*, vol 81, pp. 718–723, 2014, https://doi.org/10.1016/j.proeng.2014.10.066.

[[19] H. Kim, J. Dykeman, A. Samant, and Cliff Hoschouer, "Prediction and Reduction of Edge Cracking in Forming Advanced High-Strength Steels", *2017 Great Designs in Steel*, sponsored by American Iron and Steel Institute.

[20] R. Wiedenmann, P. Sartkulvanich, and T. Altan, "Finite element analysis on the effect of sheared edge quality in blanking upon hole expansion of advanced high strength steel," *International Deep Drawing Research Group (IDDRG) Conference*, 2009, Golden, CO, USA.

17 Finite Element Analysis of AA6061/AA5754 Bimetallic Sheets by Constrained Groove Pressing

Vedat Taşdemir, Ömer Seçgin, and Nuri Şen

17.1 INTRODUCTION

Severe plastic deformation (SPD) can be described as an ultrafine grain (UFG) and nano-structured metal forming method based on to the extreme plastic deformation of the materials (Głuchowski et al., 2011; Solhjoei et al., 2014; Nazari and Honarpisheh, 2018). SPD reduces the material grain size to nano (smaller than 100 nm) and ultrafine (100 nm–1 µm) grains, which increases the strength of the material (Solhjoei et al., 2014; Shahmirzaloo et al., 2018). In UFG structured materials, SPD causes superplastic behavior with high deformation rates at very low temperatures (Güral et al., 2012; Yadav et al., 2016).

The main purpose of SPD is to produce high-strength, lightweight materials that are compatible with the environment (Nazari and Honarpisheh, 2018). In traditional metal forming methods such as rolling, forging, and extrusion, materials typically undergo no or only one plastic deformation because it is not appropriate to use materials that have undergone too much thinning in forming processes such as multistep (transitive) rolling, drawing, or extrusion (Taşdemir, 2020). Researchers have conducted extensive investigations of SPD methods of producing flawless, homogeneous products from parts subjected to very large deformations.

Ultrafine-grained materials produced by SPD very superior properties (Valiev et al., 2016). Some of the methods used in forming sheet materials can be listed as follows: accumulative roll bonding (Tamimi et al., 2014), cone–cone (Bouaziz, Estrin, and Kim, 2009), CGP (Lee and Park, 2002), friction stir processing (Ma, 2008), equal channel angular rolling (Chen, Cheng, and Xia, 2007), repetitive corrugation and straightening (RCS) (Sunil, 2015), RCS by rolling (Thangapandian and Balasivanandha Prabu, 2017), asymmetric rolling (Ucuncuoglu et al., 2014), differential speed rolling (Taşdemir, 2022), continuous frictional angular extrusion (Huang and Prangnell, 2007), and continuous cyclic bending (Takayama et al., 2004).

DOI: 10.1201/9781003441755-17

Among these methods, CGP and RCS are often confused with each other. The main difference between CGP and RCS is that in CGP the sheet material is constrained both in width and length. RCS, on the other hand, is usually free along the width or the length or both. This affects both the mechanical and dimensional properties of the part to be shaped.

Many researchers have studied and continue to work on the mechanical and microstructural properties of sheet materials shaped using SPD. The purpose of this study is to analyze the effects of different groove angles, friction coefficients, and contact types between bimetallic materials on effective plastic deformation with the help of finite element analysis (FEA).

17.2 CONSTRAINED GROOVE PRESSING

CGP is a severe plastic deformation method used to obtain ultrafine-grained or nanostructured sheet materials. In this method, the sheet material is first taken between the molds with grooves on it (Figure 17.1(a)) and pressed (Figure 17.1(b)). Then, the pressed sheet material is flattened by being pressed between flat dies (Figure 17.1(c)). The sheet material that has been flattened and turned 180° (Figure 17.1(d)) is pressed a second time between the grooves (Figure 17.1(e)) and flattened again (Figure 17.1(f)). This second pressing and flattening ensures the deformation of the undeformed regions and homogeneous plastic deformation at every point. This process is repeated depending on the grain structure and strength desired to be obtained. Figure 17.1 shows the schematic of the CGP method.

Theoretical equations for determining effective strain and shear strain can be derived using Figure 17.2 (for one pressing). A single pressing causes shear strain as follows (Wang et al., 2015):

$$\gamma_{xy} = \frac{h}{w} = \tan\theta \qquad (17.1)$$

Here, w, h, and θ are the width, height, and angle of the groove, respectively.
Shear strain is calculated as

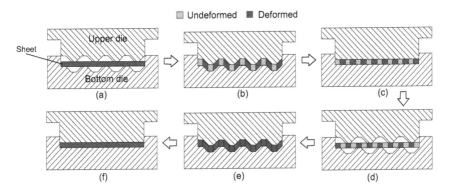

FIGURE 17.1 Schematic representation of constrained groove pressing (CGP) process

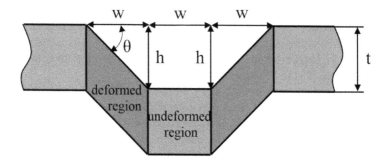

FIGURE 17.2 Deformation in a pressed material: t, material thickness; w, groove width; h, groove height; θ, groove angle.

$$\varepsilon_{xy} = \frac{\gamma_{xy}}{2} \tag{17.2}$$

Since CGP assumes shear deformation under the condition of planar strain (Shirdel, Khajeh, and Moshksar, 2010), the corresponding effective strain is as follows:

$$\varepsilon_{eff} = \sqrt{\frac{2}{9}\left[\left(\varepsilon_x - \varepsilon_y\right)^2 + \left(\varepsilon_y - \varepsilon_z\right)^2 + \left(\varepsilon_z - \varepsilon_x\right)^2\right] + \frac{4}{3}\left[\left(\varepsilon_{xy}\right)^2 + \left(\varepsilon_{yz}\right)^2 + \left(\varepsilon_{zx}\right)^2\right]} \tag{17.3}$$

$$\varepsilon_x = \varepsilon_y = \varepsilon_z = \varepsilon_{yz} = \varepsilon_{zx} = 0$$

$$\varepsilon_{eff} = \sqrt{\frac{4\varepsilon_{xy}^2}{3}} = \frac{\gamma_{xy}}{\sqrt{3}} = \frac{\tan\theta}{\sqrt{3}} \tag{17.4}$$

The total effective strain accumulated in the CGP sample as a result of N passes is calculated as

$$\varepsilon_t = N\frac{2\tan\theta}{\sqrt{3}} \tag{17.5}$$

17.3   FINITE ELEMENT ANALYSIS

FEA is a method of more effectively explaining deformation behaviors. For the study for this chapter, we conducted analyses using Simufact FE software and two alloys: AA6061 with a thickness of 0.8 mm (sheet 1, upper sheet) and AA5754 with a thickness of 1.2 mm (sheet 2, lower sheet); we took the properties of the material used from the software's library of materials. We performed the analyses at room temperature using a punch speed of 0.1 mm/s. We designed the top die, bottom die, and sheet materials in three dimensions but the analyses in two dimensions. The process parameters and material constants used in the FE analysis are given in Table 17.1, and the die set is given in Figure 17.3.

TABLE 17.1

CGP Process Parameters for Finite Element Analysis

Parameters	Values
Material	AA6061 and AA5754
Object type	Material: Elastoplastic
	Die: Rigid
	Punch: Rigid
Mesh properties	Mesher: Advancing Front Quad
	Element type: Quads, plane strain
	Element size: 0.3 mm
	Mesh created: Yes
Friction	Friction law: Coulomb
	Coefficient (μ): 0.05, 0.1, 0.2
Contact type	Touching (free surface), Glued
(between material surface)	(fixed surface)
Punch speed (mm/s)	0.1
Groove angle, (°)	30, 45, 60

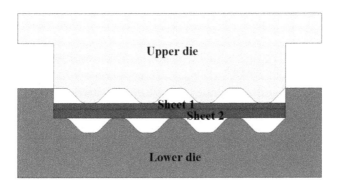

FIGURE 17.3   Cross-sectional view of the experimental setup used in the analysis.

17.4 RESULTS AND DISCUSSION

Plastic strain has an important effect on both internal microstructure homogeneity and hardness (Faraji et al., 2012); higher plastic strain provides stronger, finer-grained materials. Figure 17.4 gives the effective plastic strain for one pressing; the figure shows that the plastic deformation increases as the groove angle increases. The type of contact between materials and the groove angle also clearly affect the amount of deformation: At a 30° grove angle, the maximum plastic deformation was 0.39 with glued contact and 0.65 with touching contact. The findings were similar situation for forming with dies at 45° and 60° groove angles. In addition, it can be said that the die-formed material with a 60° groove angle exhibits a less uniform strain behavior than the others (Mendoza-Cuesta et al., 2021).

Additionally, whereas there is independent movement between the two materials with touching contact, there is an inseparable boundary between the two material surfaces with glued contact (Figure 17.5), and this also affects the results.

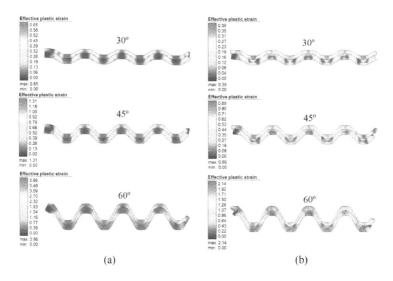

(a)   (b)

FIGURE 17.4   Effective plastic strain at all groove angles for one pressing: a) glued contact; b) touching contact (friction: 0.1).

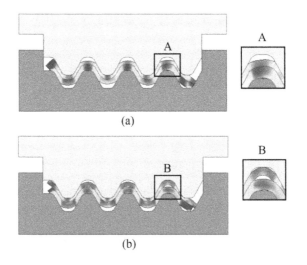

FIGURE 17.5   Views of the forming process: a) glued contact; b) touching contact (60° groove angle, 0.1 friction).

With touching contact, the deformation difference between the upper and lower sheets increases as the channel angle increases; when parts are shaped with a 60° die, significant deformation occurs in the upper part with a thickness of 0.8 mm. In this case, it can be said that as the groove angle increases, the deformation to the thin parts increases.

Figure 17.6 displays the effective plastic strain occurring at all friction coefficients for one pressing (Pressing 1); the figure graphically demonstrates that the type of contact is very important: While the lowest effective plastic strain was found at a friction coefficient of 0.05 with glued contact, it was obtained at a friction coefficient of 0.2 in the with touching contact. Because there is no movement between the sheets in glued contact, there was no significant change in plastic deformation. However, with touching contact, the sheets can move independently of each other. This may have resulted in both a lower and a more heterogeneous plastic deformation.

Figure 17.7 gives the results of the effective plastic strain resulting from one pass; it is evident that the effective plastic strain deformation increases as the number of pressings increases. The figure also shows that following the second flattening, there are very slightly deformed parts in the lower part. We considered that this was because the upper and lower sheets are two independent parts and their interfaces slide over each other during shaping. The figure shows that the upper sheet, with a thickness of 0.8 mm and located close to the movable upper die, undergoes more deformation. We consider that the independent upper

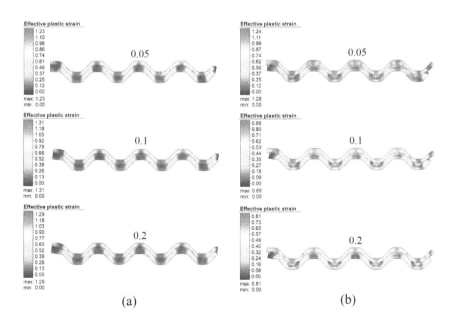

(a)   (b)

FIGURE 17.6   Effective plastic strain at all frictions for one pressing: a) glued contact; b) touching contact (groove angle: 45°).

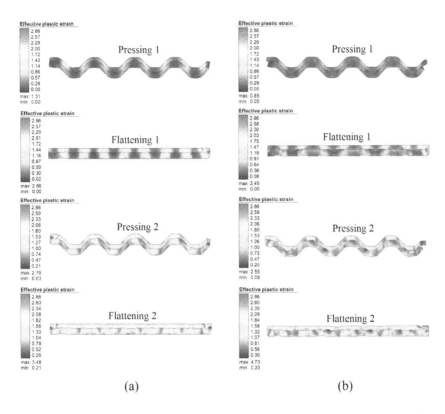

FIGURE 17.7   Effective plastic strain obtained for one pass: a) glued contact; b) touching contact (45° groove angle, 0.1 friction).

sheet prevents the lower sheet from undergoing further deformation. As the number of passes increases, the tensile strength, yield strength, and hardness of the parts increase, but their ductility/elongation decreases (Mohammadi Aghbolagh, Alimirzaloo, and Khamedi, 2020).

17.5   CONCLUSIONS

In this study, we performed finite element analysis of bimetallic sheet material pairs using constrained groove pressing, a severe plastic deformation method. We identified the following results. CGP increases strength by reducing the grain structure of sheet materials, and its usage is becoming increasingly widespread. We identified that the contact type between the bimetal sheet pair formed is an important factor. The sliding of the upper and lower sheets against each other during forming directly affects the results: The effective plastic strain increased as the groove angle increased. However, the effective plastic strain became less uniform as the groove angle increased. We found the lowest effective plastic strain at a friction coefficient of 0.2 with touching contact, but the coefficient was 0.05

with glued contact. Additionally, glued contact gave more consistent results than touching contact.

REFERENCES

Bouaziz, O., Estrin, Y. and Kim, H. S. (2009) 'A new technique for severe plastic deformation: The cone-cone method', *Advanced Engineering Materials*, 11(12), pp. 982–985. doi: 10.1002/adem.200900217.

Chen, Z. H., Cheng, Y. Q. and Xia, W. J. (2007) 'Effect of equal-channel angular rolling pass on microstructure and properties of magnesium alloy sheets', *Materials and Manufacturing Processes*, 22(1), pp. 51–56. doi: 10.1080/10426910601015907.

Faraji, G. *et al.* (2012) 'Deformation behavior in the Tubular Channel Angular Pressing (TCAP) as a noble SPD method for cylindrical tubes', *Applied Physics A: Materials Science and Processing*, 107(4), pp. 819–827. doi: 10.1007/s00339-012-6809-6.

Głuchowski, W. *et al.* (2011) 'Ultrafine grained copper alloys processed by continuous repetitive corrugation and straightening method', *Materials Science Forum*, 674, pp. 177–188. doi: 10.4028/www.scientific.net/MSF.674.177.

Güral, A. *et al.* (2012) 'Microstructural characterization of 7075 aluminum alloy severe deformed by Equal Channel-Angular Pressing (ECAP)', *Journal of the Faculty of Engineering and Architecture of Gazi University*, 27(4), pp. 807–812.

Huang, Y. and Prangnell, P. B. (2007) 'Continuous frictional angular extrusion and its application in the production of ultrafine-grained sheet metals', *Scripta Materialia*, 56(5), pp. 333–336. doi: 10.1016/j.scriptamat.2006.11.011.

Lee, J. W. and Park, J. J. (2002) 'Numerical and experimental investigations of constrained groove pressing and rolling for grain refinement', *Journal of Materials Processing Technology*, 130–131, pp. 208–213. doi: 10.1016/S0924-0136(02)00722-7.

Ma, Z. Y. (2008) 'Friction stir processing technology: A review', *Metallurgical and Materials Transactions A: Physical Metallurgy and Materials Science*, 39A(3), pp. 642–658. doi: 10.1007/s11661-007-9459-0.

Mendoza-Cuesta, A. *et al.* (2021) 'Finite element analysis of constrained groove pressing on strain behavior of Armco Iron Sheets', *Materials Engineering*, 23(2), p. e21611262. doi: 10.25100/iyc.v23i2.11262.

Mohammadi Aghbolagh, V., Alimirzaloo, V. and Khamedi, R. (2020) 'Constrained groove pressing process of Al/Cu bimetal sheet', *Materials and Manufacturing Processes*, 35(2), pp. 130–141. doi: 10.1080/10426914.2019.1692351.

Nazari, F. and Honarpisheh, M. (2018) 'Analytical model to estimate force of constrained groove pressing process', *Journal of Manufacturing Processes*, 32, pp. 11–19. doi: 10.1016/j.jmapro.2018.01.015.

Shahmirzaloo, A. *et al.* (2018) 'Interface sheet-constrained groove pressing as a modified severe plastic deformation process', *Materials Science and Technology (United Kingdom)*, 34(14), pp. 1669–1678. doi: 10.1080/02670836.2018.1471379.

Shirdel, A., Khajeh, A. and Moshksar, M. M. (2010) 'Experimental and finite element investigation of semi-constrained groove pressing process', *Materials and Design*, 31(2), pp. 946–950. doi: 10.1016/j.matdes.2009.07.035.

Solhjoei, N. *et al.* (2014) 'A comparative study to evaluate the efficiency of Rcs and Cgp processes', *Indian Journal of Scientific Research*, 1(2), pp. 563–572.

Sunil, B. R. (2015) 'Repetitive corrugation and straightening of sheet metals', *Materials and Manufacturing Processes*, 30(10), pp. 1262–1271. doi: 10.1080/10426914.2014.973600.

Takayama, Y. *et al.* (2004) 'Microstructural and textural evolution by continuous cyclic bending and annealing in a high purity titanium', *Materials Transactions*, 45(9), pp. 2826–2831. doi: 10.2320/matertrans.45.2826.

Tamimi, S. *et al.* (2014) 'Accumulative roll bonding of pure copper and IF steel', *International Journal of Metals*, 2014, pp. 1–9. doi: 10.1155/2014/179723.

Taşdemir, V. (2020) 'Severe plastic deformation analysis of tubular parts via Double Pass Parallel Tubular Channel Angular Pressing (DP-PTCAP) method', *Fırat University Journal of Engineering Science*, 32(2), pp. 313–324.

Taşdemir, V. (2022) 'Finite element analysis of the springback behavior after V bending process of sheet materials obtained by Differential Speed Rolling (DSR) method', *Revista de Metalurgia*, 58(1), pp. 1–9. doi: 10.3989/REVMETALM.219.

Thangapandian, N. and Balasivanandha Prabu, S. (2017) 'Effect of combined repetitive corrugation and straightening and rolling on the microstructure and mechanical properties of pure aluminum', *Metallography, Microstructure, and Analysis*, 6(6), pp. 481–488. doi: 10.1007/s13632-017-0400-7.

Ucuncuoglu, S. *et al.* (2014) 'Effect of asymmetric rolling process on the microstructure, mechanical properties and texture of AZ31 magnesium alloys sheets produced by twin roll casting technique', *Journal of Magnesium and Alloys*, 2(1), pp. 92–98. doi: 10.1016/j.jma.2014.02.001.

Valiev, R. Z. *et al.* (2016) 'Producing bulk ultrafine-grained materials by severe plastic deformation: Ten years later', *Jom*, 68(4), pp. 1216–1226. doi: 10.1007/s11837-016-1820-6.

Wang, Z. S. *et al.* (2015) 'Influences of die structure on constrained groove pressing of commercially pure Ni sheets', *Journal of Materials Processing Technology*, 215(1), pp. 205–218. doi: 10.1016/j.jmatprotec.2014.08.018.

Yadav, P. C. *et al.* (2016) 'Microstructural inhomogeneity in constrained groove pressed Cu-Zn alloy sheet', *Journal of Materials Engineering and Performance*, 25(7), pp. 2604–2614. doi: 10.1007/s11665-016-2142-0.

18 Fracture Formation Limits in Sheet Metal Forming

Hitesh Dnyaneshwar Mhatre and Amrut Mulay

18.1 INTRODUCTION

Incremental sheet forming (ISF) represents an adaptable approach to dieless sheet metal shaping, finding applications across diverse sectors such as the automotive, aerospace, and biomedical industries and enabling the production of intricate sheet metal parts [1], [2]. The shaping tool movement causes the central area of maximum strain in ISF to undergo continuous displacement. This dynamic aspect retards the localization of strain, extending the capacity for overall deformation before eventual fracture.

Failure in ISF results from the interplay between material behavior and the stress conditions it exposes [3]. ISF subjects the deforming specimen to multiaxial stress, and these loading conditions and strain patterns vary within the material depending on the component's shape under production. This phenomenon can be better understood by monitoring how strain evolves within this contact region.

Unlike in conventional forming processes that often result in local necking, ISF is characterized by fracture; hence, it relies on fracture forming limit (FFL) curves rather than the traditional forming limit curves (FLC) to define its deformation limits [4], [5]. Despite the enhanced formability of ISF compared with conventional techniques, it is limited by gradual thinning and eventual fracture [5], [6]. An examination of failed sections reveals ductile characteristics within the cracks located in the ISF region [7]. In a 2D strain plot, the standard FFL is a straight line, varying from uniaxial to equibiaxial tension modes, similar to FLC but with a constant slope [6].

Accurately anticipating fractures in ISF holds paramount importance for effective process planning. In the realm of material mechanics, the utilization of continuum damage models proves highly effective in providing a theoretical estimation of critical fracture thresholds [8]. These damage models are commonly classified as coupled or uncoupled [9]. In coupled models, the damage variable influences plasticity relations, whereas in uncoupled models like the modified Mohr–Coulomb [10] or Hosford–Coulomb [11] models, damage progression occurs without direct interaction with the stress–strain relationship. When the cumulative damage unpredictably influences a predefined significant level in uncoupled models, failure initiation is determined, making them more computationally tractable and increasingly preferred [12].

To assess material damage, Martins et al. [13] employed a fracture criterion focused on hydrostatic stress, whereas Oyane et al. [14] applied the Oyane fracture criterion to gauge damage within magnesium ISF. In a finite element analysis of a simple cone, Malhotra et al. [15] used a damage technique that revealed that damage accumulates mostly in locations characterized by local bend over within the sheet during deformation. Mirnia et al. [16] embraced the modified Mohr–Coulomb fracture criterion, and their statistical assessments underscored comparable outcomes. They highlighted notable damage accumulation within the intermediate region connecting the deformation and contact zones.

18.2   THEORETICAL ASPECT

In the lead-up to fracture, the strain patterns observed in truncated tapering and pyramidal sections manufactured through single-point incremental forming (SPIF), along with those derived from in-plane torsion shear tests, demonstrate a noticeable proportionality ($\beta = d\ \varepsilon 2/d\ \varepsilon 1 = \varepsilon 2/\varepsilon 1$). This quality enables the evaluation of formability limits using experimental data related to the significant and minor in-plane stresses next to outset of fracture. Furthermore, it simplifies the establishment of connections with fundamental principles in plasticity, damage mechanics, and ductile fracture mechanics [17].

18.2.1   FRACTURE FORMING LIMIT LINE

Isik et al. [17] introduced a fresh perspective on the formation limits of fracture in sheet metal forming under tension among in-plane shear. Figure 18.1(a) visually represents the FFL on the primary strain diagram, depicted as a straight line with a negative slope of −1 descending from left to right:

$$\varepsilon_{1f} + \varepsilon_{2f} = C \tag{18.1}$$

where

$$C = \frac{3}{2}\left(\ln\frac{l}{2r}\right)$$

Equation 18.1 defines C about common microstructural features like the spacing between inter-hole l (interparticle inclusions) and the radius of the hole particles, r [18].

In metal forming, a slope of −1 on the FFL indicates fracture failure, a critical level of thickness reduction, denoted as $R_f = (t_0\_t_f)/t_0$, that remains constant regardless of the strain loading conditions:

$$\varepsilon_{1f} + \varepsilon_{2f} = -\ln(1 - R_f) \tag{18.2}$$

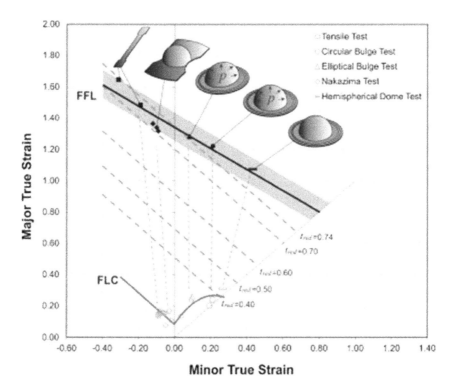

FIGURE 18.1 Forming limits of aluminum AA1050-H111 constructed from several sheet formability tests in the principal strain space [17].

Here, t_0 represents the original, i.e., undeformed sheet thickness, while t_f represents the sheet thickness at fracture. This intersection of the FFL with the y-axis occurs within one primary in-plane strain plot:

$$C = -\ln\left(1 - R_f\right) \qquad (18.3)$$

where $C = \varepsilon_{1f}$ under conditions of plane strain (with $\varepsilon_{2f} = 0$) signifies the fundamental level of primary in-plane strain at the initiation of fracture.

Martins et al. [19] subsequently illustrated that fracture border specified by Eq. 18.1 may be expressed as follows:

$$D_{crit} = \frac{1+r}{3}\left[\varepsilon_1 + \varepsilon_2\right] \qquad (18.4)$$

The anisotropy coefficient is denoted as r, and the critical degree of ductile damage is symbolized as D_{crit}, which is derived from McClintock's [20] non-coupled void growth damage-based criterion. Additionally, Martins et al. [19]

established that equation (18.4) should be revised by incorporating the subsequent term:

$$D_{crit} = \frac{1+r}{3}\left[\varepsilon_1 + \varepsilon_2 - \left(1+\beta\right)\varepsilon_o\right] \qquad (18.5)$$

where $\qquad \beta = \dfrac{d\varepsilon_2}{d\varepsilon_1}.$

In materials where the onset of damage accumulation occurs only after surpassing a defined threshold strain, εo, the FFL displays an upward extension, making it inappropriate for a linear regression. In these instances, a curve should be employed instead of a straight line.

Variations in critical damage at the place of fracture, anisotropic characteristics, and threshold strain levels are indicative of the material influence on the FFL. FFL is unaffected by the process because ductile strength and fracture strength are basic material characteristics, unlike FLC, which is a nonmaterial quality and is dependent on the strain route [21].

Martins et al. [19] extensively investigate the intricate relationships among the FFL, ductile damage, fracture toughness, and crack initiation under tensile stresses. Unlike the FLC, which has a well-established experimental determination approach through Nakajima or Marciniak formability tests [22], [23], and ISO 12004–2:2021. Currently, there is no standardized experimental procedure for determining the FFL in sheet metal forging. When comparing fracture stress data from different research projects, the lack of internationally recognized and controlled procedures for measuring the FFL in sheet metal forming causes obstacles and uncertainty.

This study describes an experimental technique for deciding the fracture formation limit in sheet metal forming. The study describes a step-by-step approach, including exact measurement sites and the reasoning for estimating gauge length strains, that indicates the stresses at which specimens shatter during testing [21]. While the major focus is on identifying the FFL, the suggested approach may also be used to determine the FFL under in-plane shear situations. According to the findings of Martins et al. [19], the visual representation of the initiation of fracture under in a plane shear is a proportional decline from the left with a slope of +1 that goes perpendicular to the direction of fracture within the main strained area.

18.3   EXPERIMENTAL METHOD AND PROCEDURE

18.3.1   DETERMINATION OF FFL

To identify the FFL and encompass multiple strain loading scenarios within the primary strain range, it is necessary to create test samples with different geometries and subject them to varied loading circumstances. The outlines of these specimens, which were determined from mechanical, fracture, and formability characterization tests, are shown in Figure 18.1 [21]. Prior to initiating the tests, it is crucial to prepare the specimens by employing electrochemical etching to create a lattice of circles with diameter d. The selection of the appropriate electrolyte, current, voltage, and neutralizer should be tailored to the specific sheet material in use and should align with the requirements of the marking equipment.

Contrary to FLC, the FFL cannot be easily determined from circular lattice in-plane strain values. This limitation arises because the resultant values cannot accurately describe fracture stresses even when very tiny circles are positioned near the fracture locations. To obtain the fracture strain ε_{3f} in the direction of thickness and estimate the thickness of the sheet at fracture, tf, it is essential to section the samples close to the fractures, t_f [21].

To calculate the thickness of the sheet at fracture point, t_f, an average was computed as $t_f = \sum_{i=1}^{n_m} \left(\dfrac{t_f}{n_m} \right)$, based on multiple individual measurements (t_{fi}) taken along the crack. These measurements were obtained employing a digital camera-equipped optical microscope with an accuracy of ± 0.0001 mm [21]. Figure 18.2(a) provides a schematic representation of this technique, and the determination of ε_{3f} is as follows:

$$\varepsilon_{3f} = \ln \left(\frac{t_f}{t_o} \right) \tag{18.6}$$

By examining Figure 18.2(b) and calculating the minor axis, b, of the ellipse produced due to the plastic distortion and circle fracture during the test, it is possible to determine the corresponding fracture strain, ε_{2f}, on the specimen surface [21]:

$$\varepsilon_{2f} = \ln \left(\frac{b}{d} \right) \tag{18.7}$$

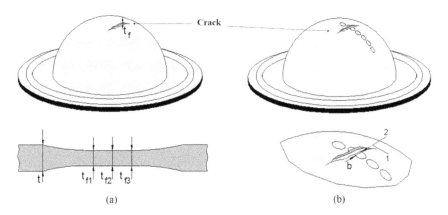

(a)   (b)

FIGURE 18.2   Assessing fracture stresses in sheet specimens: (a) Schematic showing measurement points; (b) schematic of the minor axis b of the ellipse on the specimen fractured plane [21].

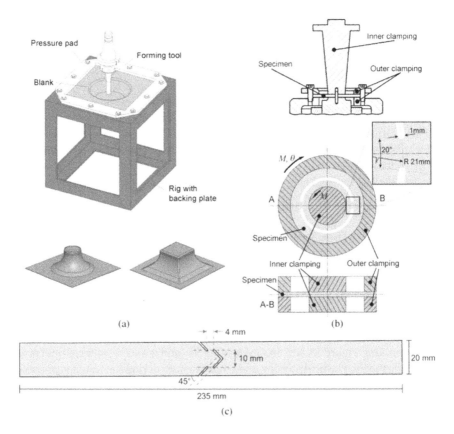

FIGURE 18.3 Schematic representation of the experimental setups that were utilized in the (a) SPIF and (b) in-plane torsion tests together with the (c) geometrical details of the plane shear test specimens that were performed in a universal testing machine [17].

Finally, incompressibility principles are used to determine the surface residual fracture strain, ε_{1f}:

$$\varepsilon_{1f} = -\left(\varepsilon_{2f} + \varepsilon_{3f}\right) \tag{18.8}$$

The experiments were conducted using 1 mm thick sheets of aluminum AA1050-H111. The material's mechanical properties were assessed through tensile tests performed on specimens cut from the supplied sheets at angles of 0, 45, and 90 degrees relative to the rolling direction.

In Figure 18.3, various SPIF experiments used different geometries, including truncated cone and pyramid, with varying drawing angles. SPIF was conducted using a hemispherical tip tool with an 8-mm diameter.

18.4   RESULTS

Figure 18.4 depicts the FFL estimated using the suggested approach, which includes measuring and calculating fracture strains (ε_{1f}, ε_{2f}) across the surface of every test specimen. The straight line shown in the figure may be used to approximate the FFL. The solid black marks indicate the stresses at fracture, while the dashed line labeled FFL represents the previously established fracture locus from standard sheet formability testing. The inset images provide a visual of crack details for different types of test specimens. This validates the SPIF test. [21].

$$\varepsilon_1 + 0.68\varepsilon_2 = 1.34 \tag{18.9}$$

According to equation 18.1, the measured slope of −0.68 closely approximates the theoretic −1 slope, in line with a crucial decrease in thickness R_f at the side of the fracture site. The small difference may result from assumptions in Martins et al. theoretical model [19], including the use of McClintock's noncoupled ductile

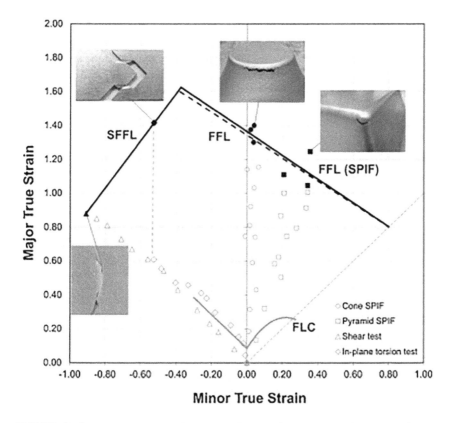

FIGURE 18.4 Determining the fracture loci (FFL) from the experimental strains at fracture that were obtained from the SPIF and shear tests [17].

damage model [20], Hill's plasticity criterion [24], and inherent experimental errors in fracture strain estimation.

For the specimens shown in Figure 18.1, Figure 18.4 illustrates strain pairs that occur at the initiation of necking. This dataset enabled the development of the FLC surrounded by the major strain plot, enabling comparative analysis. Figure 18.4 also shows strain loading paths from SPIF tests.

18.5 CONCLUSION

This method for pinpointing fracture formation limits in the primary strain plot uses circular grids, a common technique for assessing the FLC at necking initiation, but in this novel approach, fracture thickness and the minor axis of the formed ellipses are merged to establish the gauge length, defining the fracture strains. This differs from the usual method of using both minor and major ellipse axes formed by the plastic deformation of electrochemically etched circles in the specimen necked region [21].

The experimental methodologies for identifying fracture loci comprise a limited set of incremental sheet forming and in-plane shear tests, coupled with specific techniques for discerning stresses at fracture. An alternative yet successful approach involves establishing the forming limit line through SPIF experiments employing linear loading pathways until fracture, as opposed to the conventional necking-based methods used in tensile and traditional sheet forming tests. This proposed technique is entirely grounded in sheet metal testing, given the prevalent plane-stress flow conditions in sheet metal forming processes, distinguishing it from alternative methods found in the literature.

REFERENCES

[1] J. Jeswiet, F. Micari, G. Hirt, A. Bramley, J. Duflou, and J. Allwood, "Asymmetric single point incremental forming of sheet metal," *Cirp Ann. Technol.*, vol. 54, no. 2, pp. 88–114, Jan. 2005, doi: 10.1016/S0007-8506(07)60021-3.

[2] T. McAnulty, J. Jeswiet, and M. Doolan, "Formability in single point incremental forming: A comparative analysis of the state of the art," *CIRP J. Manuf. Sci. Technol.*, vol. 16, pp. 43–54, 2017, doi: 10.1016/j.cirpj.2016.07.003.

[3] T. L. Anderson, *Fracture Mechanics: Fundamentals and Applications, Fourth Edition*, Fourth Edi. CRC Press, 2017. doi: doi:10.1201/9781315370293.

[4] J. M. Allwood, D. R. Shouler, and A. E. Tekkaya, "The increased forming limits of incremental sheet forming processes," *Key Eng. Mater.*, vol. 344, no. July 2007, pp. 621–628, 2007, doi: 10.4028/www.scientific.net/kem.344.621.

[5] K. Prasad, H. Krisnaswamy, D. Banerjee, and U. Chakkingal, "An investigation into the influence of interrupted loading in improving the stretch-flangeability of dual phase steel," in *Tribology in Manufacturing Processes*, in Defect and Diffusion Forum, vol. 414. Trans Tech Publications Ltd, 2022, pp. 81–87. doi: 10.4028/p-gi07rp.

[6] M. B. Silva, M. Skjoedt, A. G. Atkins, N. Bay, and P. A. F. Martins, "Single-point incremental forming and formability—Failure diagrams," *J. Strain Anal. Eng. Des.*, vol. 43, no. 1, pp. 15–35, 2008, doi: 10.1243/03093247JSA340.

[7] P. Gupta and J. Jeswiet, "Observations on heat generated in single point incremental forming," *Procedia Eng.*, vol. 183, pp. 161–167, 2017, doi: 10.1016/j.proeng.2017.04.060.

[8] G. Hirt, J. Ames, M. Bambach, R. Kopp, and R. Kopp, "Forming strategies and process modelling for CNC incremental sheet forming," *CIRP Ann.*, vol. 53, no. 1, pp. 203–206, Jan. 2004, doi: 10.1016/S0007-8506(07)60679-9.

[9] V. K. Barnwal, S.-Y. Y. Lee, J. Choi, J.-H. H. Kim, and F. Barlat, "Performance review of various uncoupled fracture criteria for TRIP steel sheet," *Int. J. Mech. Sci.*, vol. 195, p. 106269, Apr. 2021, doi: https://doi.org/10.1016/j.ijmecsci.2021.106269.

[10] K. Pack and D. Mohr, "Combined necking & fracture model to predict ductile failure with shell finite elements," *Eng. Fract. Mech.*, vol. 182, pp. 32–51, Sep. 2017, doi: https://doi.org/10.1016/j.engfracmech.2017.06.025.

[11] D. Mohr and S. J. Marcadet, "Micromechanically-motivated phenomenological Hosford—Coulomb model for predicting ductile fracture initiation at low stress triaxialities," *Int. J. Solids Struct.*, vol. 67–68, pp. 40–55, Aug. 2015, doi: https://doi.org/10.1016/j.ijsolstr.2015.02.024.

[12] S. Bharti, A. Gupta, H. Krishnaswamy, S. K. Panigrahi, and M.-G. G. Lee, "Evaluation of uncoupled ductile damage models for fracture prediction in incremental sheet metal forming," *CIRP J. Manuf. Sci. Technol.*, vol. 37, pp. 499–517, May 2022, doi: https://doi.org/10.1016/j.cirpj.2022.02.023.

[13] P. A. F. F. Martins, N. Bay, M. Skjoedt, and M. B. Silva, "Theory of single point incremental forming," *CIRP Ann.*, vol. 57, no. 1, pp. 247–252, Jan. 2008, doi: https://doi.org/10.1016/j.cirp.2008.03.047.

[14] M. Oyane, T. Sato, K. Okimoto, and S. Shima, "Criteria for ductile fracture and their applications," *J. Mech. Work. Technol.*, vol. 4, no. 1, pp. 65–81, Apr. 1980, doi: https://doi.org/10.1016/0378-3804(80)90006-6.

[15] R. Malhotra, L. Xue, T. Belytschko, and J. Cao, "Mechanics of fracture in single point incremental forming," *J. Mater. Process. Technol.*, vol. 212, no. 7, pp. 1573–1590, Jul. 2012, doi: 10.1016/j.jmatprotec.2012.02.021.

[16] M. J. Mirnia and M. Shamsari, "Numerical prediction of failure in single point incremental forming using a phenomenological ductile fracture criterion," *J. Mater. Process. Technol.*, vol. 244, pp. 17–43, Jun. 2017, doi: https://doi.org/10.1016/j.jmatprotec.2017.01.029.

[17] K. Isik, M. B. Silva, A. E. Tekkaya, and P. A. F. F. Martins, "Formability limits by fracture in sheet metal forming," *J. Mater. Process. Technol.*, vol. 214, no. 8, pp. 1557–1565, Aug. 2014, doi: 10.1016/j.jmatprotec.2014.02.026.

[18] A. G. Atkins, "Fracture in forming," *J. Mater. Process. Technol.*, vol. 56, no. 1, pp. 609–618, Jan. 1996, doi: https://doi.org/10.1016/0924-0136(95)01875-1.

[19] P. A. F. F. Martins, N. Bay, A. E. Tekkaya, and A. G. Atkins, "Characterization of fracture loci in metal forming," *Int. J. Mech. Sci.*, vol. 83, pp. 112–123, Jun. 2014, doi: https://doi.org/10.1016/j.ijmecsci.2014.04.003.

[20] F. ~A. McClintock, "A criterion for ductile fracture by the growth of holes," *J. Appl. Mech.*, vol. 35, no. 2, p. 363, Jan. 1968, doi: 10.1115/1.3601204.

[21] V. A. M. Cristino, M. B. Silva, P. K. Wong, and P. A. F. Martins, "Determining the fracture forming limits in sheet metal forming: A technical note," *J. Strain Anal. Eng. Des.*, vol. 52, no. 8, pp. 467–471, 2017, doi: 10.1177/0309324717727443.

[22] G. Centeno, A. J. Martínez-Donaire, D. Morales-Palma, C. Vallellano, M. B. Silva, and P. A. F. Martins, *Novel Experimental Techniques for the Determination of the Forming Limits at Necking and Fracture*, vol. 2, no. 2008. Elsevier Ltd., 2015. doi: 10.1016/B978-0-85709-483-4.00001-6.

[23] S. Bruschi *et al.*, "Testing and modelling of material behaviour and formability in sheet metal forming," *CIRP Ann.*, vol. 63, no. 2, pp. 727–749, Jan. 2014, doi: 10.1016/J.CIRP.2014.05.005.

[24] S. Xu and K. J. Weinmann, "Prediction of forming limit curves of sheet metals using Hill's 1993 user-friendly yield criterion of anisotropic materials," *Int. J. Mech. Sci.*, vol. 40, no. 9, pp. 913–925, 1998, doi: 10.1016/S0020-7403(97)00145-8.

19 The Role of Lubrication in Sheet Metal Forming Processes

*Vishal Bhojak, Jinesh Kumar Jain,
and Tejendra Singh Singhal*

19.1 INTRODUCTION

The tribological aspects of sheet metal forming are influenced by a multitude of factors encompassing material properties, surface finish, temperature, sliding velocity, contact pressure, and lubricant characteristics. In Figure 19.1, a schematic representation illustrates the key parameters governing the phenomena of friction, wear, and lubrication [1]. The efficacy of a lubricating agent and the coefficient of friction (COF) is contingent on the variations in these specified variables.

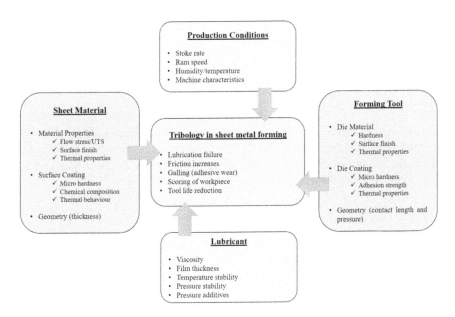

FIGURE 19.1 Influential parameters impacting tribological aspects in the sheet metal forming process

DOI: 10.1201/9781003441755-19

Notably, the frictional attributes operative at the interface between the die and workpiece bear direct consequence on the flow of material within the die cavity. The judicious selection of lubricants and the attainment of high-quality sheet metal components necessitate a comprehensive understanding of the manifold determinants underpinning frictional behavior [2]. The magnitude and spatial distribution of frictional forces within the context of sheet metal forming exert a profound influence on the dynamics of metal flow, the manifestation of part defects, overall product quality, as well as the wear and cost considerations associated with tooling.

19.2   TYPES OF LUBRICATION

The Stribeck curve, schematically represented in Figure 19.2, delineates how the initiation of distinct lubrication mechanisms is contingent on lubricant viscosity (η), sliding velocity (v), and normal pressure (p) [3],[4]. Metal forming processes offer diverse methods for the provision of lubrication.

19.2.1   DRY CONDITION

In this mode, as illustrated in Figure 19.2, no lubrication is present at the mating surfaces, resulting in a significant level of friction. The extent of friction is contingent upon the coefficient of friction of the mating materials and is subject to variation accordingly. Dry conditions are typically employed when material formability is adequate to produce simple-geometry parts without the need for lubrication, or when the impact of friction on part quality is negligible, as exemplified in air bending, V-die bending, and U-die bending processes without stretching. Specific metal-forming operations, such as the hot rolling of plates or slabs and nonlubricated extrusion of aluminium alloys, are conducive to dry conditions.

19.2.2   BOUNDARY

As depicted in Figure 19.2, boundary lubrication occurs when solid surfaces are in proximity, enabling surface interactions between lubricant films and the solid asperities to dominate the contact. Boundary lubrication is the prevailing lubrication mode in metal forming processes.

19.2.3   MIXED FILM

In this variant of lubrication, depicted in Figure 19.2, boundary lubrication conditions are observed on the micropeaks of the metal surface, with the lubricant filling the microvalleys. Mixed-film lubrication is frequently encountered in sheet metal forming.

19.2.4   HYDRODYNAMIC

Figure 19.2 also illustrates hydrodynamic lubrication, which arises in specific sheet metal forming activities such as high-speed sheet rolling characterized by high velocities at the material–tool interface [3], [4].

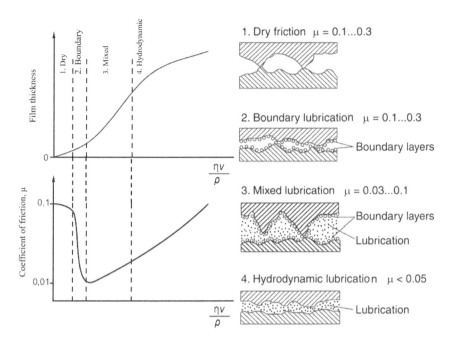

FIGURE 19.2 Diverse lubrication mechanisms as delineated by the Stribeck curve, encompassing four distinct regimes.

19.3 FRICTION LAWS

In the domain of metal forming, two prevalent friction models, Coulomb's friction model (Eq. 1) and the shear friction model (Eq. 2), assume prominent roles in elucidating the frictional dynamics. These models serve as valuable tools for encapsulating and comprehending the manifold contact phenomena illustrated collectively in Figure 19.2. Both Coulomb's friction model and the shear friction model offer a unifying framework by amalgamating these diverse contact phenomena into a singular nondimensional coefficient or factor, thereby facilitating the quantification of interface friction [5]:

$$\tau_f = \mu p \qquad\qquad \text{Eq. 19.1}$$

In the context of metal forming analyses, the frictional shear stress, denoted as τ_f, is linked to the normal pressure, represented by P, through the COF, symbolized as μ. It is noteworthy that in several metal-forming techniques, the interface pressure, p, can attain levels surpassing multiples of the yield strength of the material in question. Under such circumstances, the shear stress, τ_f, must adhere to the constraint imposed by the shear strength, denoted as k, of the material undergoing deformation within the workpiece.

This constraint implies that the conventional linear relationship between τ_f and p, as expressed by Eq. 19.1, ceases to be valid at elevated contact pressure regimes. Therefore, when the product of μ and p surpasses the magnitude of τ_f the practical utility of the coefficient of friction becomes compromised. In response to this limitation inherent in Coulomb's model, Orowan introduced the shear friction model, as articulated in Eq. 19.2, to address and circumvent this specific drawback [5]:

$$\tau_f = f\bar{\sigma} = m\frac{\bar{\sigma}}{\sqrt{3}} = mk \qquad \text{Eq. 19.2}$$

In the context of this study, it is imperative to define key parameters: k represents the shear strength, f denotes the friction, m signifies the shear, with its allowable range between 0 and 1, and $\bar{\sigma}$ characterizes the flow stress exhibited by the material undergoing deformation. Within the conceptual framework presented in Figure 19.3 [1], the frictional shear stress, denoted as τ_f, is intricately linked to the normal pressure anticipated by Coulomb's model under conditions of low pressure. Notably, at elevated levels of interface pressure denoted as p, τ_f converges to the value of k.

In Eq. 19.2, the parameter m exhibits binary behavior: It assumes a value of zero when friction is absent and a value of unity in the presence of adhering friction. When two ostensibly flat surfaces come into contact, surface irregularities give rise to discrete contact patches, and the actual contact area, denoted as A_r, is an aggregation of these discrete contact areas. Usually it represents only a fraction of the apparent contact area, denoted as A_a.

The real contact area ratio, designated as α, represents the relationship between the authentic contact area (A_r) and the apparent contact area (A_a). Researchers such as the authors of [6], [7] have formulated a comprehensive friction model

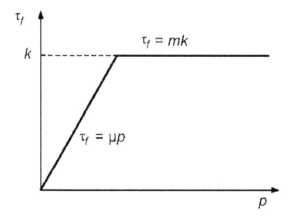

FIGURE 19.3   The correlation between shear stress due to friction and normal pressure in the context of varying interface pressures at both lower and higher magnitudes.

that incorporates the pivotal influence of the true contact area ratio α on frictional behavior:

$$\tau_f = f'\alpha k = m_r \frac{\bar{\sigma}}{\sqrt{3}}$$

Eq. 19.3

Within this theoretical framework, the following key parameters are established: A_r/A_a is the real contact area ratio α; f' represents the modified friction; m_r is the modified shear, which is contingent on α; and $\bar{\sigma}$ represents the flow stress. The frictional shear stress, denoted as τ_f, within this conceptual framework is also contingent on α. Notably, the model proposed by Bay and Wanheim [7] lacks consideration for lubricant behaviours. To account for the influence of lubricants, Bowden and Tabor [8] introduced a comprehensive model for friction, particularly addressing boundary and mixed-film lubrication regimes occurring at the interface between the tool and workpiece. Frictional shear stress τ_f is expressed as

$$\tau_f = \alpha\tau_a + (1-\alpha)\tau_b$$

Eq. 19.4

where α represents the genuine contact area ratio, τ_a signifies the average shear stress at the asperity peaks in contact, and τ_b denotes the average shear stress, specifically at the lower stress levels within the lubricating pockets. It is noteworthy that this model establishes a lucid relationship between α and τ_a, as well as the impact of lubricant properties, encapsulated in τ_b. The latter is influenced by factors such as viscosity, pressure, sliding velocity, and film thickness. In accordance with Eq. 19.4, τ_f is contingent upon the actual contact area ratio and τ_b. Notably, τ_b will assume a value of zero in cases where no lubricant is present at the interface between the tool and workpiece.

In order to assess the influence of lubricant behavior on friction, an artificial lubrication film was postulated at the interface between the tool and workpiece. The analysis incorporated the Reynolds equation, which is rooted in fluid mechanics theory, to elucidate how the film thickness varies, consequently defining the fluctuations in friction [9]. This approach offers a granular representation of lubricant behavior. Nonetheless, it is important to acknowledge that implementing this model in practical metal forming scenarios poses considerable challenges.

19.4   LUBRICANTS IN SHEET METAL FORMING

In the process of selecting lubricants for sheet metal forming, several critical considerations come to the forefront: the techniques employed for lubricant application, the types of additives incorporated, measures for corrosion control, cleanliness and subsequent removal methods, compatibility with preexisting lubricants and pre-applied oils, the implications for post-metal forming operations (such as welding and adhesive joining), and the overarching considerations of environmental safety and recyclability [10], [11].

19.4.1 Classification of Lubricants

The frictional force conditions vary across distinct metal forming processes, which calls for different lubricants. Following are some common categories of lubricants:

- **Oils:** Petroleum-based oils are commonly utilized as lubricants in light-duty sheet metal forming processes, including coining, blanking, and stamping. Oils can significantly enhance the manufacturing process. This category encompasses paraffinic oils and naphthenic oils, among others.
- **Soluble Oils:** Soluble oils contain emulsifiers, facilitating their dilution in water. These oils are typically referred to as preformed emulsions. Commonly, soluble oils are diluted with water at ratios ranging from 10% to 50%.
- **Semi-Synthetics:** Semi-synthetic lubricants exhibit greater miscibility with water than soluble oils due to their lower mineral oil content, typically constituting less than 30% of the total concentrate volume.
- **Synthetics:** Synthetic lubricants fall into two primary categories: hydrocarbon-based and water-based. These lubricants often exhibit a milky or hazy appearance.
- **Dry-Film Lubricants:** This category can be further divided into water-soluble and water-free ("hot melt") dry-film lubricants. Water-soluble dry-film lubricants arc applied at rolling mills in concentrations ranging from 0.5 to 1.5 g/m^2 [12]. They adhere to the sheet metal surface and provide adequate corrosion protection; however, they may not be compatible with many adhesives used in vehicle body assembly. Water-free dry-film lubricants are also sparingly applied at the rolling mill. These lubricants offer superior drawing performance compared with oil-based alternatives and are renowned for their compatibility with a wide array of commonly used adhesives, making them highly advantageous in sheet metal forming processes [13].

19.4.2 Methods of Lubricant Application

To ensure optimal lubrication performance, minimize lubricant wastage, and mitigate environmental risks, it is imperative that lubricants are applied with precision and care [10]. The predominant techniques for lubricant application include

- **Drip Method:** This straightforward and cost-effective approach involves applying the lubricant by dripping it onto the panel or sheet blank. However, achieving precise control over the quantity of lubricant applied can be challenging.
- **Roll Coating:** In this method, lubrication is administered to a moving blank positioned between two rollers operating under controlled

pressure. This technique allows for accurate regulation of the lubricant quantity.

- **Electro-Deposition:** Primarily suited for high-speed applications, electro-deposition minimizes lubricant wastage but necessitates a substantial capital investment, which may pose challenges for smaller stamping facilities. Lubricant is deposited onto the panel surface with an electric charge.
- **Airless Spraying:** While boasting a low lubricant wastage rate, airless spraying may not be compatible with high-viscosity lubricants. Nevertheless, it excels in delivering precise amounts of lubricant to targeted areas.
- **Mops and Sponges:** Several smaller stamping operations continue to employ this cost-effective technique for lubricant application to panels. However, it often results in excessive lubricant wastage, suboptimal control over lubricant distribution, and a disorderly work environment.

19.4.3 ADDITIVES IN LUBRICANTS

Various additives are strategically employed to enhance the functional characteristics of lubricants. In heavy-duty metal stamping, the utilization of extreme pressure (EP) additives is a common practice. These additives can be categorized into two distinct groups: temperature-activated and non-temperature-activated. Specifically, temperature-activated EP additives, such as chlorine, phosphorus, and sulfur, initiate chemical reactions and form a protective layer on the metal surface as the temperature at the interface elevates. This chemical layer plays a pivotal role in preventing direct metal-to-metal contact during stamping processes [14]. It is noteworthy that the effectiveness of EP additives varies within specific temperature ranges, with phosphorus exhibiting its efficacy up to 205 °C, chlorine coming into play between 205 °C to 700 °C, and sulfur being effective within the temperature range of 700 °C to 960 °C [10].

Lubricants enriched with EP additives exhibit a notable characteristic: They tend to become less viscous and are susceptible to combustion when subjected to elevated temperatures resulting from the deformation process involving the tool and workpiece. Conversely, lubricants infused with extreme temperature (ET) additives exhibit the opposite behavior: As the temperature increases, these lubricants tend to thicken, rendering them particularly effective in high-temperature environments. They have the capacity to adhere to the warm workpiece surface and establish a protective film barrier that effectively reduces friction between the tool and workpiece. Additives such as phosphorus, chlorine, and sulfur are harnessed as ET additives.

19.4.4 CORROSION CONTROL IN SHEET METAL STAMPING

Lubricants can also control corrosion and avert oxidation on stamped components for postprocessing. These lubricants are tested according to the ASTM

(a) Process with oil-based (wet) lubricant

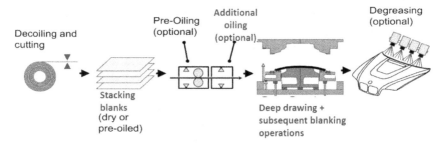

(b) Process with dry-film lubricant

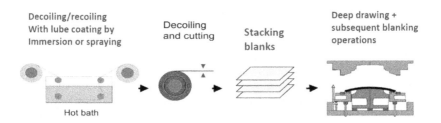

FIGURE 19.4   Configurations of the stamping process.

D130 standard corrosion test [15]. This test involves immersing a sheet sample in the lubricant for a specified duration to assess its corrosion resistance properties. Figure 19.4 depicts configurations of stamping with oil-based and dry-film lubricant.

19.4.5   CLEANING PROCEDURES IN SHEET METAL STAMPING

Ensuring the complete removal of lubricants assumes paramount importance prior to painting and electro cathodic coating (E-coating) of stamped components with primers. The presence of surface contaminants can significantly impair the quality of painting and E-coating applications. One method for the elimination of straight oils is using vapor degreasers, a solvent-based cleaning technology. However, vapor degreasers are not extensively employed due to safety and environmental considerations. Alternatively, alkaline cleaners can effectively substitute for vapor degreasers. These cleaners can be applied to the part through impingement spraying or immersion in water. Notably, oil-containing lubricants that have been diluted with water respond favorably to this cleaning process. Many practitioners opt for multistage washers to achieve the comprehensive removal of oil residues.

19.4.6 PRE-LUBRICANTS IN CONTEMPORARY STAMPING PROCESSES

In the various stages of modern sheet metal stamping processes, a diverse array of lubricants and washer oils are routinely employed. A thin layer of pre-lubricant is a common application practice, often administered to sheet metal directly at the rolling machine. Such pre-lubricants play a crucial role in facilitating the efficient progress of sheet metal through the stamping process.

19.4.7 POST-METAL-FORMING OPERATIONS IN SHEET METAL PROCESSING

The incorporation of lubricants into sheet metal forming processes predominantly occurs prior to the press forming stage. However, it is imperative to recognize that the influence of lubricants extends beyond the initial forming phase and significantly impacts subsequent post-metal forming procedures, including welding, adhesive bonding, and painting. In order to ensure that the stamped component is entirely free from residual oils, lubricants must possess not only effective lubrication properties but also the capacity for facile removal from the manufactured panels.

Consequently, it is necessary to comprehensively consider the advantages and drawbacks associated with these lubricants [16]. For instance, wet versus dry-film lubricants can have consequential implications, as elucidated in Figure 19.5 [14].

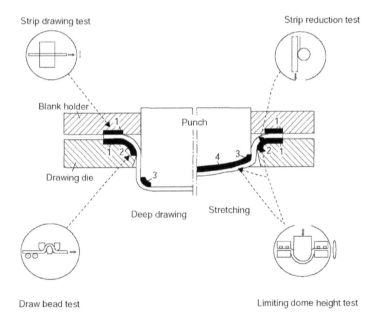

FIGURE 19.5 Various tribo-tests are conducted to assess stamping lubricants. Point 1 is employed to assess flange deformation, Point 2 is utilized to analyze bending and unbending properties, Point 3 is employed for evaluating bending and stretching characteristics, and Point 4 is employed to examine high-friction scenarios with minimal deformation [17], [18].

19.5 THE TRIBOLOGICAL ASSESSMENT OF LUBRICANTS IN SHEET METAL FORMING

The analysis of lubrication behavior within diverse sheet metal forming processes is conducted through laboratory scale test, deep drawing and ironing, and warm and hot stamping. Each testing procedure for lubrication discussed in subsequent sections.

19.5.1 LABORATORY-SCALE TESTS

In the quest to evaluate the effectiveness of lubricants employed in sheet metal forming processes, laboratory-scale investigations play a pivotal role. These assessments involve the utilization of a diverse array of tribological tests that have evolved and been refined over time. Notable among these tests are the draw bead test (Figure 19.5) and the strip drawing test (SDT) (Figure 19.6).

The limiting dome height (LDH) test also assesses the performance of stamping lubricants [17]. It is particularly sensitive to the frictional interaction between the punch and the sheet material, exerting a pronounced influence on the precise location of the fracture point within the test specimen. This examination closely simulates the frictional conditions encountered at the die or punch shoulders during the stamping process, as it entails a predominantly stretching action with minimal sliding, as depicted in Figure 19.5 [18]. Furthermore, the strip reduction test [19], occupies a prominent position in the evaluation of lubricants, particularly for stainless steels undergoing ironing processes. In this rigorous evaluation, the metal strip undergoes substantial thickness reduction as it is drawn through a roller element (Figure 19.6) [20], [21]. This test proves invaluable for the assessment of stamping lubricants and tool coatings, relying on measurements of galling either on the roller or the strip to draw meaningful conclusions regarding lubricant performance. In laboratories, the twist compression test (TCT) is also a common method for evaluating lubricants and additives. It particularly determines the COF of stamping lubricants. The TCT entails recording pressure and torque measurements while a rotating annular tool exerts force on a stationary sheet metal specimen (Figure 19.7).

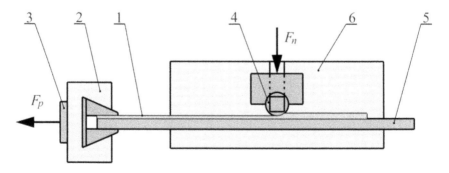

FIGURE 19.6 The components of testing strip reduction in sheet metal forming: 1—strip, 2—clamping jaws, 3—utilized piezoelectric transducer, 4—implementing tool pin, 5—tempered steel rod, and 6—upright guiding structure.

The COF between the tool and the specimen is subsequently derived through analytical calculations [22]:

$$\mu = \frac{T}{r \times P \times A}$$

Eq. 19.5

In this context, let A symbolize the contact area between the tool and the sample, P represent the applied pressure on the tool, r denote the mean radius of the tool, and μ stand for the COF. The SDT (shown in Figure 19.8) [23] was devised to comprehensively examine and assess the impact of various parameters such as die radii, material coatings, and lubricants. This test protocol was primarily developed to evaluate the performance of advanced high-strength steels, including but not limited to DP 780, TRIP 780, and DP 980 grades. The SDT serves as a simplified adaptation of the traditional round cup deep drawing test (Figure 19.8), aiming to provide more nuanced insights into material behavior and lubrication effects.

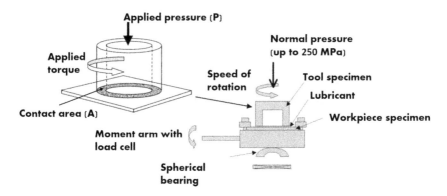

FIGURE 19.7   Schematic representation of twist compression test; normal and enlarged view [22].

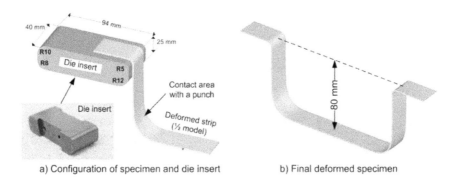

FIGURE 19.8   Sheet metal strip drawing testing: (a) the arrangement of the test specimen and die insert, encompassing radii of R5, R8, R10, and R12 at the peripheries; (b) the ultimate deformation of the test specimen.

Notably, during testing, the stresses and strains can vary during deformation, and these discrepancies can alter the COF [23].

It is imperative to acknowledge the inherent limitations of laboratory testing in replicating the complex process conditions encountered in real-world stamping operations, encompassing factors such as temperature, contact pressure, and speed. The evaluation of lubricants under authentic production conditions presents formidable challenges, often accompanied by substantial costs. Therefore, the availability of a robust laboratory-based tribological test that can faithfully emulate the environmental and operational conditions prevalent in stamping operations emerges as a highly valuable and practical asset.

19.5.2 Deep Drawing and Ironing

Several European automobile manufacturers have adopted the deep drawing test (Figure 19.9) to evaluate the performance of lubricants within production-like scenarios [16]. In these experiments, deep drawing procedures are conducted to closely mimic the conditions prevalent in actual stamping operations. By modulating the Blank Holder Force (BHF), it becomes feasible to emulate both (a) the sheet-die interface pressures encountered in production and (b) the punch velocities characteristic of mechanical stamping presses.

The die shoulder and flange regions are particularly prone to friction during deep drawing, as evidenced in Figure 19.9. The lubrication state within these regions exerts a profound influence on two critical aspects: (a) the potential for sidewall thinning or failure in the drawn cup and (b) the draw-in length, denoted as L_d in Figure 19.9, of the flange. According to Coulomb's law, an increase in blank holder pressure, P_b, corresponds to a concurrent elevation in frictional stress, τ, as articulated in Eq. 19.1. Consequently, evaluating lubricants in the context of deep drawing entails determining the maximum permissible BHF that can be applied without inducing cup wall rupture.

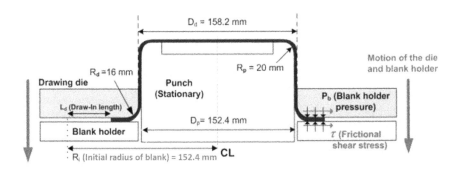

FIGURE 19.9 The experimental arrangement for deep drawing examination [23].

Qualitative and quantitative assessments of lubricants consider the following parameters:

- Maximum punching force (where lower force implies superior lubrication).
- Maximum permissible BHF (higher BHF indicates enhanced lubricant performance without inducing fracture in the drawn cup).
- Visual inspection for zinc powdering and galling.
- Measurement of L_d within the flange (greater lubrication yields longer draw-in lengths).
- Circumference within the flange region (increased lubrication results in reduced perimeter dimensions).

The deep drawing test has been employed for the analysis of numerous stamping lubricants in conjunction with DP600 steel. Specifically, at a substantial BHF of 70 tons, Figure 19.10 offers a comparative illustration of load-stroke curves, highlighting distinctions between fully drawn cups and cups that have experienced fracture.

Automotive stamping uses a diverse array of lubricants that can interact to result in suboptimal performance of the stamping lubricant. Consequently, testing only individual lubricants can yield misleading results. The deep drawing test allows for evaluating the effectiveness of these intricate lubrication systems under

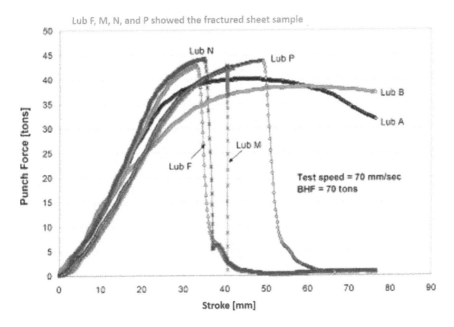

FIGURE 19.10 Load-stroke curves were acquired for multiple lubricants under examination at an elevated BHF [16].

real-world production conditions [24]. This examination facilitates a comprehensive understanding of the implications associated with the integration of mixed lubrication systems on stamping performance.

19.5.3 WARM AND HOT STAMPING

Producing lightweight materials such as aluminum, magnesium alloys, and boron-alloyed steels is challenging at room temperature, and therefore, fabricating them requires high temperatures. It is therefore necessary to test lubricants for their effectiveness in warm and hot stamping as well as for room temperature forming [25]. The ironing test is common for evaluating warm and hot stamping lubricants. Figure 19.11 depicts the conduct of ironing experiments to evaluate the performance of five distinct lubricants in the warm forming of AISI 1008 CR steel at a temperature of 100 °C employing a heated ironing die. Among the lubricants assessed, lubricant B emerged as the most effective, while lubricants C and D exhibited the poorest performance.

To replicate the process conditions inherent to hot stamping, the authors of [27] introduced modifications to the cup drawing test, as visually depicted in Figure 19.12. This adapted test configuration features punch and die corner radii measuring 10 mm, with punch and die diameters of 50 mm and 59 mm, respectively. Load cells are strategically incorporated within the cup drawing test apparatus to facilitate the precise measurement of punch force and BHF.

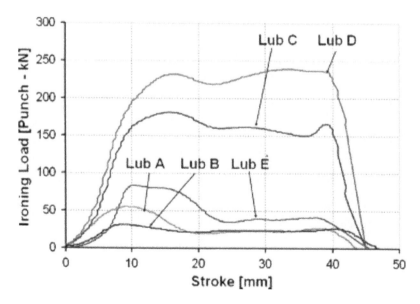

FIGURE 19.11 Load-stroke curves were acquired for a range of lubricants, designated as "Lub", under examination at a temperature of 100 °C [26].

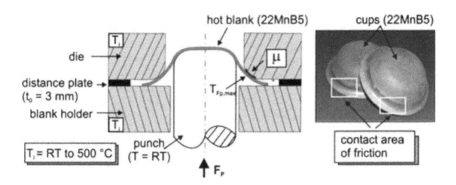

FIGURE 19.12   Schematic representation of the cup drawing test and two formed cups at elevated temperatures: t_0, initial sheet thickness; T, temperature; T_i, initial temperature; RT, room temperature [27].

The thermal environment of the testing apparatus is meticulously controlled through the individual heating cartridges and cooling units integrated into the tools. Notably, these tools have the capacity to reach temperatures of up to 650 °C. To optimize the formability of the blank, the punch's temperature is maintained at room temperature. However, the temperatures of both the die and the blank holder are systematically manipulated to investigate the impact of tool temperature on friction during the test. To ensure the stability of the process conditions, a distance plate is strategically positioned between the die and the blank holder. Moreover, BHF is deliberately not applied to prevent premature cooling or quenching of the blank prior to the forming operation.

19.5.4   Punching and Blanking

Punching and blanking processes, in comparison with other sheet metal forming techniques, entail significantly more contact pressure and higher temperatures at the tool/workpiece interface. A notable challenge encountered in these operations is the sustained lubrication of the interface. This challenge arises due to the continuous shearing process, wherein the punch continually interfaces with new surfaces. Additionally, the presence of oxygen on the punch surface can lead to the adhesion and hardening of torn sheet fragments, resulting in a phenomenon known as galling.

Punching tests were executed using Wn.1.4401 and Wn.1.4301 stainless-steel sheet materials at a rate of two strokes per minute, conducted under both dry and lubricated conditions [28]. The lubricants assessed in these tests included plain mineral oil and chlorinated paraffin oil. Backstroke force was meticulously evaluated employing a circular punch. With an increasing number of strokes, the backstroke force demonstrated an upward trend, concomitant with an escalation in the accumulation of sheet material on the punch surface [29]. Scanning electron microscopy revealed that the sheet material pickup initially commenced in

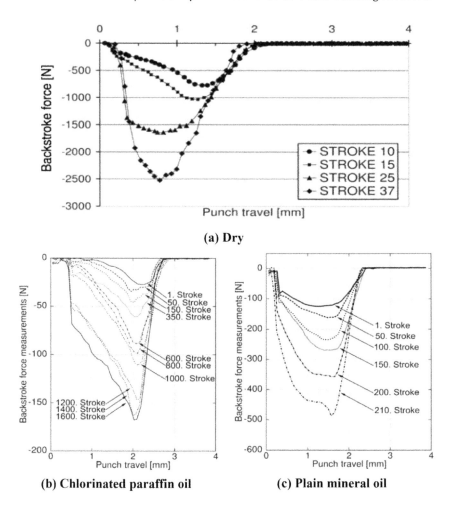

FIGURE 19.13   Backstroke force versus punch travel for (a) dry condition, (b) chlorinated paraffin oil, and (c) plain mineral oil.

the axial direction and subsequently extended radially until it either reached the die surface or filled the space between the die and punch [26]. Notably, when compared with chlorinated paraffin oil, mineral oil exhibited a higher propensity for inducing galling, as indicated by the greater backstroke force, as illustrated in Figure 19.13 [29].

19.6   SUMMARY

The field of tribology plays a crucial and essential part in the study of metal forming processes since it involves the investigation of surface phenomena such as contact, friction, lubrication, and wear, which all directly impact the quality of components produced, and sometimes their cost. This chapter undertakes a complete investigation of lubricating processes in sheet metal forming methods, including investigating the coefficient of friction. The chapter provides a detailed analysis of lubricant testing procedures. The utilization of tribology in mechanical systems shows potential for reducing global carbon emissions by minimizing wear, decreasing friction, and prolonging the lifespan of resources.

REFERENCES

[1] J. A. Schey, "Tribology in Metalworking : Friction, Lubrication, and Wear," *Am. Soc. Met.*, p. 736, 1983. www.worldcat.org/title/472157575 (accessed Sep. 12, 2023).

[2] V. Bhojak and J. K. Jain, "Tribological Behavior in Bulk and Sheet Forming Processes," *Met. Form. Process.*, pp. 121–133, Aug. 2022, doi: 10.1201/9781003226703-7.

[3] B. Bhushan, *Introduction to Tribology*. John Wiley and Sons, 2002. www.wiley.com/en-sg/Introduction+to+Tribology%2C+2nd+Edition-p-9781119944539 (accessed Sep. 12, 2023).

[4] L. F. Folle *et al.*, "The Role of Friction on Metal Forming Processes," *Tribol. Mach. Elem.—Fundam. Appl.*, Feb. 2022, doi: 10.5772/INTECHOPEN.101387.

[5] D. E. Orowan, "The Calculation of Roll Pressure in Hot and Cold Flat Rolling," *Proc. Inst. Mech. Eng.*, vol. 150, no. 1, pp. 140–167, Nov. 2016, doi: 10.1243/PIME_PROC_1943_150_025_02.

[6] T. Wanheim, N. Bay, and A. S. Petersen, "A Theoretically Determined Model for Friction in Metal Working Processes," *Wear*, vol. 28, no. 2, pp. 251–258, 1974, doi: 10.1016/0043–1648(74)90165-3.

[7] N. Bay, T. Wanheim, and A. S. Petersen, "Ra and the Average Effective Strain of Surface Asperities Deformed in Metal-Working Processes," *Wear*, vol. 34, no. 1, pp. 77–84, 1975, doi: 10.1016/0043-1648(75)90310-5.

[8] F. P. Bowden and D. Tabor, *The Friction and Lubrication of Solids*. Clarendon Press, Methuen and Co. Ltd, 2001. https://global.oup.com/academic/product/the-friction-and-lubrication-of-solids-9780198507772 (accessed Sep. 12, 2023).

[9] S. Patra and L. K. Singhal, "Influence of Hot Band Annealing and Cold Rolling on Texture and Ridging of 430 Stainless Steel Containing Aluminum," *Mater. Sci. Appl.*, vol. 4, no. 1, pp. 70–76, 2013, doi: 10.4236/MSA.2013.41009.

[10] J. P. Byers, *Metalworking Fluids (Manufacturing Engineering and Materials Processing)*. CRC Press, 2017. www.routledge.com/Metalworking-Fluids/Byers/p/book/9781498722223 (accessed Sep. 12, 2023).

[11] V. Bhojak, J. Lade, J. K. Jain, A. Patnaik, and K. K. Saxena, "Investigation of Annealing on CR-2 Grade Steel Using Taguchi and Taguchi Based Gray Relational Analysis," *Adv. Mater. Process. Technol.*, pp. 1–16, Feb. 2022, doi: 10.1080/2374068X.2022.2037878.

[12] N. Bay *et al.*, "Environmentally Benign Tribo-Systems for Metal Forming," *CIRP Ann.*, vol. 59, no. 2, pp. 760–780, Jan. 2010, doi: 10.1016/J.CIRP.2010.05.007.

[13] W. R. D. Wilson, T. C. Hsu, and X. B. Huang, "A Realistic Friction Model for Computer Simulation of Sheet Metal Forming Processes," *J. Eng. Ind.*, vol. 117, no. 2, pp. 202–209, May 1995, doi: 10.1115/1.2803295.

[14] M. Tolazzi, M. Meiler, and M. Merklein, "Tribological Investigations on Coated Steel Sheets Using the Dry Film Lubricant Drylube E1," *Adv. Mater. Res.*, vol. 6–8, pp. 565–572, May 2005, doi: 10.4028/WWW.SCIENTIFIC.NET/AMR.6-8.565.

[15] M. Meiler and H. Jaschke, "Lubrication of Aluminium Sheet Metal within the Automotive Industry," *Adv. Mater. Res.*, vol. 6–8, pp. 551–558, May 2005, doi: 10.4028/WWW.SCIENTIFIC.NET/AMR.6-8.551.

[16] S. Subramonian, N. Kardes, Y. Demiralp, M. Jurich, and T. Altan, "Evaluation of Stamping Lubricants in Forming Galvannealed Steels for Industrial Application," *J. Manuf. Sci. Eng.*, vol. 133, no. 6, Dec. 2011, doi: 10.1115/1.4003948/439918.

[17] G. M. Dalton and J. A. Schey, "Effect of Bead Finish Orientation on Friction and Galling in the Drawbead Test," *SAE Tech. Pap.*, Feb. 1992, doi: 10.4271/920632.

[18] M. Vermeulen and J. Scheers, "Micro-Hydrodynamic Effects in EBT Textured Steel Sheet," *Int. J. Mach. Tools Manuf.*, vol. 41, no. 13–14, pp. 1941–1951, Oct. 2001, doi: 10.1016/S0890-6955(01)00059-1.

[19] J. L. Andreasen, N. Bay, and L. De Chiffre, "Quantification of Galling in Sheet Metal Forming By Surface Topography Characterisation," *Int. J. Mach. Tools Manuf.*, vol. 38, no. 5–6, pp. 503–510, 1998, doi: 10.1016/S0890-6955(97)00095-3.

[20] D. D. Olsson, N. Bay, and J. L. Andreasen, "Prediction of Limits of Lubrication in Strip Reduction Testing," *CIRP Ann.*, vol. 53, no. 1, pp. 231–234, Jan. 2004, doi: 10.1016/S0007-8506(07)60686-6.

[21] T. Trzepiecinski and H. G. Lemu, "Recent Developments and Trends in the Friction Testing for Conventional Sheet Metal Forming and Incremental Sheet Forming," *Met. 2020*, vol. 10, no. 1, p. 47, Dec. 2019, doi: 10.3390/MET10010047.

[22] H. Kim, J. Sung, F. E. Goodwin, and T. Altan, "Investigation of Galling in Forming Galvanized Advanced High Strength Steels (AHSSs) Using the Twist Compression Test (TCT)," *J. Mater. Process. Technol.*, vol. 205, no. 1–3, pp. 459–468, Aug. 2008, doi: 10.1016/J.JMATPROTEC.2007.11.281.

[23] H. Kim, T. Altan, and Q. Yan, "Evaluation of Stamping Lubricants in Forming Advanced High Strength Steels (AHSS) Using Deep Drawing and Ironing Tests," *J. Mater. Process. Technol.*, vol. 209, no. 8, pp. 4122–4133, Apr. 2009, doi: 10.1016/J.JMATPROTEC.2008.10.007.

[24] M. H. Sulaiman, P. Christiansen, and N. Bay, "The Influence of Tool Texture on Friction and Lubrication in Strip Reduction Testing," *Lubricants*, vol. 5, no. 1, p. 3, Jan. 2017, doi: 10.3390/LUBRICANTS5010003.

[25] A. Yanagida and A. Azushima, "Evaluation of Coefficients of Friction in Hot Stamping By Hot Flat Drawing Test," *CIRP Ann.*, vol. 58, no. 1, pp. 247–250, Jan. 2009, doi: 10.1016/J.CIRP.2009.03.091.

[26] S. Chandrasekharan, H. Palaniswamy, N. Jain, G. Ngaile, and T. Altan, "Evaluation of Stamping Lubricants at Various Temperature Levels Using the Ironing Test," *Int. J. Mach. Tools Manuf.*, vol. 45, no. 4–5, pp. 379–388, Apr. 2005, doi: 10.1016/J.IJMACHTOOLS.2004.09.014.

[27] M. Geiger, M. Merklein, and J. Lechler, "Determination of Tribological Conditions within Hot Stamping," *Prod. Eng.*, vol. 2, no. 3, pp. 269–276, 2008, doi: 10.1007/S11740-008-0110-8.

[28] C. Z. F. Klocke, T. Massmann, F. M. R. A. Schmidt, and J. Schulz, "Fineblanking with Non-Chlorinated Lubricants," *Tribologie and Schmierungstechnik*, 2008. www.researchgate.net/publication/288169086_Fineblanking_with_Non-chlorinated_Lubricants (accessed Sep. 12, 2023).

[29] D. D. Olsson, N. Bay, and J. L. Andreasen, "Analysis of Pick-Up Development in Punching," *CIRP Ann.*, vol. 51, no. 1, pp. 185–190, Jan. 2002, doi: 10.1016/S0007-8506(07)61496-6.

Index

Note: Page numbers in *italics* indicate a figure and page numbers in **bold** indicate a table on the corresponding page.

For Product Safety Concerns and Information please contact our EU
representative GPSR@taylorandfrancis.com
Taylor & Francis Verlag GmbH, Kaufingerstraße 24, 80331 München, Germany

www.ingramcontent.com/pod-product-compliance
Ingram Content Group UK Ltd.
Pitfield, Milton Keynes, MK11 3LW, UK
UKHW021113180425
457613UK00005B/73